MEASUREMENT OF NUCLEAR RADIATION WITH SEMICONDUCTOR DETECTORS

MEASUREMENT OF NUCLEAR RADIATION WITH SEMICONDUCTOR DETECTORS

BY

D. N. POENARU

N. VÎLCOV

CHEMICAL PUBLISHING COMPANY, INC.

NEW YORK

1969

Measurement of Nuclear Radiation with Semiconductor Detectors

ISBN: 978-0-8206-0103-8

Chemical Publishing Company:
www.chemical-publishing.com
www.chemicalpublishing.net

First American Edition: Chemical Publishing New York 1969

Printed in the United States of America

Translated by

MAGDALENA GABRIELA SEZON

PREFACE

In the last years the semiconductor detectors (SD) have been more and more frequently used in nuclear physics due to their manifold advantages as compared to other types of detectors, the high energy resolution being the most important one. We dare to say that in a not too distant future such detectors will be used in most applications, as transistors have succeeded at the present time in replacing vacuum tubes.

The purpose of this book is to present existing achievements in this field as well as the physical basis necessary for understanding subsequent developments. It consists of seven chapters the first two being a brief reviewing of some general notions of nuclear physics and solid state physics required for a better understanding of the following chapters. Thus, Chapter I reviews the main physical processes encountered when various types of nuclear radiations penetrate through matter, a series of data concerning silicon and germanium, of particular interest for SD, being presented.

The elements of semiconductor physics required for the understanding of the working principles of semiconductor devices are the object of Chapter II. The elementary processes which take place under the action of nuclear radiations and give rise to electric charge pulses delivered by SD are analysed in Chapter III. Chapter IV deals with electrical and spectrometric characteristics, of interest equally for device constructors and research workers in the field. The description of a great number of semiconductor devices constructed up to the present and the illustration of their specific applications are given in Chapter V. The putting into evidence of the excellent spectrometric performances of SD cannot be made without a suitable electronic equipment. The working principles of such an equipment, its particularities and conclusions pertaining to the optimization of the spectrometer system are presented in Chapter VI. The principal types of vacuum tubes and transistor electronic circuits used in SD spectrometers are dealt with in the last chapter.

As compared to the Romanian edition, the English version has been completed by taking account of several papers recently published.

The book is intended to research workers, electronic engineers and post-graduate students interested in modern techniques for the detection of nuclear radiations either from the standpoint of the experimenter, or of those who manufacture electronic nuclear equipment.

The extraordinary advances which have been made recently in this field have raised difficult problems in the writing of this book, whose solutions may perhaps differ from the optimum ones.

The authors would like to express their thanks to Iliffe Books Ltd, Institute of Physics, the Institute of Electrical and Electronics Engineers, International Atomic Energy Agency, North-Holland Publishing Company, European Nuclear Energy Agency, as well as to G. Dearnaley, E. Gatti, F.S. Goulding, G.C. Huth, F.A. Tove and others, who lent us the permission to reproduce figures from their papers.

Thanks are also due to the Foreign Language Editorial Office of the Publishing House of the Romanian Academy for the help given in the publication of this book and in advance to all those who will be kind enough to send their critical observations and suggestions for the improvement of this book.

CONTENTS

CHAPTER I

INTERACTION OF THE NUCLEAR RADIATION WITH MATTER

CHAPTER II

FUNDAMENTAL PROCESSES IN SEMICONDUCTORS AND METALS

CHAPTER III

WORKING PRINCIPLES OF NUCLEAR RADIATION SEMICONDUCTOR DETECTORS

CHAPTER IV

CHARACTERISTICS OF SEMICONDUCTOR DETECTORS

CHAPTER V

SEMICONDUCTOR DETECTOR TYPES

CHAPTER VI

AMPLIFICATION OF SEMICONDUCTOR DETECTOR ELECTRIC PULSES

CHAPTER VII

SEMICONDUCTOR DETECTOR ASSOCIATED ELECTRONICS

MAIN NOTATIONS

a — lattice constant
a — acceleration
$\mathcal{A}$ — amplification factor
A — atomic mass
A_q — charge sensitivity
B — frequency band width
c — light velocity in vacuum (2.997925×10^8 m/s)
C — electric capacity
d — depletion layer thickness
D — diffusion constant
e — electron charge (1.60210×10^{-19} C)
$\in$ — mean energy necessary for an electron-hole pair production
E — total energy
E_a — energy of acceptor levels
E_b — energy of Bose level
E_c — the bottom of the conduction band
E_d — energy of donor levels
E_f — energy of Fermi level
E_g — energy gap
E_i — energy of the intrinsic level
E_v — the top of the valency band
$\vec{\mathcal{E}}$ — electric field intensity
f — Fano factor
f — frequency
$\vec{F}$ — force
g_m — mutual conductance
G — generation rate of charge carriers
$\hbar$ — Planck constant ($1.05450 \cdot 10^{-34}$ J.s)
$\vec{H}$ — magnetic field intensity
i — $\sqrt{-1}$
i — electric current intensity
$\mathcal{I}$ — intensity of a radiation beam
I — mean excitation potential of atoms
I_M — current pulse amplitude
$\vec{J}$ — electric current density
k — Boltzmann constant ($1.38054 \cdot 10^{-23}$ J/°K)
$\vec{k}$ — wave vector
$\vec{K}$ — reciprocal lattice vector

$\mathcal{K}$ — chemical potential
l — orbital quantum number
L — diffusion length
L_D — Debye length
L_n^l — associated Laguerre polynomial of degree n and order l
m — mass
m — magnetic quantum number
m_0 — rest mass of the free electron ($9.1091 \cdot 10^{-31}$ kg)
m^* — effective mass
M — mass of a heavy particle
n — principal quantum number
n — charge-carrier concentration
$\vec{n}$ — crystalline lattice vector
N — Avogadro's number ($6.00252 \cdot 10^{23}$ mol^{-1})
N — total particle number
N — bound charge concentration
N_p — ion pair number
$N(\omega)$ — noise spectrum density
p — pressure
p — complex frequency
$\vec{p}$ — momentum
P_l^m — associated Legendre polynomial of degree l and order m
Q — electric charge
$\vec{r}$ — position vector
R — particle range
R — electric resistance
$\mathcal{R}$ — volume recombination rate
R — radial wave function
s — spin quantum number
s — surface recombination rate
S — entropy
S — slope of a vacuum tube
t — time
t_r — time of a transient response
T — absolute temperature
u — electric voltage
U — potential energy
$\vec{v}$ — velocity
$\mathcal{V}$ — volume
V — electric potential
w — distribution function
W — probability of penetration through a potential barrier
x — distance
Y — angular wave function
z — number of quantum states
z — number of elementary charges of a particle
Z — electric charge of the nucleus (in e-units)

α — traggling parameter

β — $\frac{v}{c}$ ratio

β — reaction coefficient

δ — absolute density

δ — delta function (Dirac)

ε — relative dielectric constant

ζ — neutron or gamma ray absorption coefficient

η — charge collection efficiency

θ — polar angle

$\varkappa$ — wave number

λ — wave length

μ — electrochemical potential

μ — charge carrier mobility

μ — amplification factor of vacuum tubes

ν — number of particles in the energy interval unit

ν — number of defects yielded by crystal irradiation

ρ — electric resistivity

ρ — state (level) density

ρ_v — volume density of the electric charge

σ — electric conductivity

σ — cross-section of an atomic process

σ^2 — mean square deviation

Σ — area of a surface

τ — time constant

φ — azimuth angle

Φ — the volume in the phase space

χ — distribution function of energy losses (Landau)

Ψ — wave function

ω — angular frequency

CHAPTER I

INTERACTION OF THE NUCLEAR RADIATION WITH MATTER

The nuclear particles and the quanta of electromagnetic energy are not directly perceptible quantities. For recording the presence of nuclear particles and obtaining quantitative information about their energy and number the following two types of detectors are used:

- passive detector, storing information during exposure to radiations and requiring a subsequent processing of data (e.g. photographic emulsions),
- active detectors, radiation to electric signal converters (e.g. gaseous detectors, scintillation counters, or semiconductor detectors).

Indifferently of the manner in which the detector acts, the ionizing radiation must dissipate part of its energy within the detector sensitive region.

This energy, initially absorbed by a limited number of primary particles, is distributed afterwards inside an important volume of the detector and creates, at a macroscopical scale, noticeable events.

In the case of *SD**) the useful event generates the ionization while the collection of positive and negative ions by an electric field leads to the appearance of an electric signal which can be subsequently amplified and recorded. Nuclear radiations are of three types : charged particles (electrons, alpha particles, protons, a.s.o.), neutral particles (neutrons, neutrinos, a.s.o.) and electromagnetic radiations (gamma-rays and X-rays). Ionizations can be produced only by interaction processes of charged particles with matter; neutrons and gamma-rays must firstly produce charged particles enabling the detection. We begin by reviewing the main interaction processes between matter and different radiation types.

---

*) *SD* — semiconductor detector

1.1. INTERACTION OF HEAVY CHARGED PARTICLES WITH MATTER

"Heavy particles" are usually considered to be those having a mass much greater than the electron mass (mesons, protons, deuterons, alpha particles). During the passage of heavy charged particles through matter, different processes can occur [T3] such as:

— inelastic collisions with the atoms and molecules of the medium, in which the energy lost by the particle is used for exciting or ionizing the atoms and molecules,

— inelastic collisions with atomic nuclei leading to "bremsstrahlung" radiation emission, nuclei excitation or producing nuclear reactions,

— elastic collisions with the nucleus (or the atom as a whole),

— elastic collisions with free electrons,

— Vavilov-Cerenkov effect, i.e., light emission during the particle passage through matter with velocities exceeding the light wave group velocity.

The inelastic collisions with electrons [S26] and the elastic collisions with the nuclei (or the atoms) of the stopping material are usually preponderant in comparison with the other processes [S10], [3].

The relative weight of the collisions with electrons and atoms in attenuating the beam intensity and slowing down the heavy charged particles depends on their charge and velocity.

There are two limiting cases:

— incident particles with small charge ($z \simeq 1$) and high velocity ($v \gg cz\ (137)^{-1}$),

— incident particles with great charge ($z \gg 1$) and low velocity $\left(v < \dfrac{cz}{137}\right)$.

In the first case, the energy is lost at small rates, especially due to inelastic collisions with electrons. Because of its great mass a heavy particle knocking an electron is very slightly deflected from its initial direction.

Particles with the same energy, cover practically the same range in matter before being stopped. When the particle velocity decreases to a value comparable with that of the outer electrons in the material atoms ($v \sim cz(137)^{-1}$), their rôle in stopping the particles is diminished. Finally when the velocity decreases further, the energy loss due to inelastic collisions with electrons becomes negligible and the elastic collisions with atoms become preponderant.

New theories concerning the stopping power and some other details are given in the excellent paper of Bergstrom and Domeij [B15].

For particles with great charge and low velocity, the effective charge decreases abruptly with the velocity, but specific phenomena are observed on the final part of the path for particles with small charge.

1.1.1. *Stopping power of heavy charged particles.* The mean energy-loss per unit length of the path is called "stopping power": $\mathrm{d}E/\mathrm{d}x$.

Usually, instead of length one considers the surface unit mass, and introduces the "mass stopping power": $\frac{1}{\delta}\cdot\frac{\mathrm{d}E}{\mathrm{d}x}$ (δ — the density of the material).

The Bethe-Bloch formula [3] may be used for calculating the stopping power in MeV g^{-1} cm^2 units

$$-\frac{\mathrm{d}E}{\delta\cdot\mathrm{d}x}=\frac{z^2Z}{A}F(\beta)\left[f(\beta)-\ln I-\frac{\sum_i C_i}{Z}-\frac{\Delta\delta}{2}\right] \tag{1.1}$$

where:

δ = density of the material, ze = charge of the incident particle, Z = the atomic number of the material, A = the atomic weight of the material, β = ratio of incident particle velocity to light velocity (v/c), C_i = correction coefficient to the electron binding energy in the i-shell; $\Delta\delta$ = density correction at high energies; I = mean excitation potential of the material atoms (a constant of the material) which, as suggested by Bloch [B34] can be calculated from the relation $I = \alpha Z$, where [W22] $\alpha = 11.5$ eV.

$F(\beta)$ and $f(\beta)$ are given by the relationships:

$$F(\beta)=\frac{4\pi e^4 N}{m_0 c^2\beta^2}=\frac{4\pi m_0 c^2 r_0^2 N}{\beta^2}, \tag{1.2}$$

$$f(\beta)=\ln\frac{2m_0c^2\beta^2}{1-\beta^2}-\beta^2, \tag{1.3}$$

where

$$r_0^2 = 7.9403 \times 10^{-26} \text{ cm}^2$$

e = the electron charge, $N = 6.02252 \times 10^{23}$ = Avogadro's number, m_0 = rest mass of the electrons, $m_0c^2 = 0.510976$ MeV*).

The values of the coefficients $F(\beta)$ and $f(\beta)$ as functions of the proton energy E (MeV) are given in a table in [3].

They are valid for any charged particle with velocity $\beta = v/c$ and mass $M \gg m_0$.

From the Bethe-Bloch formula one can see that the mean energy loss by ionization in a given material depends only on the particle charge and velocity and not on its mass: $-\frac{\mathrm{d}E}{\mathrm{d}x} \sim \frac{z^2}{v^2}$. On an average the protons with energy E lose by ionization the same energy as the deuterons having the equivalent energy $M_d(M_p)^{-1}E$, or the tritons with energy $M_t(M_p)^{-1}E$. The alpha particles with energies $M_\alpha(M_p)^{-1}E$ passing through the same layer of material have an energy loss four times greater. For instance, the stopping power of alpha particles with 8 MeV energy is four times the stopping power of the protons with 2 MeV energy; the equivalences are valid for $E_\alpha > 2$ MeV.

It has been checked up experimentally that the energy losses by ionization do not depend on the incident particle masses, are proportional to their squared charges and increase abruptly when the velocity decreases (at low velocity, the time available for interaction is longer). Fig. 1.1 gives the stopping power dependence of the proton energy for silicon and germanium [D11]. For a rough estimation of the stopping power in different materials one can use the curves from [A 15] but for a more accurate calculation we suggest Sternheimer's works [S 28], [S 29], [S 30] and the more recent tables of Williamson *et al.* [W 21]

1.1.2. *Statistics of energy losses.* Due to the statistical fluctuations of the collision number and to the energy loss fluctuations in different collisions, a monoenergetic particle beam, having passed through a homogeneous layer of material, contains an energy spectrum.

The function describing the energy loss dispersion generally depends on the particle type, its energy and the thickness of the layer. The heavy charged particles penetrating a material layer of thickness x_0

*) The elementary particle energy is measured in electron-volt units (symbol — eV) and its multiples 1 eV $= 1.602 \times 10^{-19}$ J; 1 MeV $= 10^6$ eV.

have an energy dispersion about the mean value, that can be fitted by a Gaussian curve [F 9] with a full width at half-maximum (*FWHM*) Γ

$$(\Delta E_{FWHM})^2 = \Gamma^2 = P x_0. \tag{1.4}$$

According to Bohr, the quantity P is not a function of the particle velocity and can be obtained from the relation

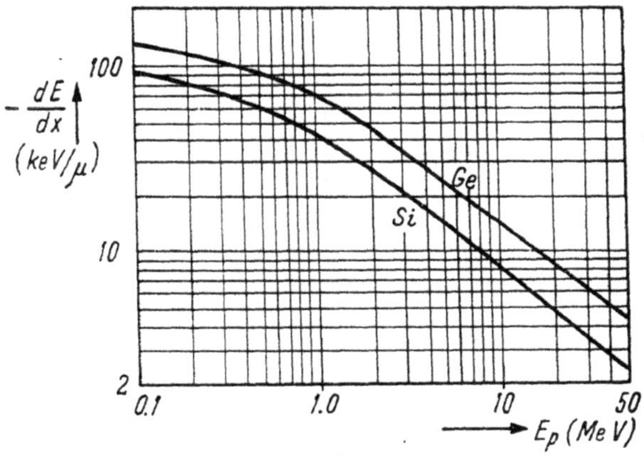

Fig. 1.1 — Proton stopping power in Si and Ge [D 11].

$$P = 4\pi z^2 e^4 n = 4 \pi z^2 e^4 NZ, \tag{1.5}$$

where n is the number of electrons and N the number of atoms per 1 cm^3 of material.

A more rigorous formula is given by Livingston and Bethe

$$P = 4 \pi z^2 e^4 n B, \tag{1.6}$$

where B takes into account the electron binding energies in atoms [S 26] (it differs from unity, especially at low velocities of the particles).

In the range of very high particle energies, the energy loss fluctuations are very different from those given by the Gaussian distribution, the energy dispersion is asymmetrical relative to the mean value and the most probable energy loss has a maximum which does not coincide with the mean value given by the Bethe-Bloch formula. Fig. 1.2. shows an example of the probability variation regarding the energy loss of 730 MeV protons passing through a silicon layer of 0.48 g/cm^2 thickness [M 1]. The theoretical study of this problem was carried out by Williams, Landau, Vavilov, etc. [S 26], [M 1].

According to Landau a particle, when passing through a material layer, has the probability of losing an energy within ΔE and $\Delta E + \mathrm{d}\,\Delta E$ equal to

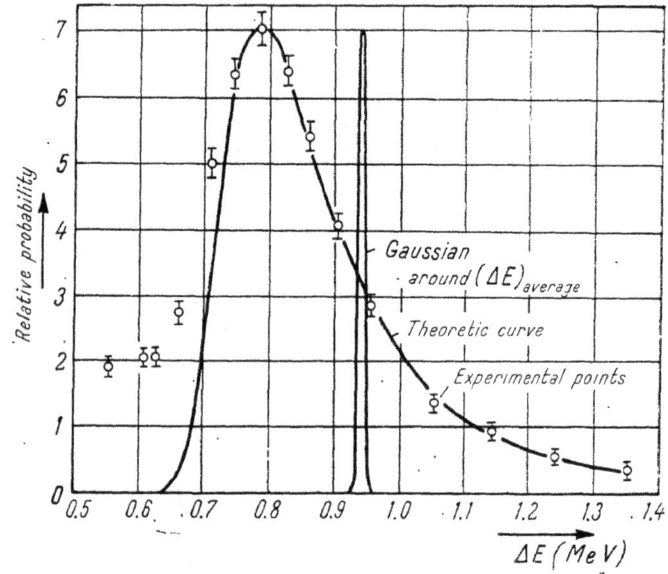

Fig. 1.2 — Fluctuations of 730 MeV proton energy losses in silicon of 0.48 g/cm² thickness [M 1].

$$f(\Delta E,\ x_0) = \chi\left(\frac{\Delta E - \Delta E_0}{E_\xi}\right)\cdot \mathrm{d}\left(\frac{\Delta E - \Delta E_0}{E_\xi}\right), \tag{1.7}$$

where

$$E_\xi = \frac{2\,\pi\, z^2\, e^4}{m\, v^2}\, n\, x_0 = \frac{1.53 \times 10^5 \times \delta\,(\mathrm{g/cm^3}) \times x_0\,(\mathrm{g/cm^2})}{\beta\, A z^{-2} Z^{-1}} \tag{1.8}$$

$E_\xi =$ the energy loss in a collision occurring only once on the path x_0, ΔE_0 — represents the most probable energy loss (different from the mean energy loss) and is given by the expression

$$\Delta E_0 = E_\xi\left[\ln \frac{2m_0 v^2 E_\xi}{I^2(1-\beta^2)} - \beta^2 + 0.373\right]. \tag{1.9}$$

Unlike the mean energy loss, ΔE_0 is not a linear function of the absorber thickness. The function giving the energy loss distribution inside thin absorbers is plotted in Fig. 1.3.

It has a $FWHM = 3.98\,E$ and is normalized so that the surface under the curve should be equal to unity. The results are valid, if the energy loss ΔE, when penetrating the material, is small in comparison with the initial energy of the particle and if $E_\xi > 1$. In the case of low energies, the energy loss fluctuations (straggling) have a Gaussian distribution.

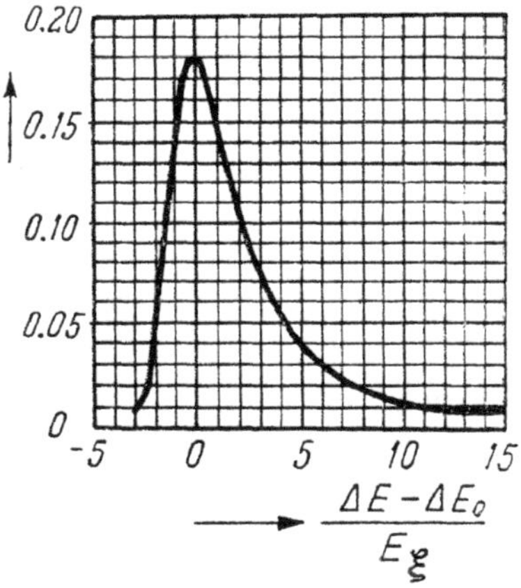

Fig. 1.3 — Distribution function of energy losses in thin absorbers [L 3].

1.1.3. *The range-energy relations.* That part of the path where the energy of the particle with mass M and charge ze decreases from E_1 to E_2 can be calculated with the Bethe-Bloch formula

$$\Delta R = \int_{E_2}^{E_1} \frac{\mathrm{d}E}{-\left(\frac{\mathrm{d}E}{\mathrm{d}x}\right)} = \frac{M}{z^2} \cdot \frac{m_0}{4\pi e^4 n} F(v), \qquad (1.10)$$

where $F(v)$ is a function depending on the particle velocity at the beginning and at the end of the path.

Because the Bethe-Bloch formula is not adequate at low energies of the particles, the integration limit E_2 cannot be $E_2 = 0$. The mean range of a particle with energy E is usually calculated from

$$R(E) = R(E_0) + \int_{E_0}^{E} \frac{\mathrm{d}E}{-\frac{1}{\delta} \cdot \frac{\mathrm{d}E}{\mathrm{d}x}} \qquad (1.11)$$

where $R(E_0)$ is the experimental range value at a particle energy E_0 when the electron screening of the particle charge has a negligible influence. One may take $E_0 \simeq 1.5$ MeV for protons and $E_0 \simeq 5$ MeV for alpha particles.

The range value R_{zM} of a particle with mass M and charge ze can be expressed as a function of the proton range in the same material

$$R_{zM}(E) = \frac{1}{z^2} \cdot \frac{M}{M_p} R_p\left(\frac{M_p}{M} E\right) \tag{1.12}$$

where $R_p\left(\frac{M_p}{M} E\right)$ is the proton range at equivalent energy $M_p (M)^{-1} \cdot E$.

For instance, the ranges of the hydrogen and helium isotopes are related by the relationships:

$$R_d(E) \simeq 2 R_p\left(\frac{E}{2}\right), \tag{1.13}$$

$$R_t(E) \simeq 3 R_p\left(\frac{E}{3}\right), \tag{1.14}$$

$$R_{He^3}(E) \simeq \frac{3}{4} R_\alpha\left(\frac{4}{3} E\right). \tag{1.15}$$

For mesons

$$R_\mu(E) = 0.114\, R_p\,(8.8\, E), \tag{1.16}$$

$$R_\pi(E) = 0.149\, R_p\,(6.7\, E), \tag{1.17}$$

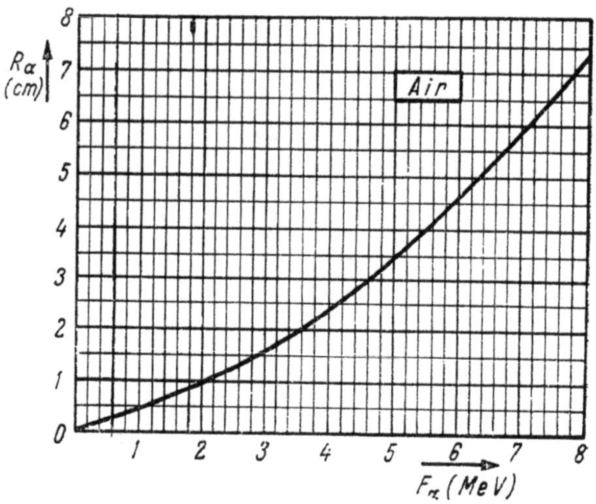

Fig. 1.4 – Alpha-particle range in air [B 18].

$$R_\pi(E) = 1.31\, R_\mu\left(\frac{E}{1.31}\right). \qquad (1.18)$$

For particles with different charges we must take into account the phenomena occurring at low velocities

$$R_p(E) = 1.007 R_\alpha(3.972\, E) - C_{p\alpha}. \qquad (1.19)$$

For air

$$C_{p\alpha} = \begin{cases} -\ 0.2 \text{ cm} \quad \text{when} & E > 0.5 \text{ MeV}, \\ -\ 0.13 \text{ cm} & E = 0.1 \text{ MeV}, \\ -\ 0.02 \text{ cm} & E = 6.7 \text{ KeV}. \end{cases}$$

Fig. 1.4 gives the range — energy relation for alpha particles in air [B 18].

In detecting the nuclear particles with semiconductor detectors we are interested in the range — energy relations of different particles in silicon and germanium. These relations are plotted in Figs. 1.5 and 1.6 respectively [G 24]; Fig. 1.7 shows the energy — range relation for fission fragments in silicon [G 24].

The difference between the path length and the range is due to multiple scatterings; the Coulombian interaction of the fast charged particles with the atomic nuclei of the material scatters the particles that will go a path with detours, longer than the foil thickness.

In addition, the statistical fluctuations of the energy losses by ionization lead to a statistical distribution of heavy charged particle ranges.

The range dispersion is characterized by the mean square values of the fluctuations $(X-R)$ about the mean value R from relations (1.10)—(1.19).

The variance is

$$\sigma_R^2 = \langle (X - R)^2 \rangle. \qquad (1.20)$$

If the range fluctuation is described by the Gaussian distribution

$$P(X)\mathrm{d}X = \frac{1}{\alpha\sqrt{\pi}} \cdot \exp\left[-\frac{(X-R)^2}{\alpha^2}\right] \cdot \mathrm{d}X \qquad (1.21)$$

the integral and differential transmission curves of the charged particles have the form shown in Fig. 1.8. The range extrapolated from the curve is

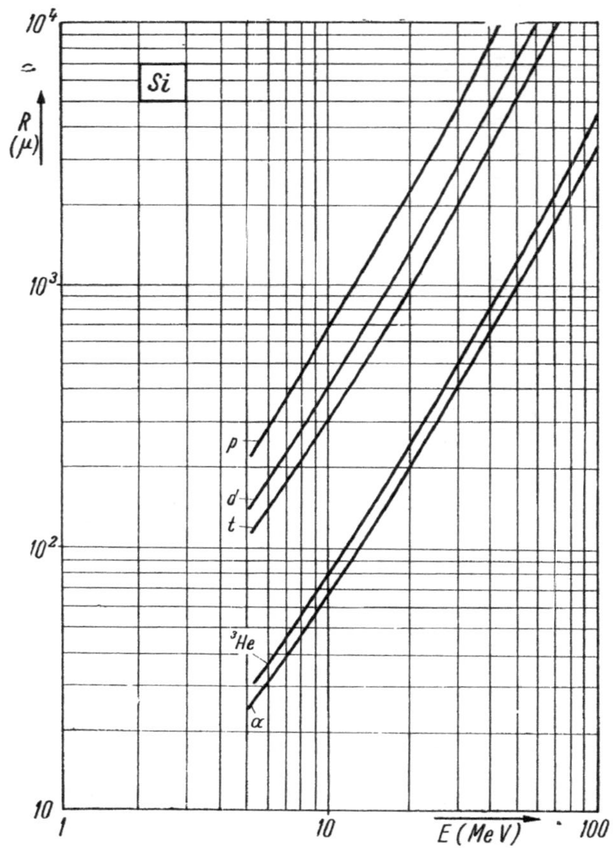

Fig. 1.5 — Heavy charged particle range — energy curve for silicon [G 24].

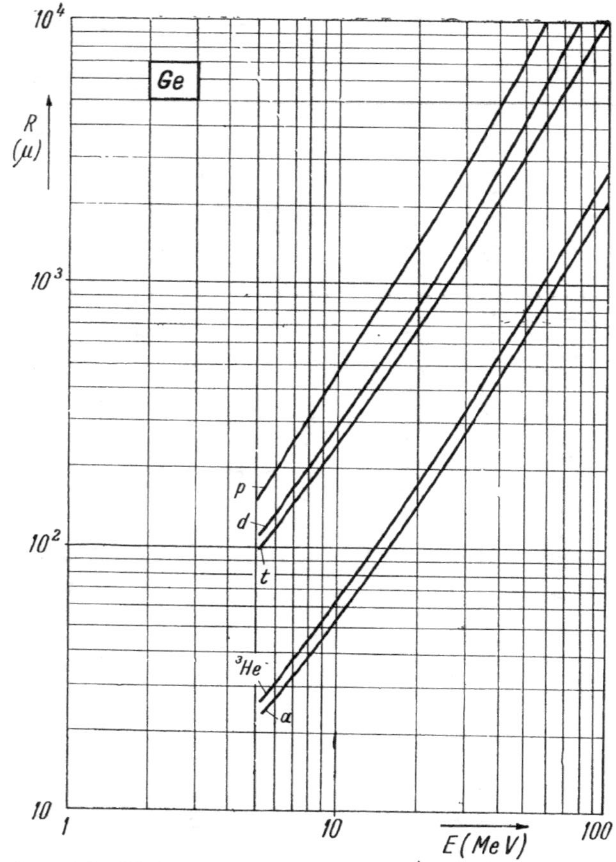

Fig. 1.6—Heavy charged particle range — energy curve for germanium [G 24].

$$R_n = R + \frac{\sqrt{\pi}}{2} \cdot \alpha = R + S \,, \tag{1.22}$$

where

$$S = \frac{\sqrt{\pi}}{2} \cdot \alpha = \sqrt{\frac{\pi}{2}} \cdot \sigma_R \,, \tag{1.23}$$

where α is the "straggling" parameter linked to the dispersion σ_E by the relation

$$\alpha = \sqrt{2}\, \sigma_R$$

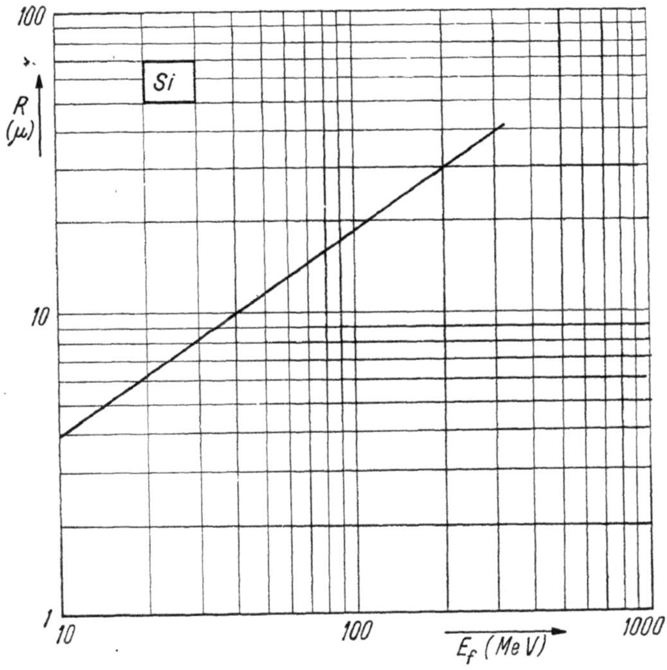

Fig. 1.7—Fission fragment range—energy curve for silicon [G 24].

The alpha particles of Po^{210} ($E_\alpha = 5.3$ MeV) have in air $R = 3.842$ cm and $R_n = 3.879$ cm.

Empirical expressions can be used to calculate the alpha-particle mean range in silicon and germanium [G 10]

$$R_{\alpha\,\mathrm{Si}}[\mu] = 2.13\ E^{1.45}\ \mathrm{MeV} + 2.2, \tag{1.24}$$

$$R_{\alpha\,\mathrm{Ge}}[\mu] = 1.95\ E_{\alpha}^{1.34}\ \mathrm{MeV} + 1.72. \tag{1.25}$$

These relations give an accuracy better than 10%, for alpha particle energies ranging within 1—5 MeV.

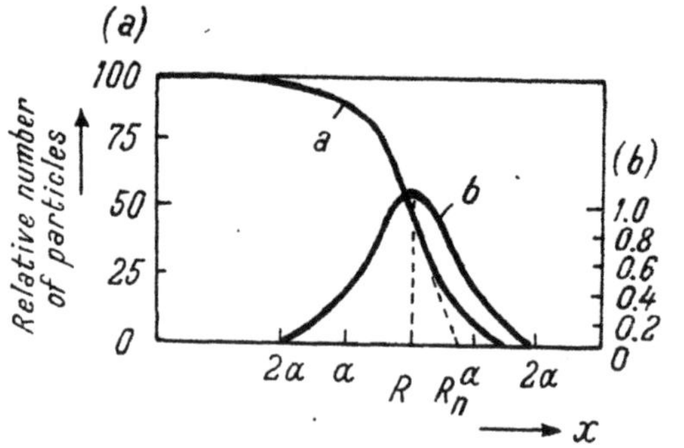

Fig. 1.8 — Integral (*a*) and differential (*b*) transmission curve of charged particles.

1.1.4. *The relation energy-loss – ionization.* Unlike $\mathrm{d}E/\mathrm{d}x$, the value of the specific ionization $\mathrm{d}N_p/\mathrm{d}x$ (the ratio of the number of ion pairs to unit length of the ionizing particle path) cannot be inferred theoretically [S 26]. The existing relation of proportionality between $\mathrm{d}N_p/\mathrm{d}x$ and $\mathrm{d}E/\mathrm{d}x$ has been established experimentally for a large energy range of the incident particle.

The primary ionization is given by the number of ion pairs formed directly as a result of the particle collision with the material atoms. The electrons yielded in this way (or the recoiled atoms in the case of elastic collisions) can have sufficient energy for a new ionization of the material atoms. The maximum energy transferred to the electron by the heavy particle is $4\frac{m_0}{M}E$ and the maximum velocity $2v$, where M, E, v are the mass, energy and velocity of the incident particle. Thus an alpha particle of 5 MeV can give to a single electron a maximum energy of 3 keV. The electrons due to ionization have on average an energy of 200 eV. This energy is sufficient to produce a secondary ionization representing an important part of the total ionization.

A secondary electron having enough energy, and being yielded by an incident particle is known as delta radiation. A large fraction out of the incident particle energy used for ionizing the penetrated material appears under the form of delta radiations (this fraction amounts to 50%).

A very important experimental fact is the nearly total independence of the energy ϵ lost for the formation of an ion pair of the initial energy of the particle which produces the ionization. This is of a particular importance for gaseous and solid state detector in spectroscopic measurements and in stating the particle range-energy relationship. For air e.g. $\epsilon_{air} = 34.7$ eV

$$\epsilon_{air} = 34.7 \text{ eV}. \tag{1.26}$$

In the case of recoil atoms and fission fragments when the velocity decreases ϵ becomes dependent on the ionization efficiency. A "ionization defect" is introduced

$$\Delta E = E - \epsilon' \cdot N_p , \tag{1.27}$$

with E = energy of the incident particle, N_p = number of ion pairs produced, $\epsilon' = \dfrac{\mathrm{d}E}{\mathrm{d}N_p}$ = the particle energy loss for an ion pair formation.

1.1.5. *Fano fluctuations.* The energy resolution of the detectors used in spectrometry and the accuracy of particle energy measurements after the ionization produced by them depend on the fluctuations of the number of ion pairs yielded by particles with the same initial energy. This problem has been theoretically investigated by Fano in 1947 [F 9]. According to his calculations, the mean square deviation of the number of ion pairs created by charged particles with identical energies is :

$$\sigma_N^2 = \langle (N_p - \overline{N}_p)^2 \rangle = f N_p , \tag{1.28}$$

where $\overline{N}_p$ = the average number of ion pairs and f = Fano factor.

The theory can establish only the upper limit of f as $f < \dfrac{3}{4}$ for gases. For semiconductor detectors the calculations [R 8] [A 5] and the experiments [A 11], [M 10] give a value $f < 0.15$.

1.1.6. *Identification of fast charged particles by* $E \cdot \left(\frac{\mathrm{d}E}{\mathrm{d}x}\right)$ *measurements.* The product $E \cdot \left(\frac{\mathrm{d}E}{\mathrm{d}x}\right)$ is proportional to z^2M for charged unrelativistic particles when neglecting the influence of the logarithmic term in the expression of the mean energy loss. The simultaneous measuring of the energy E and of $\frac{\mathrm{d}E}{\mathrm{d}x}$ permits us, after performing the product $E \cdot \frac{\mathrm{d}E}{\mathrm{d}x}$, to separate the particles, according to their mass, within a sufficiently large energy range. For instance, for particle energies ranging within 5 — 25 MeV, the product $E \cdot \frac{\mathrm{d}E}{\mathrm{d}x}$ in several materials is proportional to $1:1.8:2.6$ for protons, deuterons and tritons, respectively.

This method is used to study nuclear reaction products.

1.2. PASSAGE OF ELECTRONS THROUGH MATTER

As in the case of heavy charged particles, the electron passage through matter is accompanied by elastic and inelastic collisions with the nuclei and atoms of the stopping material. Yet unlike the case of heavy particles, for electrons, these collisions lead to comparatively large energy losses and significant changes of the moving electron direction. Consequently, the electron beam intensity in heavy material decreases almost exponentially at the increase of the absorber thickness. For the same reason the electron path in the material is generally a complicated zigzagged curve and the range of electrons with the same energy has a large dispersion. For an electron energy $E_\beta > 0.5$ MeV another difference lies in the increasing rôle of the losses through radiation due to "bremsstrahlung" emission occurring at the electron slowing-down in the nuclear field of the stopping material. At electron energies lower than 1 MeV these losses can be neglected in most practical cases. The principal energy loss mechanism in this range is the inelastic scattering on the bound electrons of the material, leading to ionization and excitation of atoms.

For electrons and protons, having the same velocity, the energy losses by ionization per unit path length are almost identical.

Fig. 1.9 gives the electron stopping power in silicon [D 11]. The stopping power passes through a flat minimum at about 1 MeV. After all, this behaviour is specific for all charged particles and is due to the relativistic contraction of the particle Coulombian field along its direction of motion. The protons have their domain of "minimum ionization" at $\sim$ 1300 MeV and the μ mesons at $\sim$ 200 MeV.

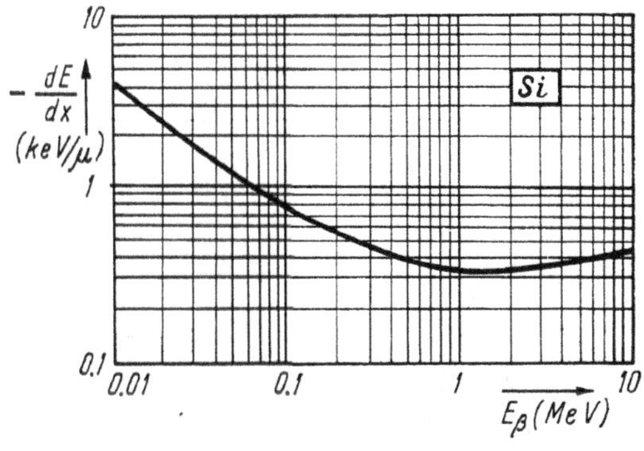

Fig. 1.9—Electron stopping power in silicon [D 11].

The form of the monoenergetic electron beam absorption curve depends on the atomic number Z of the absorber and the geometry of detection [C 12].

The range of the electron beam with continuous spectrum, agrees with that of a monoenergetic beam, the energy of which is equal to the upper limit of the spectrum. Generally, the absorber thickness, for a 50% attenuation of the beam intensity, is linked to the maximum range R_{Max} [S 10] through the relation

$$X_{1/2} \simeq \frac{1}{7} R_{Max}. \quad (1.29)$$

Fig. 1.10 is a plot of the range–energy curves for electrons in silicon and germanium [S 24].

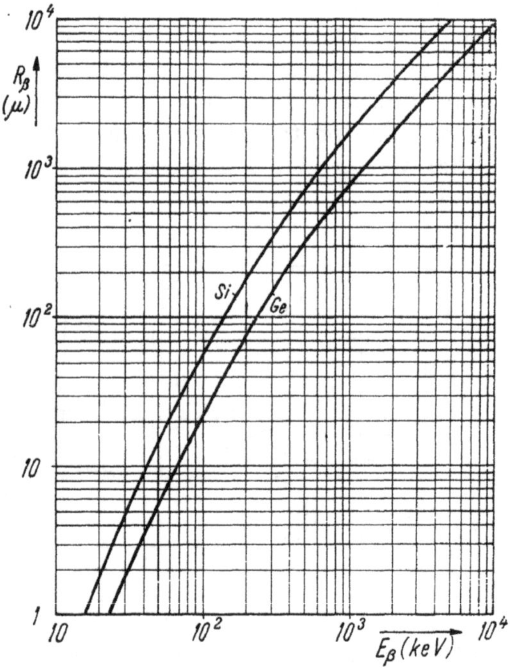

Fig. 1.10 — Electron ranges in silicon and germanium [G 24].

1.3. INTERACTION PROCESSES OF GAMMA AND X-RAYS

The electromagnetic radiations penetrating a material are able to produce the following types of interaction :

— Compton effect,
— photoelectric effect,
— formation of electron-positron pairs,
— photonuclear reactions (much less probable).

1.3.1. *The Compton effect.* As illustrated in Fig. 1.11, the Compton effect consists in an interaction during which the electron with energy E is ejected at an angle φ as against the direction of the incident γ-quantum with energy E_γ while a quantum with lower energy E'_γ is scattered under an angle θ.

Through this effect, the corpuscular nature of the electromagnetic radiation is made obvious. On considering the energy and momentum conservation in the collision photon-electron, the energy of the resulted electron is [S 10]

$$E = E_\gamma - E'_\gamma, \tag{1.30}$$

where

$$E'_\gamma = \frac{m_0c^2E_\gamma}{m_0c^2 + E_\gamma(1 - \cos\theta)}, \tag{1.31}$$

m_0 = the rest mass of the electron and c = the light velocity.

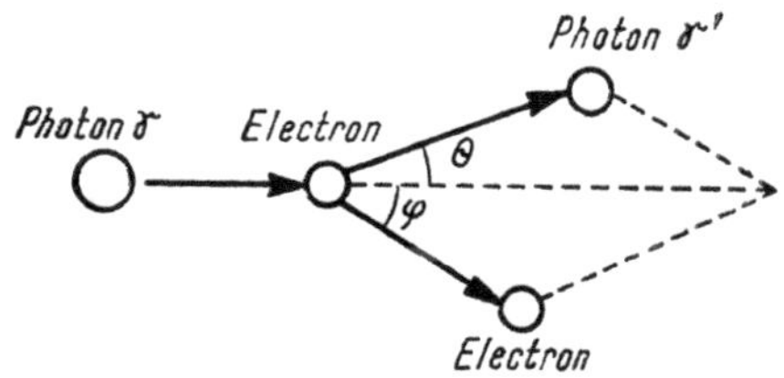

Fig. 1.11 — Compton diffusion.

Replacing $m_0c^2 \simeq 0.5$ MeV and expressing the energy in MeV, it results

$$E'_\gamma \simeq \frac{E_\gamma}{1 + 2E_\gamma(1 - \cos\theta)}. \tag{1.32}$$

The electrons of maximum energy are ejected forward and have the energy

$$E_{Max} \simeq \frac{4E_\gamma^2}{1 + 4E_\gamma}. \tag{1.33}$$

The cross-section of the Compton scattering behaves as $ZE_\gamma^{-1/2}$, Z being the atomic number of the scattering material. After a first scattering, the secondary quantum can undergo a new scattering and accordingly, in the gamma-ray absorption multiple scatterings must be taken into account.

1.3.2. *The photoelectric effect.* The hard X-rays and gamma-rays can be captured in an electronic shell of the atom with subsequent emission of an electron of energy $E_\gamma - E_p$ where E_p is the binding energy of the electron in the p shell (usually K).

The atom, left in an excited state, comes back to its ground state emitting a new gamma radiation which can be also captured through photoelectric effect by the external shells and releasing an electron of energy $E_\gamma - E_p - E'_p$, a.s.o.

The total energy of the emitted electrons is $\sim E_\gamma$. The probability of the photoelectric effect is even higher when the photon energy is closer to the binding energy of the ejected electron.

The photoelectric absorption probability varies approximately as $Z^5 . E_\gamma^{-7/2}$ and for $E_\gamma \leqslant 100$ keV the photoelectron emission is prevailing.

1.3.3. *The electron-positron pair production.* The electron-positron pair creation arises from the interaction between the nuclear field and a gamma photon with an energy higher than 1.02 MeV. An electron and a positron with a total kinetic energy $E_\gamma - 1.02$ MeV are created, the particles being generally emitted in the forward direction. When the positron slows down, it is annihilated by an electron. At the annihilation of the particle-antiparticle pair two gamma photons appear, each of them with an energy $E'_\gamma = 0.511$ MeV.

The probability of pair production behaves [S 10] as $(E_\gamma - 1.02)^{2.1} \times Z^2$.

1.3.4. *The energy distribution of the electrons arising from interaction.* Considering a monoenergetic photon beam and a material consisting of a single type of atoms, the following energy distribution for electrons arising from interaction results:

— photoelectrons with energy E_γ (giving a spectrum line);

— the Compton electrons represent a continuous spectrum with energies within 0 and E_{Max} having a characteristic form (the yield of electrons with low and high energies is greater than that of electrons with mean energies);

— the electron-positron pairs have a continuous energy spectrum within 0 and $E_\gamma - 1.02$ MeV.

As an example, the distribution of the electrons yielded in the interaction of the gamma radiation of 2 MeV with water [C 12] is shown in Fig. 1.12.

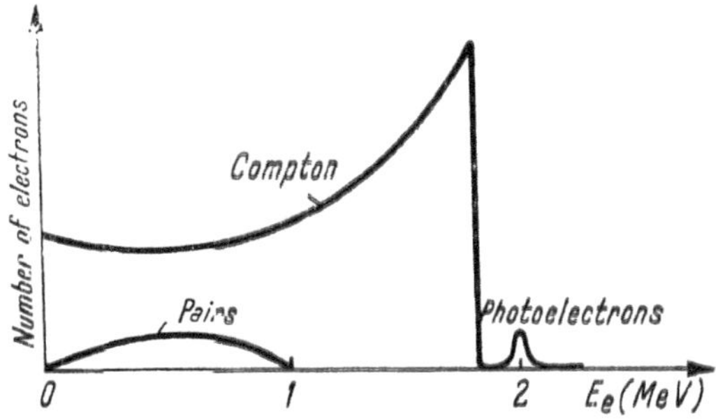

Fig. 1.12 — Energy distribution of electrons yielded in the interaction of 2 MeV gamma-rays with water [C 12].

1.3.5. *Attenuation of photon beams.* The probability of a gamma-ray being captured by an atom with atomic number Z may be expressed in terms of the cross-section for a certain type of interaction. The cross-section of Compton scattering is:

$$\sigma_c \simeq 0.2 \times Z E^{-1/2} \times 10^{-24} \text{ cm}^2 . \qquad (1.34)$$

In practice, as cross-section units the barn 1 bn $= 10^{-24}$ cm^2 and its subunit, the milibarn 1 mbn $= 10^{-3}$ bn are used. The cross-section of the photoelectric absorption is

$$\sigma_{ph} \simeq 3 \cdot 10^{-9} Z^5 E^{-\frac{7}{2}} \text{ bn}, \qquad (1.35)$$

and for the pair formation

$$\sigma_p = 6.2 \cdot 10^{-5} (E - 1.02)^{2.1} Z^2 \text{ bn} \qquad (1.36)$$

A gamma-ray beam having an intensity $\mathcal{J}_0$ before impinging on a material layer of thickness x, will have, when going out, an intensity given by an exponential law

$$\mathcal{J} = \mathcal{J}_0 \cdot \exp(-\zeta x), \qquad (1.37)$$

where

$$\zeta \simeq N(\sigma_c + \sigma_{ph} + \sigma_p), \quad [\text{cm}^{-1}] \qquad (1.38)$$

ζ = the absorption coefficient (the inverse of the thickness for which the beam intensity reduces to $(2{,}718)^{-1}$) while N is the number of atoms per cm^3 in the same material.

In Fig. 1.13 the variation of the gamma-ray absorption coefficient components in silicon and germanium [E 12] is given. It is

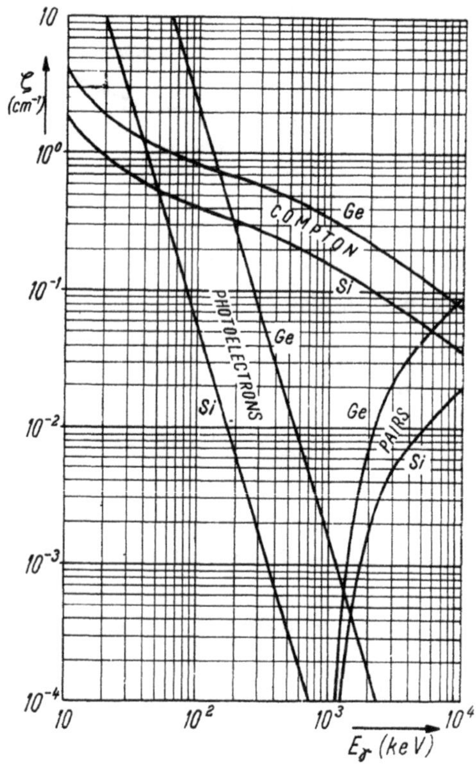

Fig. 1.13 — Gamma-ray absorption coefficient for photoelectric effect, Compton diffusion and pair production in silicon and germanium [E 12].

obvious that, in the interesting region for the nuclear spectroscopy of charged particles (from 200 KeV to 3 MeV), the Compton effect is predominant.

1.4. INTERACTION PROCESSES OF NEUTRONS

In contrast with the electromagnetic radiations interacting particularly with the electron shells of the atoms, the neutrons have a high probability of interaction with the atomic nuclei.

The nuclear radiations produced by neutrons are of several types :

— elastic and inelastic scattering ; the incident neutron is scattered by the nucleus, to which it transfers or not an excitation energy, the first case representing the inelastic scattering and the second the elastic scattering ;

— nuclear reactions. The neutron knocked nucleus emits other neutrons or charged particles : protons (n, p reactions), deuterons (n, d reactions), alpha particles (n, α reactions), combinations of them (n, np reactions) a.s.o., or goes to an excited state from which it comes back with gamma-quanta emission (n, γ reactions). A particular case of nuclear reaction is the nucleus fission into two fragments of about equal masses.

At low neutron velocities, the (n, γ) reactions and sometimes the fission are predominant.

The scattering and the nuclear reactions become dominant for fast neutrons with energies $E_n > 100$ keV.

The simplest process of interaction is the neutron scattering by protons. The energy of the recoiled proton being a function of the emission angle φ, the proton may have any energy within 0 and E_n

$$E_p = E_n . \cos^2 \varphi. \tag{1.39}$$

The cross-section of this process is great, it changes slightly with the energy E_n and is well known within a wide range of neutron energies, but there is a disadvantage for the nuclear spectroscopy : if the angle φ is not clearly delimited, there is no univocal relation between E_n and E_p. The angle delimitation reduces much of the detection efficiency.

For the neutron detection, the following exothermic reactions, yielding charged particles, are also important

$$^{6}\mathrm{Li} \ (n, \alpha) \ ^{3}\mathrm{H}, \ Q = 4.78 \ \mathrm{MeV}$$
$$^{3}\mathrm{He} \ (n, p) \ ^{3}\mathrm{H}, \ Q = 0.76 \ \mathrm{MeV}$$
$$^{10}\mathrm{B} \ (n, \alpha) \ ^{7}\mathrm{Li}, \ Q = 2.80 \ \mathrm{MeV}$$

Light nuclei are preferred due to their lower Coulombian barriers for charged particles emission.

In contrast with the charged particles, the neutron (as well as gamma-rays) is extracted from the beam by a single process of interaction, hence the neutron beam intensity decreases exponentially :

$$\mathcal{J} = \mathcal{J}_0 \cdot \exp\ (-\zeta x),$$

where $\zeta = N\sigma$ is the absorption coefficient ; N = the number of the absorber nuclei per cm^3 ; σ = the cross-section. Fig. 1.14 gives the cross-section for some important processes in the neutron detection.

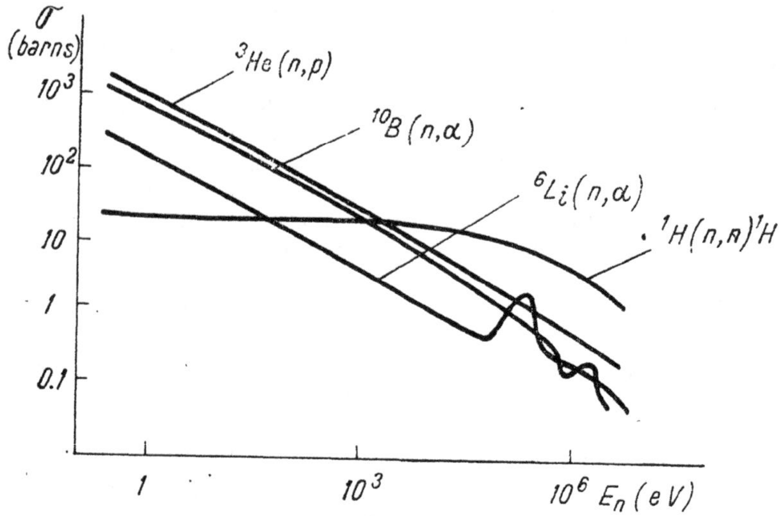

Fig. 1.14 — Reaction cross-section vs energy for different nuclear reactions induced by neutrons [D 3].

Good results have been obtained for ^{3}He in the fast neutron detection with semiconductor detectors because the cross-section of the process, in this case, has no resonances. We get also a constant efficiency of detection at neutron energies up to 10 MeV (the efficiency being higher than for the other reactions).

1.5. CONCLUSIONS

The ionization is the main interaction process of the nuclear radiations with matter. The matter ionization is direct in the heavy charged particle case and indirect for gamma-rays and neutrons.

The gamma-rays release photoelectrons, Compton electrons and positron-electron pairs able to induce the ionization of the material traversed by gamma-rays.

As regards the fast neutrons, the recoiled protons, ejected as a result of elastic collisions, produce the matter ionization and the slow neutrons give rise to gamma-rays and charged particles through the (n, α) and (n, p) reactions.

In solids, the yield of free charged carriers by ionizing particle interaction with outer electrons is a process similar to the gas ionization. Yet in this respect there are two essential differences between gases and solids :

— in gases free electrons and positive ions are released, while in solids electron-hole pairs are created ;
— the energy required for the creation of a single electron-hole pair in solids is lower than the energy needed for a single electron-pair in gases.

This leads to smaller statistical fluctuations in the number of the created carriers and consequently to more accurate estimates of the incident particle energy.

For example, the mean energy required for an electron-hole pair production in silicon is 3.5 ± 0.07 eV and in germanium 2.94 ± 0.15 eV in comparison with 30 eV in gases.

Estimations made for gamma-rays, electrons, protons, alpha particles, heavy ions and fission fragments have shown that ϵ does not depend on the particle energy and type. The values agree with those calculated by Shockley.

A particle with 5 MeV energy, stopped in silicon, yields 1.43×10^6 electron-hole pairs with a statistical fluctuation (in the case of a Gaussian distribution) of $\pm 1.2 \times 10^3$ namely 0.08%. For an ionization chamber the statistical fluctuation is in this case 0.23% and for scintillation detectors 0.8%.

Though in practice both the detectors and the associated apparatus noise as well as the inefficient charge collection prevent the achievement of theoretical resolution, the practical performances of the semiconductor detectors are higher than of any other device assuming ionization.

There are two types of radiation beam attenuation : exponential intensity attenuation, characterized by the attenuation coefficient, and energy losses with range, specified by the respective range. The exponential attenuation relies on monoact mechanism (the particles are removed out of the beam after a single act of interaction) while the attenuation with range on a continuous energy loss mechanism at small rates. An intermediate case is the electron beam attenuation.

The form of the attenuation curve may be specified by the range R to the half thickness $X_{1/2}$ ratio. An interaction is considered to be strong when it is characterized by a great attenuation coefficient (small $X_{1/2}$), in the case of exponential attenuation, or by a great energy loss (a short range) in the case of energy losses.

Otherwise the interaction is weak. Examples are given in table 1.1. [C 12]. In solids the range of these radiations is much smaller (see Figs. 1.5, 1.6 and 1.7) because several interaction processes occur in the same volume.

Table 1.1

Nuclear radiation penetration in air [C 12]

Radiation type	Maximum energy (MeV)	Range in air (cm)	$X_{1/2}$ (cm)
α	5	3.48	—
α	1	0.57	—
β	5	1900	310
β	1	370	30
neutrons	5	∞	8700
neutrons	1	∞	6300
γ	5	∞	20000
γ	1	∞	8500

Also from this point of view, the solid state detectors offer the advantage of stopping the same energy on a smaller volume than in gases.

CHAPTER II

FUNDAMENTAL PROCESSES IN SEMICONDUCTORS AND METALS

The semiconductor and metal properties are studied by the solid-state physics. The great majority of the processes occurring in metals and semiconductors, is due to the quantum nature of the phenomena taking place on atomic or molecular scale.

Consequently we begin by presenting the elementary phenomena on atomic scale. In view of the specific character of the present book, only relevant phenomena for the understanding of fundamental processes in metals and semiconductors are discussed.

2.1. SCHRÖDINGER EQUATION. THE PARTICLE INSIDE THE POTENTIAL WELL

The problem of the particle motion in classical mechanics reduces finally to the solving of Newton's equation

$$\vec{F} = m\,\vec{a} \tag{2.1}$$

expressing the proportionality between the acceleration $\vec{a}$ transmitted to the particle of mass "m", and the applied force $\vec{F}$. By knowing both the force field acting upon the particle and the initial conditions of the motion (position, velocity), by solving the Eq. (2.1), the particle trajectory and its velocity at each point are obtained

$$\vec{r} = \vec{r}\,(t)\,;\ \vec{v} = \dot{\vec{r}}\,(t). \tag{2.2}$$

Hence it is found out that the dynamical variables (position, velocity) characterizing the particle state, are uniquely related to the force field inducing the motion. In the same force field, for identical initial conditions, the particle will always travel the same trajectory and with the same velocity.

The description of the elementary particle motion requires a radical change of our point of view. The particle state can no longer be specified by its position and velocity, since the position and the velocity of a "quantum" particle cannot be measured at the same time with however high accuracy.

The accuracies in finding the position and velocity of a quantal particle condition each other according to the Heisenberg relation [S.3]

$$\Delta x \, \Delta p_x \geqslant \hbar, \tag{2.3}$$

where $\hbar$ is the Planck constant. The better is the accuracy in measuring the coordinate, the wider is the range within which varies the momentum (velocity) value.

Inequality (2.3) expresses a specific quantal property irrespective of the quality of the apparatus used for measuring the particle coordinate and momentum (velocity).

For characterizing the quantal particle state in the wave mechanics, the concept of wave function is introduced [L 4]. The wave function is the solution of Schrödinger's equation

$$-\frac{\hbar^2}{2m}\left(\frac{\partial^2\psi}{\partial x^2}+\frac{\partial^2\psi}{\partial y^2}+\frac{\partial^2\psi}{\partial z^2}\right)+U(x, y, z)\psi = E\psi, \tag{2.4}$$

where $U(x, y, z)$ represents the potential energy of the force field in which the particle of mass m moves, and E is the total energy of the particle in a state described by the wave function ψ. The product $\psi^*(x, y, z)\,\psi(x, y, z)$, where ψ^* is the complex conjugate of the wave function represents the probability of a particle being in a cube with sides equal to unity and with center coordinates (x, y, z) (in wave mechanics, wave functions are introduced, dependent on other dynamical variables — momentum, angular momentum, energy, a.s.o., and used in the estimation of the probability of a dynamical variable having a certain value in the respective state). Thus, for a given force field, Schrödinger's equation allows an immediate estimation of the probability of a particle being in different regions of the space. For example, in the case of a free particle ($U(x, y, z) = 0$) moving along a given direction, the Schrödinger equation has the solution

$$\psi = \mathcal{A} \exp\left(\mathrm{i}\frac{\sqrt{2mE}}{\hbar}x\right) = \mathcal{A} \exp(\mathrm{i}\,kx). \tag{2.5}$$

The wave function characterizes the so-called plane wave state in which the particle can be found with the same probability, wherever on the respective direction.

The problem of the particle in a potential well represents another case of immediate application of Schrödinger's equation. There the potential energy is zero in the whole space, except a region where it is negative and constant. Because in dynamics the origin of the potential energy is an arbitrary point one may evidently consider a positive potential energy, constant in the whole space, except a region where it is zero.

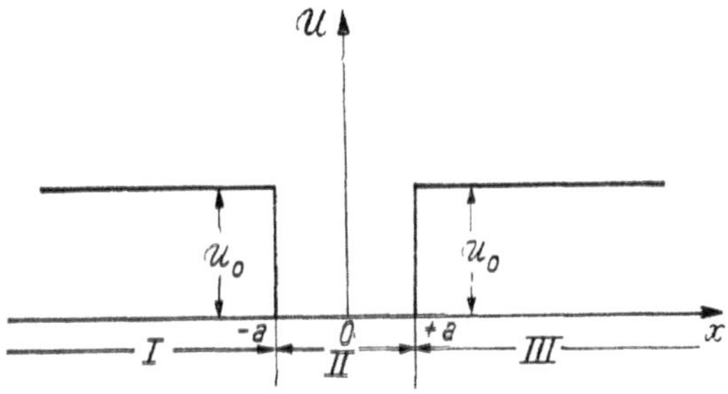

Fig. 2.1 — One-dimensional potential well.

Such a potential shape is drawn in Fig. 2.1, for the case of a particle moving along a direction "x". The potential energy expression has the form

$$\begin{cases} \mathcal{U} = \mathcal{U}_0 \text{ for } x \notin (-a,\ a), \\ \mathcal{U} = 0, \quad \text{for } x \in (-a,\ a), \end{cases}$$

and the corresponding Schrödinger's equation is

$$\begin{cases} -\dfrac{\hbar^2}{2m}\dfrac{\partial^2\psi(x)}{\partial x^2} + U_0\psi(x) = E\psi(x) \ldots x \notin (-a,\ a), \\ -\dfrac{\hbar^2}{2m}\dfrac{\partial^2\psi(x)}{\partial x^2} = E\psi(x) \qquad\qquad \ldots x \in (-a,\ a). \end{cases} \tag{2.6}$$

The potential well defines three regions of the space. Schrödinger's equation solutions for each of the three regions will be

$$\psi_I = C_1 \exp\ (Kx) + C_2 \exp\,(-Kx),$$

$$\begin{aligned}\psi_{II} &= C_3 \sin (kx) + C_4 \cos (kx), \\ \psi_{III} &= C_5 \exp (Kx) + C_6 \exp (-Kx),\end{aligned} \tag{2.7}$$

where the symbols $k = \frac{1}{\hbar}\sqrt{2mE}$; $K = \frac{1}{\hbar}\sqrt{2m(\mathcal{U}_0 - E)}$ have been introduced. The wave function must be continuous, finite and univocal in the whole space. In order that the wave function should remain finite at very great distances, the condition $C_2 = C_5 = 0$, must be accomplished. The estimation of the other four constants $-C_1$, C_3, C_4 and C_6 — is made after imposing the condition that the wave function and its derivative should be continuous at the points $x = +a$ and $x = -a$. For simplicity in the calculation we restrict ourselves to the case $\mathcal{U}_0 \to \infty$. Obviously, in such a case, the particle cannot leave the potential well because $K \to \infty$ and accordingly

$$\psi_I = 0, \; \psi_{III} = 0.$$

The values of the constants C_3 and C_4 result from the continuity condition for the wave function at the points $x = \pm a$

$$\begin{cases} C_3 \sin (ka) + C_4 \cos (ka) = 0, \\ -C_3 \sin (ka) + C_4 \cos (ka) = 0. \end{cases} \tag{2.8}$$

After adding and respectively subtracting the equations, we obtain the equivalent equations

$$\begin{cases} C_4 \cos (ka) = 0, \\ C_3 \sin (ka) = 0. \end{cases} \tag{2.9}$$

The constants C_3 and C_4 cannot be simultaneously zero, because in such a case it would also result $\psi_{II} = 0$, though the particle must be found necessarily somewhere on the direction x. At the same time, when one of the constants differs from zero, the other one must certainly be equal to zero, and thus two classes of solutions appear

$$\begin{cases} C_4 \neq 0, \; \cos (ka) = 0, \; \sin (ka) \neq 0, \; C_3 = 0, \; ka = \left(n + \frac{1}{2}\right)\pi, \\ C_3 \neq 0, \; \sin (ka) = 0, \; \cos (ka) \neq 0, \; C_1 = 0, \; ka = n\pi \end{cases} \tag{2.10}$$

where n is any positive integer. The corresponding wave functions have the form

$$\begin{cases} C_4 \neq 0, \ C_3 = 0, \ \psi_{\text{II}} = C_4 \cos\left[\left(n + \dfrac{1}{2}\right)\pi \dfrac{x}{a}\right], \\ C_3 \neq 0, \ C_4 = 0, \ \psi_{\text{II}} = C_3 \sin\left[n\pi \dfrac{x}{a}\right]. \end{cases} \tag{2.11}$$

Taking into account the definition of the parameter k and the condition imposed on the product ka (joining both classes of equations) there result the values allowed for the energy of the particle in the potential well

$$E_n = \frac{\pi^2\hbar^2}{8ma^2} n^2. \quad n = 1, 2, 3... \tag{2.12}$$

Unlike the case of the free particle, the total energy of which may take any positive value, the solution (2.5) being valid for any of those values, in the case of a particle in a potential well, the total energy takes a series of discrete values. The energy quantification turns up to be an immediate result of the particle motion limitation, that is to say in mathematical terms, a result of the boundary conditions imposed on the Schrödinger equation solutions. The total energy values allowed for a certain system form its energy spectrum.

The energy spectra are represented by groups of horizontal lines spaced proportionally to the differences of their energies.

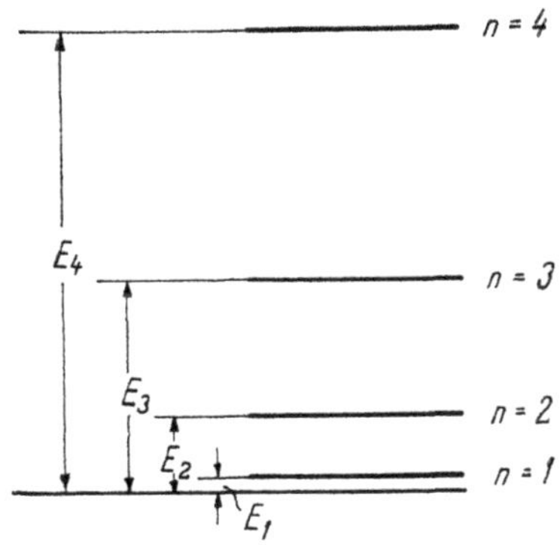

Fig. 2.2 — Energy spectrum of a particle in the potential well.

As an example, Fig. 2.2 represents the energy spectrum of the particle in the one-dimensional potential well, given by relation (2.12).

2.2. THE HYDROGEN ATOM

It is well known that the hydrogen atom contains a proton with mass M and electric charge $+e$ and an electron with mass m and electric charge $-e$, acting according to Coulomb's law. Because the proton mass is about 2000 times the electron mass we shall neglect in a first approximation the proton motion around the mass center of the system, what means to solve the problem of the electron motion around the proton (Fig. 2.3).

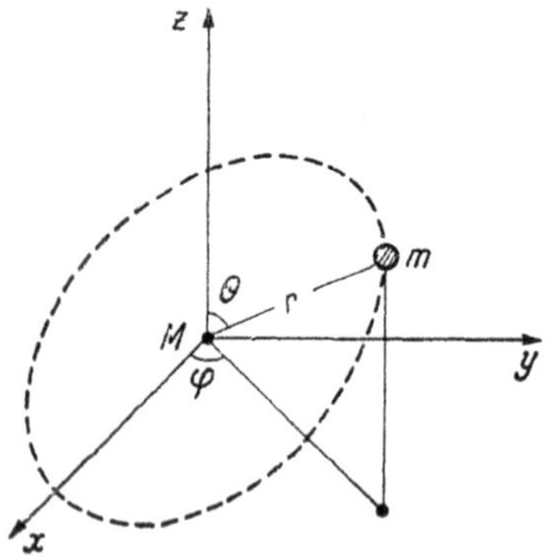

Fig. 2.3 – Structure of the hydrogen atom.

The potential energy of the system has the form

$$\mathcal{U}(x, y, z) = -\frac{e^2}{\sqrt{x^2+y^2+z^2}}, \tag{2.13}$$

where (x, y, z) are the electron coordinates, the proton being placed at the origin of the coordinate axes. We notice that the potential energy is a function only of the electron-proton distance, a property which usually defines the central fields of force. Due to the spherical symmetry of the central fields it is useful, in such cases, to introduce the spherical polar coordinates (r, θ, φ) defined by the relations

$$\begin{cases} x = r \sin\theta \cos\varphi, \\ y = r \sin\theta \sin\varphi, \\ z = r \cos\theta. \end{cases}$$

The Schrödinger equation in spherical coordinates will have the form [S.3]

$$\frac{\hbar^2}{2m}\left[\frac{1}{r^2}\frac{\partial}{\partial r}\left(r^2\frac{\partial\psi}{\partial r}\right) + \frac{1}{r^2\sin\theta}\frac{\partial}{\partial\theta}\left(\sin\theta\frac{\partial\psi}{\partial\theta}\right) + \right. \tag{2.14}$$

$$\left. + \frac{1}{r^2\sin^2\theta}\frac{\partial^2\psi}{\partial\varphi^2}\right] + \frac{e^2}{r}\psi = -E\psi.$$

For solving equation (2.14) we try a solution of the form

$$\psi(r, \theta, \varphi) = R(r)\, Y(\theta, \varphi), \tag{2.15}$$

which we insert in the equation (2.14) which becomes

$$\frac{\hbar^2}{2m}\left[Y\frac{1}{r^2}\frac{\partial}{\partial r}\left(r^2\frac{\partial R}{\partial r}\right)+\frac{R}{r^2\sin\theta}\frac{\partial}{\partial\theta}\left(\sin\theta\frac{\partial Y}{\partial\theta}\right)+\right. \tag{2.16}$$

$$\left.+\frac{R}{r^2\sin^2\theta}\frac{\partial^2 Y}{\partial\varphi^2}\right]+\frac{e^2}{r}RY = -ERY.$$

Dividing both sides of this last equation by the factor $\frac{\hbar^2 YR}{2mr}$ and removing all the terms to the left hand side, we obtain

$$\frac{1}{R}\frac{\partial}{\partial r}\left(r^2\frac{\partial R}{\partial r}\right)+\frac{1}{Y\sin\theta}\cdot\frac{\partial}{\partial\theta}\left(\sin\theta\frac{\partial Y}{\partial\theta}\right)+ \tag{2.17}$$

$$+\frac{1}{Y\sin^2\theta}\frac{\partial^2 Y}{\partial\varphi^2}+\frac{2mr^2}{\hbar^2}\left(\frac{e^2}{r}+E\right)=0$$

Equation (2.17) includes terms depending only either on the r coordinate, or on the angular (θ, φ) coordinates. Since the coordinates are independent variables, the sum of either group of terms must be constant. Making them equal to the same constant but with opposite signs, we obtain an angular equation

$$\frac{1}{Y\sin\theta}\frac{\partial}{\partial\theta}\left(\sin\theta\frac{\partial Y}{\partial\theta}\right)+\frac{1}{Y\sin^2\theta}\frac{\partial^2 Y}{\partial\varphi^2} = -\lambda, \tag{21.18}$$

and a radial equation

$$\frac{1}{R}\frac{\partial}{\partial r}\left(r^2\frac{\partial R}{\partial r}\right)+\frac{2mr^2}{\hbar^2}\left(\frac{e^2}{r}+E\right)=\lambda. \tag{2.19}$$

The angular equation (2.18) that can be also written under the form

$$\frac{1}{\sin\theta}\cdot\frac{\partial}{\partial\theta}\left(\sin\theta\frac{\partial Y}{\partial\theta}\right)+\frac{1}{\sin^2\theta}\cdot\frac{\partial^2 Y}{\partial\varphi^2}+\lambda Y = 0, \tag{2.20}$$

is valid for any central field, irrespective of the definite form of the potential. We get also the solution of equation (2.20) as a product of two factors, depending on the variables θ and φ respectively

$$Y(\theta, \varphi) = \Theta(\theta)\,\Phi(\varphi). \tag{2.21}$$

In the same way as above, we insert this solution in Eq. (2.20), make equal the sums of the terms depending on the same variable, with equal constants of opposite signs (denoted by m^2 and $-m^2$, respectively) and get the equations

$$\frac{\partial^2\Phi}{\partial\varphi^2} + m^2\Phi = 0, \tag{2.22}$$

$$\frac{\partial}{\partial\theta}\left(\sin\theta\,\frac{\partial\Theta}{\partial\theta}\right) + \left(\lambda\sin\theta - \frac{m^2}{\sin\theta}\right)\Theta = 0. \tag{2.23}$$

The solutions of Eq. (2.22) are obviously of the type

$$\Phi = C_1 \exp\,(\mathrm{i}\,m\,\varphi). \quad m = 0, \pm 1, \pm 2,... \tag{2.24}$$

The solutions of Eq. (2.23) are the associated Legendre polynomials $P_l^m(\cos\theta)$, where l satisfies the condition

$$l\,(l+1) = \lambda. \quad l = 0, 1, 2, 3...$$

The associated Legendre polynomial of degree l and order m, has the form

$$P_l^m(\cos\theta) = \frac{(1-\cos^2\theta)^{m/2}}{2^l\,l\,!}\,\frac{\mathrm{d}^{m+1}}{\mathrm{d}\,(\cos\theta)^{m+1}}\,(1-\cos^2\theta)^l, \tag{2.25}$$

where l and m are integers and $|m| \leqslant l$. In this way, the solutions of the angular Schrödinger equation are

$$Y(\theta, \varphi) = C_2 P_l^m(\cos\theta)\exp\,(\mathrm{i}\,m\,\varphi). \tag{2.26}$$

In quantum mechanics it is proved that the parameters l and m quantize the electron orbital angular momentum and its projection along a given direction (e.g. determined by an external magnetic field) chosen as the z-axis. The parameter l is the orbital quantum number and m the magnetic quantum number.

The angular distribution of the electron probability density is dependent on l and m quantum numbers which characterize therefore the "preferential" directions of the electron during its motion around the proton. Taking into account Eq. (2.26) it can be proved that the $l = 0$ and $m = 0$ states correspond to the spherical symmetry, according to which all the directions are equally preferred, the states with $m = 0$ favour the z-axis direction, the states with $m = 1$ — the plane xy, a.s.o.

For solving the radial equation (2.19) we make the above modification of parameters : $\lambda = l(l+1)$ and introduce a new parameter

$$n^2 = -\frac{me^4}{2E\hbar^2}, \tag{2.27}$$

and the variable

$$x = \frac{2me^2}{n\hbar^2} r. \tag{2.28}$$

The radial equation becomes

$$\frac{d^2R}{dx^2} + \frac{2}{x}\frac{dR}{dx} + \left(\frac{n}{x} - \frac{1}{4} - \frac{l(l+1)}{x^2}\right) R = 0. \tag{2.29}$$

Trying a solution of the form

$$R(x) = C_3\, u(x)\, x^l \exp\left(-\frac{x}{2}\right), \tag{2.30}$$

we find out that $u(x)$ must verify the differential equation for associated Laguerre polynomials

$$x\frac{d^2u}{dx^2} + [2(l+1) - x]\frac{du}{dx} + [n - (l+1)]u = 0. \tag{2.31}$$

The associated Laguerre polynomial of degree $(n + l)$ and order $(2\,l + 1)$ has the form

$$L_{n+l}^{2l+1}(x) = (-1)^{n+1}\frac{(n+1)!}{(n-l-1)!}\left\{x^{n-l-1} - \frac{(n+1)(n-l-1)}{1!}x^{n-l-2} + \right.$$

$$\left. + \frac{(n+2)(n+l-1)(n-l-1)(n-l-2)}{2!}x^{n-l-3} + \ldots\right\}. \tag{2.32}$$

The radial equation solution determined in this way

$$R_{n,l}(x) = C_3 L_{n+1}^{2l+1}(x) x^l \exp\left(-\frac{x}{2}\right), \tag{2.33}$$

is finite for the whole x range, only if $(n - l - 1)$ is a positive integer

$$n - l - 1 \geqslant 0. \tag{2.34}$$

Since l is an integer, condition (2.34) imposes that n should take only integral values.

Therefore, a quantization of the total energy values for the hydrogen atom follows, according to relation (2.27)

$$E_n = -\frac{me^4}{2\hbar^2} \cdot \frac{1}{n^2} \quad n = 1, 2, 3,... \tag{2.35}$$

The energy spectrum of the hydrogen atom can be seen in Fig. 2.4. The negative energy spectrum of the hydrogen atom can be explained considering the electron in bound states and its zero potential energy that at $r \to \infty$ (ionized atom). When $E > 0$ we can show that the total energy quantization is not necessary in solving the radial equation (2.19), the free particle assuming a continuous energy spectrum.

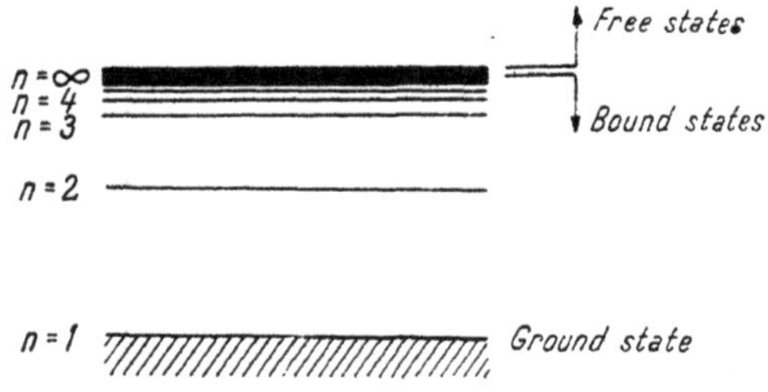

Fig. 2.4—Energy spectrum of the hydrogen atom.

The parameter n, specifying the total energy of the hydrogen atom in a certain state, is the principal quantum number. The radial distribution of the electron probability density, as resulting from Eq. (2.33), depends on the quantum numbers n and l. In a state with $n=1$, $l=0$, the electron moves in the close vicinity of the proton $\Big($at distances about the order of the first Bohr radius $a_0 = \frac{\hbar^2}{me^2} \simeq$

$\simeq 5.10^{-9}$ cm) and the total energy of the hydrogen atom has a minimum value, corresponding to its ground state. As the quantum numbers l and n take higher values, the probability of finding the electron at great distances from the proton increases, simultaneously with the total energy of the system.

Such states are called excited states of the hydrogen atom.

The quantum numbers n, l and m specify the state of the hydrogen atom. The electron states in atoms are usually denoted in the following way: the states with $l = 0, 1, 2, 3, 4, 5, \ldots$ are called $s, p, d, f, g, h, \ldots$ states; the principal quantum number is written before the respective letter, and the magnetic quantum number as a subscript (e.g. the state $n=2$, $l=1$, $m=+1$ is indicated by $2\,p_1$).

The total energy of the hydrogen atom is dependent only on the principal quantum number n. It follows that, in a given state, characterized by a set of quantum numbers (n, l, m), the quantum numbers l, m, do not influence the total energy value. The change of these two quantum numbers can lead to different states (to which will correspond, for instance, different electron trajectories) with the same total energy. It is said that there is a degeneracy in the orbital and magnetic quantum numbers. The number of states with the same total energy gives the degeneracy multiplicity. The degeneracy in the orbital quantum number is a characteristic of Coulombian fields. As regards the degeneracy in the magnetic quantum number it is a characteristic of all central force fields.

The method used above for the hydrogen atom can be extended to the ionized atoms He^+, Li^{++}, Be^{+++} a.s.o. (called "hydrogenoids"), mentioning that the potential energy will have, this time, the form

$$U(r) = -\frac{Ze^2}{r}, \tag{2.36}$$

where $+Ze$ is the nuclear charge of the hydrogenoid atom. The hydrogen and hydrogenoid atoms are the only atomic systems appropiate to an exact solution. Among the approximate methods, generally used in quantum mechanics calculations of atomic systems with two or several electrons are the perturbation and variation methods. A detailed study of these methods is beyond the framework of this book.

2.3. THEORY OF THE PERIODIC SYSTEM OF ELEMENTS

As it has been pointed out in the above section, the electron state in hydrogenoid atoms is specified by three quantum numbers (n, l, m) corresponding to the number of the degrees of freedom of the material point in the three-dimensional space. The experimental studies of energy spectra for hydrogenoid atoms fully confirmed the theoretical considerations. As the theoretical and experimental studies took into account more complex nuclei, three quantum numbers were not enough to describe the electron states in atoms.

For explaining the multiplet structure of the energy spectra of the alkaline metal atoms, Uhlenbeck and Goudsmith introduced the concept of electron spin (the intrinsic angular momentum). This is analogous to the electron having a rotational motion about an intrinsic symmetry axis, apart from its orbital motion. To this new degree of freedom the spin quantum number s has been associated. As long as the electron is in an isotropic medium, the electron spin has a random orientation. But as soon as a preferential direction appears, e.g. when applying a magnetic field, the electron spin aligns to this field vector [L.4] (taken here as the z-axis of the coordinate system as in the case of the orbital motion). The spin quantum number s can assume only two values corresponding to the two spin orientations available about the z-axis. The possible values of the quantum number set (n, l, m, s) specifying the electron states in atoms determine the structure of Mendeleev's periodic system of elements.

The exclusion principle of Pauli stating that two electrons in the same state cannot coexist in an atom, is essential for explaining the electron structure of atoms. Therefore, a given set of quantum numbers (n, l, m, s) may characterize the state of only one electron in a given atom. Thus, the principle of minimum energy conditioned by the exclusion principle (the electrons fill in states of minimum energy, as far as the exclusion principle is observed) determines the electron structure systematics of elements.

Before going further it is interesting to state the limits of the range of quantum number values. We consider a certain value n of the principal quantum number.

According to equation (2.34), to this value may correspond values of the orbital quantum number within the range

$$0 \leqslant l \leqslant n - 1.$$

Since by solving the Legendre equation we get $|m| \leqslant l$ for a given l value, the magnetic quantum number can assume the range of values

$$m = -l, \ -l+1, \ \text{------}, \ \text{------}, \ l-1, \ l$$

involving $(2l+1)$ values. Taking into account the two possible orientations of the spin, we can estimate the number of possible states for a given value of the principal quantum number

$$N_n = 2\sum_{l=0}^{n-1} 2(l+1) = 2\left[2\,\frac{n(n-1)}{2} + n\right] = 2\,n^2. \qquad (2.37)$$

The electrons of an atom, the states of which are characterized by the same principal quantum number, form a shell. The electrons of a shell having the same orbital quantum number m, form an electron orbital [D. 18].

The filling in of electron orbitals and shells can be seen in Table 2.1. The hydrogen atom has a single electron. In its ground state $n=1$, i.e. $l=0$, $m=0$ – the electron is in the $1s$ state.

The adding of another electron, also in $1s$ state, but with opposite spin orientation, leads to the He atom. Both electrons of the He atom fill in the shell $n=1$. The next shell with $n=2$ has two possible orbitals $2s$ and $2p$.

The electrons will begin to fill in that orbital corresponding to a lower total energy. We have seen that in the hydrogen atom all the states with a given n have the same total energy. In atoms with several electrons, the energy becomes dependent on the orbital quantum number too, due to the screening effect of the nuclear Coulombian field of inner electrons. In this way electrons with low orbital quantum momenta, rotating in the very neighbourhood of the nucleus, create a field opposite to the nucleus field. This produces an increase of the total energy of the electrons with higher orbital quantum momenta rotating farther from the atomic nucleus. Consequently, the total energy increases as the orbital quantum number increases. As we shall see later on, the screening effect leads finally even to the inversion in the order of the shell filling in. The $n=2$ shell filling in begins with the $2s$ orbital. After filling the shell with the 8 possible electrons for the Ne element, the filling in of the $n=3$ shell follows. The $n=3$ shell contains only the orbitals $3s$ and $3p$, which are filled in for Ar, because, due to the screening effect, the energy of the $4s$ orbital is lower than that of the $3d$ orbital. In fact, the order of

Table 2.1

Electron structure of some elements

Ele-ment	Shell	1	2		3			4			
	Orbital	1*s*	2*s*	2*p*	3*s*	3*p*	3*d*	4*s*	4*p*	4*d*	4*f*
H		1									
He		2									
Li		2	1								
Be		2	2								
B		2	2	1							
C		2	2	2							
N		2	2	3							
O		2	2	4							
F		2	2	5							
Ne		2	2	6							
Na		2	2	6	1						
Mg		2	2	6	2						
K		2	2	6	2	1					
Si		2	2	6	2	2					
P		2	2	6	2	3					
S		2	2	6	2	4					
Cl		2	2	6	2	5					
Ar		2	2	6	2	6					
K		2	2	6	2	6		1			
Ca		2	2	6	2	6		2			
Sc		2	2	6	2	6	1	2			
Ti		2	2	6	2	6	2	2			
V		2	2	6	2	6	3	2			
Cr		2	2	6	2	6	5	2			
Mn		2	2	6	2	6	5	2			
Fe		2	2	6	2	6	6	2			
Co		2	2	6	2	6	7	2			
Ni		2	2	6	2	6	8	2			
Cu		2	2	6	2	6	10	2			
Zn		2	2	6	2	6	10	2			
Ga		2	2	6	2	6	10	2	1		
Ge		2	2	6	2	6	10	2	2		

the shell filling in for all the known elements, is the following: $1s$, $2s$, $2p$, $3s$, $3p$, $4s$, $3d$, $4p$, $5s$, $4d$, $5p$, $6s$, $4f$, $5d$, $6p$, $7s$, $6d$, $5f$ [D.18].

The filling in periodicity of the electron shell of atoms, determines also the periodicity of the physical and chemical properties of those elements with electron shells just closing. All these elements form the group of noble gases and have the valency equal to zero. Similarly, the elements with an electron outside the closed shell (Li, Na, K, Rb, Cs) form the group of alkaline metals with valency equal to unity. A valency equal to unity is also proper to those elements lacking an electron to close a shell (H, F, Cl, Br) and form the halogen group. The alkaline-earth metals (Be, Mg, Ca) have two electrons outside the closed shell and the valency 2. Thus, one can envisage the explanation of this chemical property, the valency, by a mechanism according to which only the electrons outside the closed shells (just called valency electrons) determine the chemical processus of compound formation. The ionization potential, equal to the energy necessary to an electron to be extracted from the atom is also dependent on the electrons outside the closed shell, as follows from Table 2.2. The highest ionization potentials were determined for the noble gases, the lowest ones for alkaline metals, the periodicity being also present but superposed on a continuous decrease of the ionization potential when passing to heavier elements. The high ionization potential of the noble gases can be explained by the fact that the closed electron shells give rise to a very stable system with low total energy; the alkaline metals, have outside the closed shell a single electron, weakly bound in atom and thus corresponding to a low ionization potential.

Table 2.2

Ionization potentials of some elements

Element	H	He	Li	Be	B	C	N	O	F	Ne	Na
Ionization potential (eV)	13.5	24.5	5.4	9.5	8.4	11.2	14.5	13.6	16.8	21.5	5.1

The explanation of the periodicity of physical and chemical properties of the elements in terms of electron shell structure of their respective atoms has been considered as an important checking up of the quantum theory.

2.4. ELECTRONS IN CRYSTALS

The atoms having the structure described in the previous sections, join into molecules and crystals. The crystal formation is similar to that of molecules, both relying on interatomic couplings. There are several types of such couplings [D. 28].

Ionic bond. The interatomic forces result in this case from Coulombian interactions (e.g., the sodic chloride belongs to this class of crystals). The ionic bond changes significantly from one crystal to another, revealing itself the stronger, the more different are the valency values of the elements coupling together their atoms.

Covalent bond. Attraction forces appear among similar or even identical atoms, leading to the covalent bond. Such attraction forces of covalent type, explaining the building up of the hydrogen, oxygen, nitrogen and other molecules, appear, due to the use in common of the outer electrons by two or several atoms, what produces a decrease of the total energy of these atoms. Generally, the homeopolar forces are much stronger than the Coulombian forces.

The silicon and germanium atoms inside the crystal lattice are coupled through homeopolar (or covalent) bonds.

Van-der-Waals bond. This type of coupling relies on not very strong forces, arisen due to atom polarization and is specific to organic crystals.

The rigorous methods for calculating the wave functions and energy levels of electrons in crystals are very complicated. The wave functions of the constituent atoms have to be superposed inside some cells of the crystal purposely chosen and obliged to observe the boundary conditions at the neighbouring cell frontiers. Moreover, we must consider all exchange interactions appearing between neighbouring atoms of the crystal.

All the difficulties implied by a rigorous calculation have led to the formulation of crystal models, in order to reproduce their fundamental properties in a simplified mathematical form.

The periodic structure of the crystal lattice with nodes occupied by the crystal atoms is generally assimilated in these models to a sequence of equidistant identical potential wells.

The crystal lattice is usually characterized by three non-coplanar vectors, $\vec{a}$, $\vec{b}$, and $\vec{c}$ so that the lattice nodes can be represented by the vector

$$\vec{n} = n_1\vec{a} + n_2\vec{b} + n_3\vec{c}, \tag{2.38}$$

where n_1, n_2, n_3 are real integers.

The crystal lattices are often characterized by the vector of the reciprocal lattice $\vec{K}$, defined by the relation

$$\vec{K} \cdot \vec{n} = 2\pi. \tag{2.39}$$

To simplify the calculation, we reduce the problem to a one-dimensional one, when relations (2.38) and (2.39) become

$$n = n_1 a \quad (a = \text{the constant of the crystal lattice}), \tag{2.40}$$

$$Kn = 2\pi. \tag{2.41}$$

The electron motion inside the crystal lattice is described by the Schrödinger equation

$$-\frac{\hbar^2}{2m}\ \frac{\partial^2\psi(x)}{\partial x^2} + U(x)\ \psi(x) = E_n\ \psi(x), \tag{2.42}$$

where the potential energy $U(x)$ is a periodic function, with the same period as the lattice

$$U(x + a) = U(x). \tag{2.43}$$

Due to the potential energy periodicity, the wave function of the electron satisfies the relation

$$\psi(x + a) = \alpha_n\psi(x), \tag{2.44}$$

where $|\alpha_n|^2 = 1$, the electron being found with equal probabilities at the points x or $x + a$. We can introduce the quantity k, defined by

$$\alpha_n = \exp\ (ikn),$$

so that condition (2.44) may be written under the form

$$\psi(a+x) = \exp\ (ikn)\ \psi(x). \tag{2.45}$$

We notice that the k value is not uniquely determined by relation (2.45); e.g., $k' = k + K$ also verifies Eq. (2.45), as follows from the definition of K by the relation (2.41).

The general function obeying the periodicity condition (2.45) has the form

$$\psi(x) = \exp\ (ikx)\ u_k(x), \tag{2.46}$$

with $u_k(x)$ a periodic function of x, with period a. Introducing the solution (2.46) in the Schrödinger equation, we obtain the following equation for $u_k(x)$

$$\left[-\frac{\hbar^2}{2m}\left(\frac{\partial^2}{\partial x^2} + ik\right)^2 + U - E_n\right] u_k(x) = 0, \tag{2.47}$$

whence one observes that the electron energy depends both on the quantum number n and the wave number k.

The form of the wave function (2.46) suggests the analogy: motion of the electron inside the crystal-motion of the free electron (expression (2.5)), for the crystal the amplitude of the plane wave being modulated with the lattice period.

The electron state inside a crystal can be studied starting from the electron state inside an isolated atom and considering the effects of the neighbouring atom fields on that electron as a perturbation. Obviously, both the wave function and the energy levels of the electron will be changed by the presence of the other atoms in the crystal lattice. First, we shall consider the hypothesis of the strong electron coupling, according to which:

— only the atoms in the close vicinity of the atom taken into account, interact with this one:

— the shift of the energy levels resulted from the interaction with neighbouring atoms is small in comparison with the difference between the energy levels of an isolated atom.

Suppose a lattice with N identical atoms. The wave function of the electron in the j atom verifies the Schrödinger equation when there is no interaction with the other atoms

$$-\frac{\hbar^2}{2m}\frac{\partial^2 \varphi_j}{\partial x^2} + [U_j(x_j) - E(s)]\varphi_j = 0, \tag{2.48}$$

where for simplification, we supposed the electron of the isolated atom to be in an "s" state, the total energy being denoted by $E(s)$.

Because Eq. (2.48) can be written for any of the N atoms of the crystal, the energy level $E(s)$ corresponds to an N-fold degeneration. When considering the interaction, we can write the wave function in the zero-order approximation

$$\psi(x_1, x_2, \ldots x_N) = \sum_{j=1}^{N} C_j \varphi_j(x_j) \tag{2.49}$$

and the electron energy inside the crystal

$$E(k,s) = E(s) + E^{(1)}(k, s), \tag{2.50}$$

where $E^{(1)}(k, s)$ represents the perturbation (considered to be very small in the hypothesis of the strong electron coupling) induced in the presence of other atoms in the crystal lattice.

When considering the presence of the other atoms, the electron in the jth atom must satisfy the equation

$$\sum_{j=1}^{N} C_j \frac{\partial^2 \varphi_j}{\partial x_j^2} + \frac{2m}{\hbar^2}\left[E(s) + E^{(1)}(k, s) - \sum_{l=1}^{N} U_l(x_j - x_l)\right] \sum_{j=1}^{N} C_j \varphi_j = 0, \tag{2.51}$$

which can be also written as

$$\sum_{j=1}^{N} C_j \left[\frac{\partial^2 \varphi_j}{\partial x_j} + \frac{2m}{\hbar^2}(E(s) - U_j)\varphi_j\right] + \\ + \frac{2m}{\hbar^2}\left[E^{(1)}(k, s) - \sum_{\substack{l=1 \\ l \neq j}}^{N} U_l(x_j - x_l)\right] \sum_{j=1}^{N} C_j \varphi_j = 0, \tag{2.52}$$

and taking into account Eq. (2.48)

$$\sum_{j=1}^{N} C_j \left[E^{(1)}(k, s) - \sum_{\substack{l=1 \\ j \neq l}}^{N} U_l(x_j - x_l)\right] \varphi_j = 0. \tag{2.53}$$

According to our hypothesis, the influence of the other atoms on the jth atom is weak and the wave functions of the atoms do not overlap

$$\int_{\mathcal{V}} \varphi_h^+ \varphi_j \mathrm{d}\mathcal{V} = \delta_{hj} \tag{2.54}$$

where the integral (2.54) is estimated over the whole space $\mathcal{V}(\mathrm{d}\mathcal{V} = \mathrm{d}x\mathrm{d}y\mathrm{d}z)$.

Multiplying equation (2.53) by φ_h^+ and integrating over the whole space, condition (2.54) leads to

$$C_h[E^{(1)}(k, s) + \alpha] + \sum_{\substack{j=1\\ j\neq h}}^{N} \beta(x_j - x_h)C_j = 0, \tag{2.55}$$

where [L.8]

$$\begin{cases} \alpha = -\int_{\mathcal{V}} \varphi_h^+ \left[\sum_{\substack{l=1\\ l\neq j}}^{N} U_l(x_l - x_j)\right] \varphi_h \mathrm{d}\mathcal{V}, \\ \beta = -\int_{\mathcal{V}} \varphi_h^+ \left[\sum_{\substack{l=1\\ l\neq j}}^{N} U_l(x_l - x_j)\right] \varphi_j \mathrm{d}\mathcal{V}. \end{cases}$$

Because of the small overlap of the wave functions, the sum in Eq. (2.55) reduces [L 8] to terms corresponding to both neighbouring atoms of the hth atom

$$C_h[E^{(1)}(k, s) + \alpha] + \beta(x_h - x_{h-1})\, C_{h-1} + \beta\,(x_{h+1} - x_h)C_{h+1} = 0. \tag{2.56}$$

For reasons of symmetry

$$\begin{cases} x_{h+1} - x_h = x_h - x_{h-1} = a, \\ \beta(x_h - x_{h-1}) = \beta(x_{h+1} - x_h) = \beta \end{cases}$$

and so

$$C_h[E^{(1)}(k, s) + \alpha] + \beta(C_{h-1} + C_{h+1})] = 0. \tag{2.57}$$

Checking a solution of the form

$$C_h = \exp\,(\mathrm{i}\; kx_h),$$

in equation (2.57), we obtain

$$\exp\,(\mathrm{i}\,kx_h)\;[E^{(1)}(k, s) + \alpha] + \beta\,[\exp\,(\mathrm{i}\,kx_{h-1}) + \exp\,(\mathrm{i}\,kx_{h+1})] = 0, \tag{2.58}$$

and thus, the value of the perturbation is

$$E^{(1)}(k, s) = -\alpha - \beta\,\{\exp\,[-\mathrm{i}\,k\,(x_h - x_{h-1})] + \exp\,[\mathrm{i}\,k\,(x_{h+1} - x_h)] =$$
$$= -\alpha - 2\beta\,\cos k\,(x_{h+1} - x_h) = -\alpha - 2\beta\,\cos\,(ka). \tag{2.59}$$

The total electron energy in the j-th atom, when all the lattice atoms are taken into account, will be given by the expression

$$E\,(k, s) = E(s) - \alpha - 2\beta\,\cos ka. \tag{2.60}$$

Because all the atoms exert attraction forces on the electron, this one is much stronger bound in the crystal than it has been in the isolated atom

$$E(k, s) < E(s),$$

the integrals α and β being positive quantities. Assuming different values of the wave vector k, we obtain total energies of the electron under consideration within a range (Fig. 2.5) with limits

$$E(s) - \alpha - 2\beta < E(k, s) < E(s) - \alpha + 2\beta,$$

and total width

$$\Delta E(k, s) = 4\beta.$$

The appearance of an energy band involving allowed energies of the electrons inside a crystal, and resulted from the splitting of an energy level of a single atom, is not characteristic only of the s electron considered above.

Fig. 2.5 — Energy variation inside an allowed band.

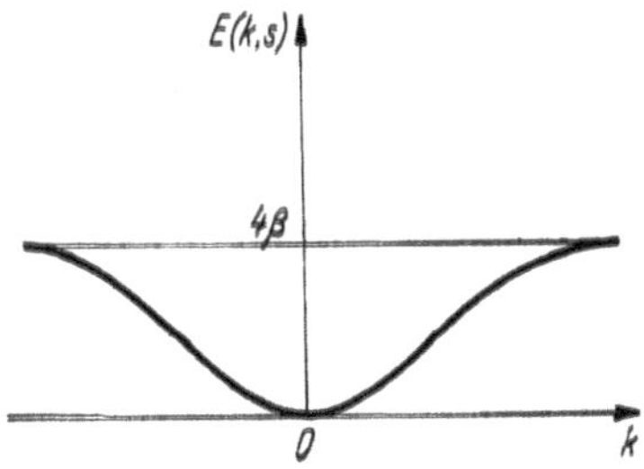

Each of the p, d, f, g states of the single atom will also give rise to a band of "allowed" energies at the splitting of the energy level with an N-fold degeneracy. These, rather narrow regions are separated by "forbidden" energy bands much wider, as we may see in Fig. 2.6.

The sequence of allowed and forbidden bands for the crystal electrons, obtained above as a result of the strong electron coupling hypothesis, represents a more general property. The strong electron coupling hypothesis holds for the inner electrons of the crystal lattice but loses its validity for electrons of the outer atomic shells.

As a result of the wave function overlapping for electrons in neighbouring atoms, the electron location inside a certain atom is a nonsense, since the electron can be found with equal probability anywhere within the crystal (hypothesis of the quasifree electron). In the hypothesis of the quasifree electron the calculations also lead to such sequences of allowed and forbidden energy bands, with the difference that, this time, the wide bands of allowed energies are separated by very narrow bands of forbidden energies. Both assump-

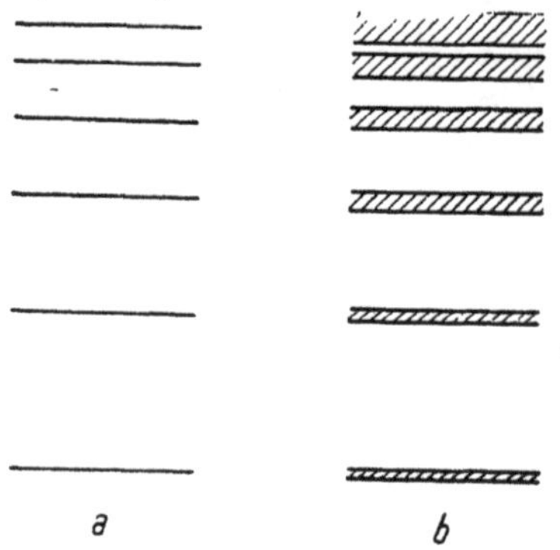

Fig. 2.6 — The crystal energy spectrum assuming strong electron coupling: a — single atom, b — crystalline lattice.

tions — of the strongly coupled and of the quasifree electron — are limit approximations of the electron states in crystals. As real cases we find intermediate situations approaching more one or the other limit, the energy spectrum of electrons inside a crystal still having in all cases the same characteristic — the presence of energy bands.

In the following we show that the presence of energy bands in the energy spectrum of a crystal is a straight consequence of the crystal structure periodicity [L.9]. For this purpose, let us assume

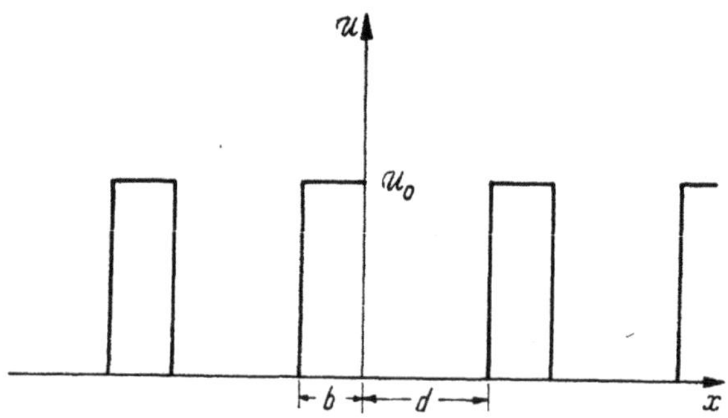

Fig. 2.7 — The Kronig–Penny potential.

the potential proposed by Kronig and Penny (Fig. 2.7) as an approach to the real potential inside a crystal. The condition of the potential periodicity

$$\mathcal{U}(x + a) = \mathcal{U}(x), \tag{2.61}$$

is obviously fulfilled, if $a = b + d$.

The Schrödinger equation takes the form

$$\begin{aligned} -\frac{\hbar^2}{2m}\frac{d^2\psi}{dx^2} &= E\psi \ldots 0 \leqslant x \leqslant d, \\ -\frac{\hbar^2}{2m}\frac{d^2\psi}{dx^2} + U_0\psi &= E\psi \ldots -b \leqslant x \leqslant 0. \end{aligned} \tag{2.62}$$

The solutions for both regions are of the type given by function (2.46). Inserting a function of this type into the first equation (2.62), we obtain the equation

$$\frac{d^2u_k}{dx^2} + 2\,ik\frac{du_k}{dx} + (\alpha^2 - k^2)u_k = 0, \left(\alpha^2 = \frac{2mE}{\hbar^2}\right), \tag{2.63}$$

with the solution

$$u_k^{(1)}(x) = C_1 \exp\,[i(\alpha - k)\,x] + C_2 \exp\,[-i(\alpha + k)\,x]. \tag{2.64}$$

In the same way, inserting the function (2.46) into the second equation (2.62), there results

$$\frac{d^2u_k}{dx^2} + 2\,i\,k\frac{du_k}{dx} - (\beta^2 + k^2)\,u_k = 0, \quad \left(\beta^2 = \frac{2m\,(U_0 - E)}{\hbar^2}\right) \tag{2.65}$$

with the solution

$$u_k^{(2)}(x) = C_3 \exp\,[i\,(\beta - ik)x] + C_4 \exp\,[-i(\beta + ik)x]. \tag{2.66}$$

The constants C_1, C_2, C_3, C_4, are estimated from the continuity conditions for the functions $u_k^{(1)}(x)$, $u_k^{(2)}(x)$ and their derivatives

$$\left\{\begin{aligned} u_k^{(1)}(0) &= u_k^{(2)}(0), \\ u_k^{(1)}(d) &= u_k^{(2)}(-b), \\ \left[\frac{du_k^{(1)}}{dx}\right]_{x=0} &= \left[\frac{du_k^{(2)}}{dx}\right]_{x=0}, \\ \left[\frac{du_k^{(1)}}{dx}\right]_{x=d} &= \left[\frac{du_k^{(2)}}{dx}\right]_{x=-b}, \end{aligned}\right. \tag{2.67}$$

Thus, we obtain a system of four homogeneous equations with four unknowns C_1, C_2, C_3, C_4

$$\begin{cases} C_1 + C_2 = C_3 + C_4, \\ C_1(\alpha - k) - C_2(\alpha + k) = C_3(\beta - ik) - C_4(\beta + ik), \\ C_1 \exp[i(\alpha - k)d] + C_2 \exp[-i(\alpha + k)d] = \\ \quad = C_3 \exp[-i(\beta - ik)b] + C_4 \exp[i(\beta + ik)b], \\ C_1(\alpha - k)\exp[i(\alpha - k)d] - C_2(\alpha + k)\exp[-i(\alpha + k)d] = \\ \quad = C_3(\beta - ik)\exp[-i(\beta - ik)b] - \\ \quad - C_4(\beta + ik)\exp[i(\beta + ik)b]. \end{cases} \tag{2.68}$$

System (2.68) is compatible when its characteristic determinant is equal to zero. Thus we obtain

$$\frac{\beta^2 - \alpha^2}{2\alpha\beta}\,\mathrm{sh}\,(\beta b)\sin(\alpha d) + \mathrm{ch}(\beta b)\cos(\alpha d) = \cos[k(b + d)]. \tag{2.69}$$

Let us assume $U_0 \to \infty$, and at the same time $b \to 0$, such that the product $U_0 b$ remains constant. Equation (2.69) takes the simple form

$$\frac{mU_0 b}{\alpha\hbar^2}\sin(\alpha d) + \cos(\alpha d) = \cos(kd) \tag{2.70}$$

or, as $d \simeq a$:

$$\frac{mU_0 b}{\alpha\hbar^2}\sin(\alpha a) + \cos(\alpha a) = \cos(ka). \tag{2.71}$$

Actually, relation (2.71) represents under an implicit form, the dependence $E(k)$ with the energy involved in the parameter α.

Because of its transcendental form, Eq. (2.71) cannot be solved, with respect to E by elementary methods. But Eq. (2.71) cannot be verified for any value of the electron energy. Hence, the electron inside a crystal cannot fill in certain energy states (if relation (2.71) is not satisfied, the system of homogeneous equations (2.68) leads to the trivial solutions $C_1 = C_2 = C_3 = C_4 = 0$, the wave functions (2.64) and (2.66) are identical to zero). This fact can be easily checked up graphically, plotting the dependence

$$\frac{mU_0 b}{\alpha \hbar^2} \sin(\alpha a) + \cos(\alpha a) = f(\alpha a),$$

shown in Fig. 2.8. Obviously equation (2.71) has solutions only in the regions where

$$\left| \frac{mU_0 b}{\alpha \hbar^2} \sin(\alpha a) + \cos(\alpha a) \right| \leqslant 1, \tag{2.72}$$

because of the cosine in the right hand side of relation (2.71). These regions are hatched in Fig. 2.8 and correspond to sets of α values

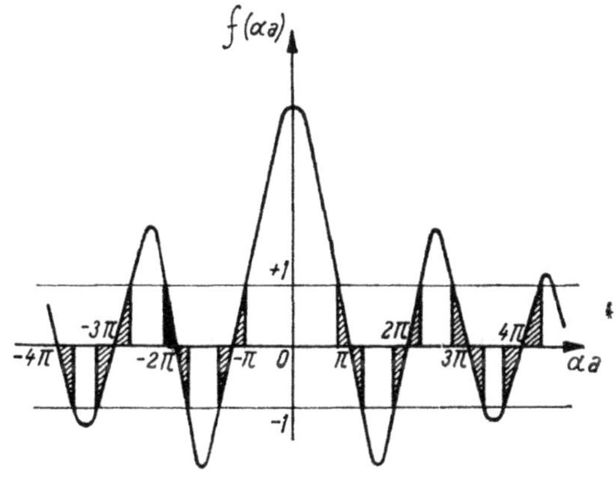

Fig. 2.8 — Plot of the inequality (2.72).

(energy bands) allowed in crystals. At the limit of such a band the condition

$$\cos ka = \pm 1, \qquad ka = n\pi, \tag{2.73}$$

is satisfied in agreement with relation (2.60). Equation (2.71) is also solved graphically with respect to E and we obtain the dependence $E(k)$ from Fig. 2.9. The energy zones of the crystal (called Brillouin zones) are also shown in this figure.

The model Kronig-Penny explains quite well the general aspect of the band energy spectrum of electrons in crystals.

The form of the curve $E(k)$ inside a given energy band must be calculated according to relation (2.60) obtained with a perturbation method which takes into account the expression of the electron wave functions.

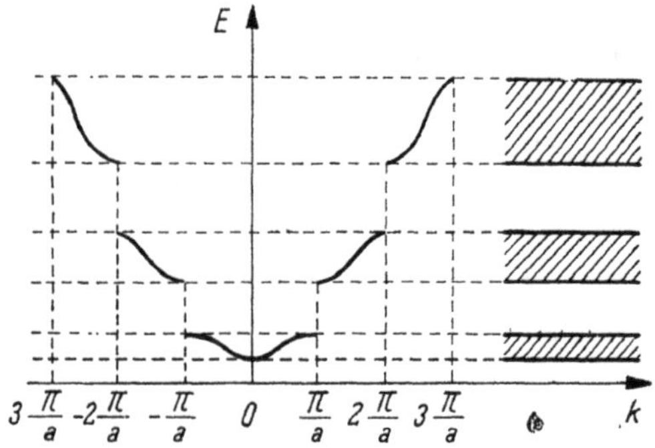

Fig. 2.9 —[Zonal structure of a crystal energy spectrum.

2.5. EFFECTIVE MASS

The electron energy dependence on the wave number k, expressed by relation (2.60), makes possible to introduce the concept of effective mass, a fundamental one in the theory of crystal energy bands.

The motion of the free electron (described by relation (2.5)) is characterized by a dependence $E(k)$ of the form

$$E = \frac{\hbar^2}{2m} k^2, \tag{2.74}$$

directly resulting from the definition of the wave number k and representing a parabola (Fig. 2.10).

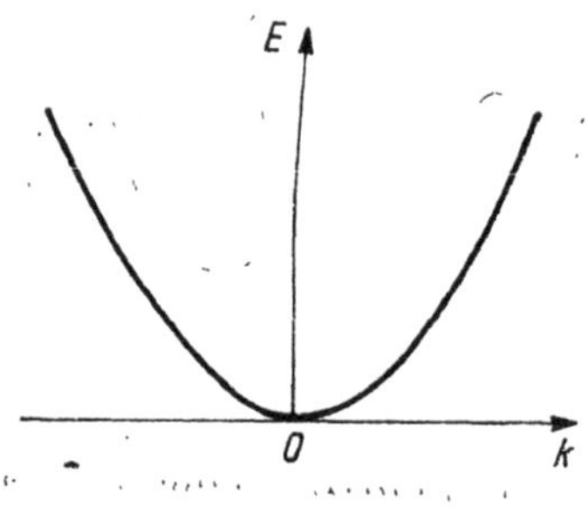

Fig. 2.10 — The dependence $E(k)$ for a free electron.

Let us differentiate two times successively relation (2.74) with respect to k. We shall obtain

$$\frac{\partial E}{\partial k} = \frac{\hbar^2}{m} k, \tag{2.75}$$

$$\frac{\partial^2 E}{\partial k^2} = \frac{\hbar^2}{m}. \tag{2.76}$$

Relation (2.76) may be considered as a definition of the free electron mass

$$m = \frac{\hbar^2}{\frac{\partial^2 E}{\partial k^2}}. \tag{2.77}$$

Evidently, for electrons inside a crystal the dependence $E(k, s)$ is no longer a parabola. For instance in the approximation of a strong electron coupling, the dependence $E(k)$ is represented by a cosine curve (in the free electron approximation, the dependence $E(k)$ remains a parabola). Nevertheless, there are variation domains of k where the dependence $E(k)$ may be approximated with a parabola. Consider, e.g., the domain $k \simeq 0$ and write under the form of series the function cos (ka), retaining only the first two terms

$$\cos (ka) \cong 1 - \frac{k^2 a^2}{2}.$$

The insertion of this function into relation (2.60) leads to

$$E(k, s) = E(s) - \alpha - 2\beta + \beta a^2 k^2 = A + Bk^2, \tag{2.78}$$

with positive constants A, B. We notice this time that the dependence $E(k)$ is parabolalike and the electron motion can be characterized by an effective mass

$$m^* = \frac{\hbar^2}{\frac{\partial^2 E}{\partial k^2}} = \frac{\hbar^2}{2\beta a^2} = m_0, \tag{2.79}$$

which is constant (with respect to k) and positive. In this way, at the lower limit of the energy band (low k values) the electron motion is similar to that of the free electron with a mass equal to the effective mass given by expression (2.79). We denote this value by m_0. In the same way we can deal with the domain

$$ka \simeq \pi,$$

corresponding to the upper limit of the energy band. Inserting the expansion

$$\cos(ka) \simeq -1 + \frac{(ka-\pi)^2}{2},$$

where we have retained only the first two terms, into (2.60) we obtain

$$E(k, s) = E(s) - \alpha + 2\beta - \beta(ka - \pi)^2. \tag{2.80}$$

The function $E(k)$ represents again a parabola. The effective mass

$$m^* = \frac{\hbar^2}{\dfrac{\partial^2 E}{\partial k^2}} = -\frac{\hbar^2}{2\beta a^2}, \tag{2.81}$$

is also constant but negative. The absolute values of both effective masses given by expressions (2.79) and (2.81) are equal. For a certain region of the energy band, the effective mass

$$m^* = \frac{\hbar^2}{\dfrac{\partial^2 E}{\partial k^2}} = \frac{\hbar^2}{2\beta a^2 \cos(ka)}, \tag{2.82}$$

is no longer a constant, but depends on the total energy value through the wave number k. When considering expression (2.82), we find again the particular cases studied beforehand

$$ka \simeq 0, \qquad m^* = m_0$$

$$ka \simeq \pi, \qquad m^* = -m_0.$$

We notice also, that according to expression (2.82), the effective mass becomes infinite at the energy satisfying the condition

$$ka = \frac{\pi}{2}.$$

The variation of the effective mass vs the wave number k is plotted in Fig. 2.11. Within every energy band, the effective mass starts from a positive value m_0, at the lower end of the band, increases to infinity, changes its sign and tends to the negative value $-m_0$, at the upper end of the energy band.

The concept of effective mass leads to difficulties that seem beyond us at first sight. Which is the physical meaning of negative

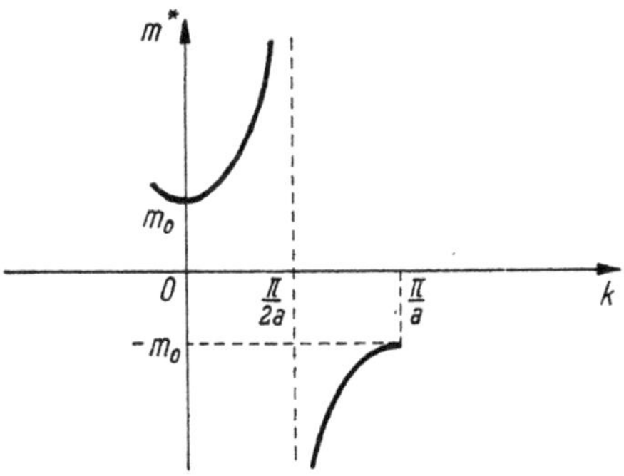

Fig. 2.11 – Effective mass variation inside an energy band.

infinite masses and generally of those masses depending on the wave number k (at low velocities, as compared to light velocity)? How can be applied the electrodynamics laws to such cases?

Actually, to understand the physical phenomena in metals and semiconductors it is enough to study the electron motion at energies towards the two ends of the energy band where the masses are approximately constant [I.3]. Thus the concept of effective mass proves its practical utility.

It results from expression (2.81) that at the upper end of the energy band the electron effective mass is negative and practically constant. The force exerted by the electric field, acting on the same direction with the electron velocity, will slow down the motion of electrons with negative mass. While the energy increases, the electron velocity $\left(\text{equal to } \frac{1}{\hbar}\frac{\partial E}{\partial k}\right)$ decreases. And then arises the question: when are such states met with in crystals?

As any physical system, the electrons in crystals tend to fill in the states of minimum energy in the absence of a perturbing energy. As in atoms, the crystal electrons fill in the quantum states according to Pauli's principle (section 2.3). If the electron number is much smaller than the number of quantum states within an energy band, the states at the lower end of this band are preferably filled.

In such states the effective mass is positive and constant and the electron motion can be studied as the motion of a free particle with mass equal to the effective mass. But, if the electron number is almost equal to the number of quantum states inside the energy band, states of maximum energy at the upper end of this band remain unoccupied. In this case, the properties of the electron ensemble are thoroughly determined by the behaviour of the unoccupied quantum states. Since every electron tends to pass into an unoccupied state of lower energy, the unoccupied states (called "holes") will be filled in by electrons coming from states of higher energies, states that remain empty in their turn. This process may be considered as a transition of the hole from the quantum state with lower energy to that with higher energy. In this way, the hole energies increase from the upper towards the lower end of the energy band inversely to the electron case. The hole, passing from state 1 to state 2 (Fig. 2.12) gives to the crystal lattice the same energy that would be given by the electron removing from state 2 to state 1. The strange behaviour of electrons with negative masses can be easily explained when using the notion of hole. Indeed, the electron stopping in an electric field aligned to the electron velocity direction represents an energy loss equivalent to an energy increase (i.e. an acceleration) of the corresponding hole. Because the holes move in the opposite direction to electrons, they need to be assigned with a positive charge so that they could be accelerated by the electric field. Thus, the hole motion inside the crystal lattice is identical to the motion of some positive charges, with absolute values equal to that of the electron charge and with effective mass

$$m^* = \frac{\hbar^2}{\left|\frac{\partial^2 E}{\partial k^2}\right|}. \tag{2.83}$$

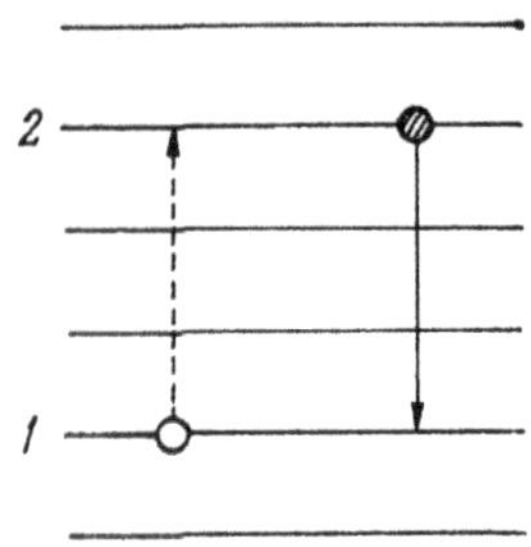

Fig. 2.12 — Equivalence of hole and electron transitions.

The introduction of the concept of hole permits to study the motion of the crystal electrons in regions with states "almost free" and "almost occupied", the problem becoming that of the free electron (hole) motion when the mass is equal to the effective one.

In spite of all simplifications introduced, the effective mass preserves a conventional characteristic that we must not forget. Let us consider, for example, the energy at which the effective mass becomes infinite according to relation (2.82). No finite force could make an electron or a hole pass through the respective state. Actually, we find that the electrons can perform transitions between any two states of the energy band and so the effective mass, as a force to acceleration ratio, loses its meaning in that region. Moreover, even towards the lower limit of the energy band, for electrons, or towards the upper limit, for holes, the experiments have shown [I 3] that the effective mass has none of the mass fundamental properties, as inertia and Newtonian attraction.

At last, the value of the effective mass depends on the electron interaction with the electric field of the crystal lattice which may be a function of the electron motion direction which makes the effective mass be a tensor [S.8] (the inertial and gravitational masses are, as we know, scalars).

2.6. ENERGY BANDS

The electron quantum state grouping into "allowed" energy bands, separated by "forbidden" energy intervals, is a characteristic of all crystals. The ways in which the crystal electrons take part in different — physical or chemical — processes are various, depending on the crystal nature. For example, the process of electric conductivity enables us to classify the crystals as metals, semiconductors or insulators. This classification takes as principle the way the energy bands split and especially the configuration of the so-called conduction bands and valency bands. We call a valency band the last band of allowed energies, totally occupied at absolute zero temperature, while the conduction band is just the upper band of allowed energies.

Let us consider first, the metals. The metal quantum states of interest in the process of electric conductivity can be grouped into two energy bands : a lower band called "of valency", involving only electron filled states and an upper band called "of conduction" involving both filled and unfilled states (Fig. 2.13. a).

Only the electrons of the conduction band ensure the electric conductivity.

The electrons of the valency band cannot take part in the process of electric conductivity. Indeed, the appearance of the electric current involves the electron acceleration and thus the change of their velocity (energy). But, since for every electron of the valency band, the neighbouring states are filled, no change of energy (or

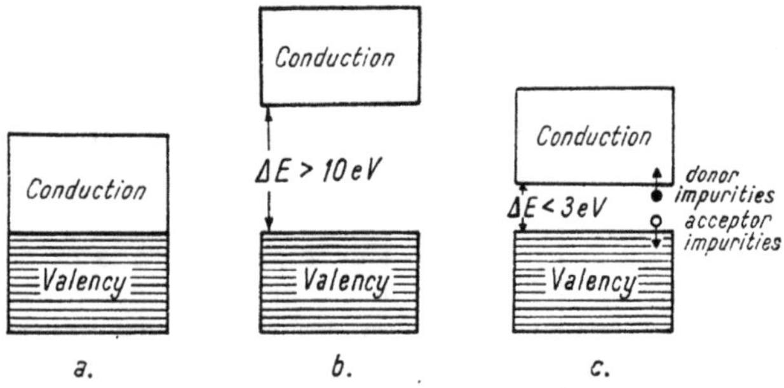

Fig. 2.13 — Energy bands in crystals : *a* — metal, *b* — insulator, *c* — semiconductor.

of any other quantum property) can occur and the passage of the electric current is not possible. We should mention the competition between the conductivity mechanisms through electrons and through holes in some metals.

The hole conductivity occurs in metals of the Zn and Cd [D 28] type where along certain directions the upper limit of the valency band in an atom comes under the lower limit of the conduction band in the next atom and the electrons from the upper levels of the valency band pass to the conduction band. The holes produced in this way enable the electric current passage through these metals.

As for insulators, the quantum states "allowed" for electrons also group into two energy bands but the number of states in the lower valency band is equal to the number of electrons, so that in the conduction band electrons are not found. Moreover, the conduction band is separated from the valency band by an "energy gap", involving no electron levels (Fig. 2.13.b).

Obviously, the electric conductivity of such a crystal is zero. Generally, an electric conductivity could appear when the electrons pass from the valency band to the conduction band, as a result of the thermal agitation.

The probability of such a transition is given by the expression

$$W = \exp\left(-\frac{\Delta E}{kT}\right), \tag{2.84}$$

with k the Boltzmann's constant, T — temperature of the crystal and ΔE — width of the forbidden energy band. Since $\Delta E > 10$ eV for most insulators — the probability W is negligible at the temperatures practically reached and the electric conductivity (proportional to the electron density inside the conduction band) remains at very low values.

The set up of the semiconductor energy bands is similar to that of insulators (Fig. 2.13.c) with the difference that the width of the forbidden band (energy gap) does not exceed 3 eV. That is why the electric conductivity due to electron transition from the valency band to the conduction band, because of the thermal agitation, is much higher than in insulators at the same temperature. This kind of conductivity, called intrinsic conductivity, is a characteristic of semiconductors in a pure state. But very often we meet semiconductors with conductivity through impurities. There are two kinds of impurities: donors and acceptors. The atoms of the donor impurities have such an electron structure that they are able to give electrons to the basic semiconductor, usually Si or Ge. If the energies of the donor atoms are within the energy gap, close to the conduction band, they are able to provide the conduction band with electrons and thus to give rise to a conductivity through electrons, conductivity of N-type, that can exceed by several orders of magnitude the intrinsic conductivity of the pure semiconductor crystal. Similarly, the acceptor atoms can receive electrons from the valency band, which leads to a conductivity through holes (conductivity of P-type). For crystals of silicon and germanium, the elements of the V-th group of the periodic system of elements behave as donors while the element of the III-rd group as acceptors. The energy differences between donor levels and the conduction band and between acceptor levels and the valency band respectively are of about 0.01 eV (see appendix AI).

Because of these low energies, all impurities are practically ionized at room temperature and thus the electric conductivity of the semiconductors is controlled by a very low percentage of impurities. Therefore we need very pure materials, produced through elaborate technical processes, to make semiconductor devices.

2.7. STATISTICAL DISTRIBUTIONS. THE FERMI LEVEL

The electron quantum states of the solid state can be empty or filled — in every state being at most one electron, according to Pauli's principle. The way, the electrons are distributed, among possible quantum states depends on their energy and the crystal temperature. The probability of a quantum state of energy E being occupied by one electron is expressed by the Fermi-Dirac statistical distribution function [T.20]

$$w_e(E) = \frac{1}{1 + \exp\left(\frac{E - E_f}{kT}\right)}, \tag{2.85}$$

where "k" is Boltzman's constant, T the absolute temperature and "E_f" the energy of the Fermi level. The form of the distribution function (2.85) depends exclusively on the crystal temperature, the other characteristics concerning the crystal type and state, being reflected in the distribution position through the Fermi level. The position of the Fermi level on the energy diagram of the crystal quantum states is decided upon by the energy diagram aspect. Let us consider, a metal for example. The statistical distribution variation vs metal temperature is shown in Fig. 2.14.

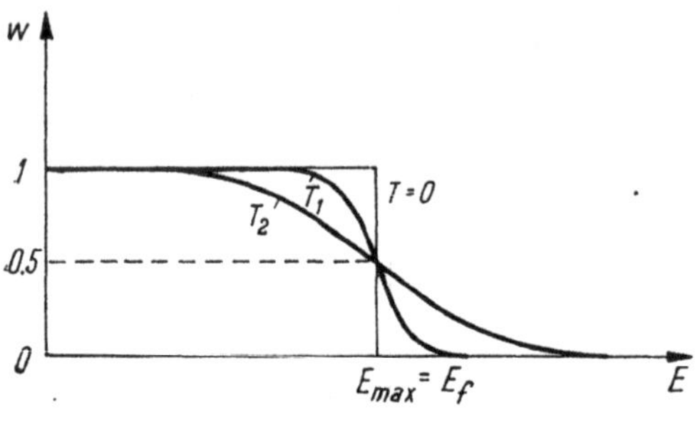

Fig. 2.14 — Fermi-Dirac statistical distribution at different temperatures ($T_2 > T_1$).

At the temperature $T = 0$, the electrons will fill in the states with the lowest possible energies. Assume E_{max} to be the energy of the highest state filled at $T = 0$. Since the electrons fill in all the states with $E \leqslant E_{max}$ and none of the states with $E > E_{max}$, the probability of occupation is 1 in the first range and 0 in the second one

$$w_e(E) = \begin{cases} 1 \ldots E \leqslant E_{max}, \\ 0 \ldots E > E_{max}. \end{cases} \quad (2.86)$$

On the other hand, when considering the expression (2.85) we observe that at $T = 0$, the exponential in the denominator is zero when $E \leqslant E_f$ and infinity when $E > E_f$. Consequently, the probability $w_e(E)$ can have only two values

$$w_e(E) = \begin{cases} 1 \ldots E \leqslant E_f, \\ 0 \ldots E > E_f. \end{cases} \quad (2.87)$$

Because the statistical distributions (2.86) and (2.87) must be identical,

$$E_f = E_{max}$$

and thus the Fermi level energy is equal to the energy of the highest level occupied by the metal electrons at the temperature of absolute zero. We notice from expression (2.85) and Fig. 2.14 that the occupation probability of the state with energy equal to the Fermi level energy is about 0.5 at $T \neq 0$. The Fermi level has equal probabilities to be occupied or unoccupied.

It can be shown for pure semiconductor crystal, at $T = 0$, that the Fermi level lies at the middle of the energy gap at equal distances from the valency and conduction bands. As concerns the problem of the position of Fermi levels, it will be considered in more detail later on.

From the electron distribution function the occupation probability of a quantum state of energy E by a hole follows easily

$$w_h = 1 - w_e = 1 - \frac{1}{1 + \exp\left(\frac{E - E_f}{kT}\right)} = \frac{1}{1 + \exp\left(\frac{E_f - E}{kT}\right)}. \quad (2.88)$$

We notice that electron and hole distribution functions are identical for a given crystal at the same temperature, provided the hole energy is measured in direction opposite to the electron energy (convention adopted above in Section 2.6). Therefore, we can say that two states equally spaced on both sides of the Fermi level have the property that the probability of one state being occupied is equal to the probability of the other being unoccupied.

The Fermi-Dirac distribution function describes not only the electron statistical distribution, but it also represents a more general property, characteristic of all elementary particles with spin $\frac{1}{2}$ called fermions (protons, neutrons, neutrinos, a.s.o.).

Besides fermions, other types of elementary particles the spins of which are integers (or zero) and called bosons interfere in the fundamental processes occurring in crystals. Bosons are considered to be the photons, the phonons (quanta of the lattice vibration energy) and other particles of less importance for the crystal physics (π-, K-mesons).

The statistical distribution of the boson energies is described by the Bose-Einstein distribution function [8]

$$w_b = \frac{1}{\exp\left(\dfrac{E - E_b}{kT}\right) - 1}, \tag{2.89}$$

where "E_b" is the Bose level energy (for photons and phonons $E_b = 0$) and the other quantities have the above meanings.

The Fermi-Dirac and Bose-Einstein distribution functions characterize the behaviour of identical particles with quantized energies. The statistical distribution of "classical" particles with continuous energy spectra is given by the Maxwell-Boltzmann distribution function

$$w_c = A \exp\left(-\frac{E}{kT}\right). \tag{2.90}$$

At sufficiently higher energies ($E - E_f \gg kT$), all the three distribution functions (2.85), (2.89) and (2.90) concur, as seen in the respective expressions.

To establish the electron structure of a crystal, it is not enough to know the electron distribution function and thus the occupation probability of quantum states with different energies. At the same time, we must also know the way in which the respective states are placed in the energy diagram of the crystal. For this purpose we introduce the concept of state density, defined as the number of states with energies different by no more than an energy unit.

If in the energy interval $(E, E + dE)$ there are dz states, the state density, corresponding to the energy E, is

$$\rho(E) = \frac{dz}{dE}. \tag{2.91}$$

Obviously, the introduction of the state density is justified only when the number of states is high enough to use a statistical model *), which happens in the case of metals and semiconductors. Roughly speaking, the study of a particle (state) ensemble within a statistical model framework reduces to the finding of the number of particles with coordinates and momenta within the intervals $(x, x + dx)$, $(y, y + dy)$, $(z, z + dz)$ and $(p_x, p_x + dp_x)$, $(p_y, p_y + dp_y)$, $(p_z, p_z + dp_z)$ grouped together in a volume element

$$d\Phi = dx\ dy\ dz\ dp_x\ dp_y\ dp_z,$$

of the 6-dimensional space (x, y, z, p_x, p_y, p_z), called the phase space. If the statistical ensemble involves fermions, the Pauli principle may be applied, provided two quantum states should be inside the volume element h^3 of the phase space (differing between them only by their spin orientation). The simplest statistical model of the crystal electrons is the degenerate Fermi gas. In the following, we consider the Fermi gas inside a limited volume $\mathcal{V}$ where the potential is constant. Choosing as the potential origin just the potential of the volume $\mathcal{V}$, the total energy of a particle will be

$$E = \frac{1}{2} \cdot \frac{p^2}{m}, \tag{2.92}$$

where p is the particle momentum in the respective state, and m the particle mass. Because the energy is not dependent on the particle coordinates, the problem can be reduced to momentum space (p_x, p_y, p_z).

In the momentum space the locus of the points representing states with the same energy E is the surface of the sphere

$$p^2 = 2\,m\,E.$$

---

*) In atomic physics, this type of statistical model is used when the differences between the state energies are comparable to the energy widths of the states (the energy widths are proportional to the deexcitation probability of the respective states).

All the points inside the sphere stand for states with energies lower or at most equal to E. The states with energies within the interval $(E,\ E + \mathrm{d}E)$ lie inside the volume

$$\mathrm{d}P = \frac{4\pi (p + \mathrm{d}p)^3}{3} - \frac{4\pi p^3}{3} \simeq 4\pi p^2 \,\mathrm{d}p,$$

enclosed between the spheres with radii p and $p + \mathrm{d}p$. The volume corresponding to it in the phase space

$$\mathrm{d}\Phi = 4\pi \mathcal{V} p^2 \,\mathrm{d}p, \tag{2.93}$$

involves, according to Pauli's principle, a number of states

$$\mathrm{d}z = \frac{\mathrm{d}\Phi}{\frac{h^3}{2}} = \frac{8\pi \mathcal{V} p^2 \,\mathrm{d}p}{h^3}. \tag{2.94}$$

Taking into account relation (2.92) one may also write

$$\mathrm{d}z = \frac{4\pi \mathcal{V} (2m)^{3/2} E^{1/2} \,\mathrm{d}E}{h^3}. \tag{2.95}$$

from which the state density corresponding to the energy E results as

$$\rho(E) = \frac{\mathrm{d}z}{\mathrm{d}E} = 4\pi \mathcal{V} \left(\frac{2m}{h^2}\right)^{3/2} E^{1/2}. \tag{2.96}$$

It may be noticed from relation (2.96) that the state density increases proportionally to the square root of the energy (Fig. 2.15).

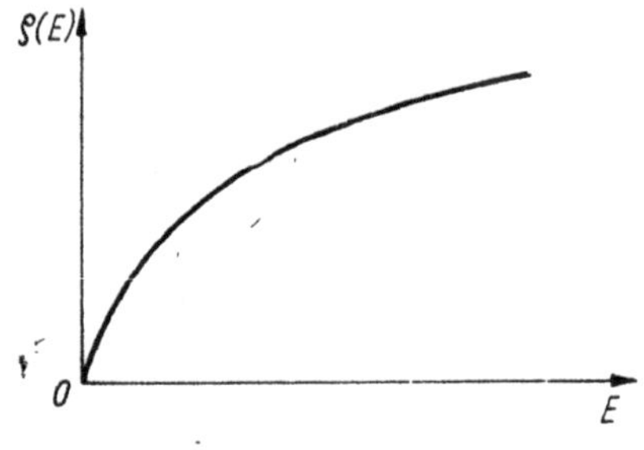

Fig. 2.15 — State density variation vs energy.

The electron energy distribution inside a crystal at a temperature T is given by the product between the state density (2.96) and the distribution function (2.85). The number of electrons with energies within the interval $(E,\ E+\mathrm{d}E)$ can be estimated from

$$\nu(E)\,\mathrm{d}E = 4\pi\mathcal{V}\left(\frac{2m^*}{\mathrm{h}^2}\right)^{3/2}\frac{E^{1/2}}{1+\exp\left(\frac{E-E_f}{kT}\right)}\,\mathrm{d}E, \qquad (2.97)$$

where m stands for the electron effective mass.

The energy distributions of electrons inside a metal at $T=0$ and $T \neq 0$ are drawn in Fig. 2.16.

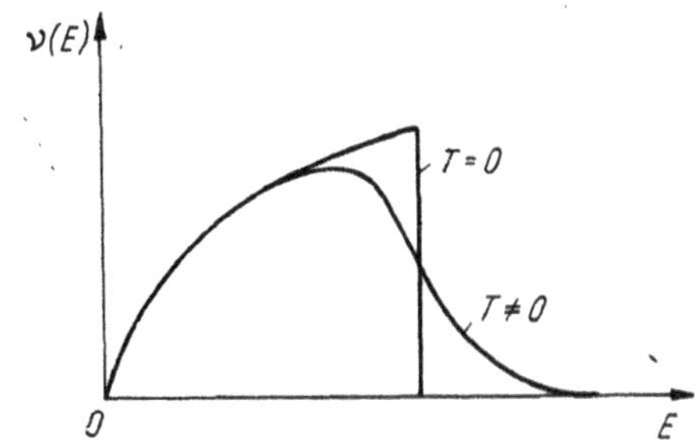

Fig. 2.16 — Electron energy distribution inside a metal.

At $T=0$, the electrons fill in all the states with energies not higher than $E=E_f$, and let all the other quantum states unoccupied. At a certain temperature T, part of the electrons receive an additional thermal energy and go to states with higher energies, letting unoccupied the states they had filled in. At moderate temperatures, the electrons in the low energy states are not affected by the temperature increase because the states with close energies are filled and for transitions to higher states, the energy given by the thermal agitation turns out not to be sufficient. At room temperature ($T=300$°K) about 1% of the metal electrons have energies exceeding the Fermi energy. Generally, states with energies approximately lying within the interval

$$|E-E_f| \leqslant kT \qquad (2.98)$$

will be partially filled.

All the other states will have occupation probabilities equal to 1 or 0 as they are on the left or the right side of the Fermi level.

From expression (2.97) it is possible to find out the Fermi level position for a certain crystal.

For this purpose, it must be supposed that the total number of electrons, estimated by integrating the expression (2.97) over the whole energy range, should be equal to the total number of the crystal electrons

$$\int_0^{\infty} \nu(E)\,\mathrm{d}E = N. \tag{2.99}$$

The integral (2.99) can be estimated in a simple way provided $T=0$, when the Fermi-Dirac distribution function takes the simple form, given by relations (2.87)

$$N = 4\pi\mathcal{V}\left(\frac{2m^*}{\mathrm{h}^2}\right)^{3/2}\int_0^{E_f} E^{1/2}\,\mathrm{d}E = \frac{8\pi}{3}\mathcal{V}\left(\frac{2m^*}{\mathrm{h}^2}\right)^{3/2} E_f^{3/2} \tag{2.100}$$

whence the Fermi energy is

$$E_f = \frac{\mathrm{h}^2}{2m^*}\left(\frac{3N}{8\pi\mathcal{V}}\right)^{2/3} = \frac{\mathrm{h}^2}{2m^*}\left(\frac{3n}{8\pi}\right)^{2/3} \tag{2.101}$$

with n-the electron concentration inside the crystal. For instance, for silver [I 3]
$n = 6\times 10^{22}$ el/cm^3 ; $m^* \simeq m_0 = 9\times 10^{-28}$ g ; $E_f = 1.44\times 10^{-11}$ erg $=$ $= 9$ eV.

The mean energy of the crystal electrons, when $T = 0$ has the value

$$\overline{E} = \frac{\int_0^{\infty} E\,\nu(E)\,\mathrm{d}E}{\int_0^{\infty} \nu(E)\,\mathrm{d}E} = \frac{4\pi\mathcal{V}\left(\frac{2m^*}{\mathrm{h}^2}\right)^{3/2}\int_0^{E_f} E^{3/2}\,\mathrm{d}E}{4\pi\mathcal{V}\left(\frac{2m}{\mathrm{h}^2}\right)^{3/2}\frac{2}{3}E_f^{3/2}} = \frac{3}{5}E_f. \tag{2.102}$$

It is interesting to compare the values of the Fermi energy (and of the mean energy) with the value of the expression kT, for the ambient temperature $T = 300°$K

$$kT = 1.38\times 10^{-16}\times 300 = 4.14\times 10^{-14}\ \text{erg} \simeq 0.026\ \text{eV}.$$

It has been found that the energy due to thermal agitation of the order of kT is much lower than the Fermi level energy and the mean energy. For this reason, the electron energy distributions are practically the same for $T = 0°K$ and $T = 300°K$ for metals and the degenerate Fermi gas model is a good approach to electrons inside metals.

Let us now consider the case of a semiconductor crystal, e.g. the silicon. For electrons at the lower limit of the conduction band (as for holes at the upper limit of the valency band) the effective mass may be considered approximately equal to the free electron mass: $m^* = m_0 = 9.1 \times 10^{-31}$ kg and the electron density of the pure silicon is $n \simeq 10^{10}$ el/cm^3 at room temperature (the industrial silicon of 5000 Ω cm has an electron density of 10^{12} el/cm^3).

In the same way, we obtain the Fermi energy

$$E_f = 0.0000006 \text{ eV}.$$

Since the Fermi energy is much lower than the thermal agitation energy, the electrons in the filled states can easily undergo transitions to any of the crystal unoccupied states, the Pauli principle restricts no longer the electron interactions and the degenerate Fermi gas becomes an ordinary gas. The Fermi-Dirac distribution function must be replaced by the Maxwell-Boltzmann distribution function specific to classical particles

$$\frac{1}{1 + \exp\left(-\frac{E_f}{kT}\right) \exp\left(\frac{E}{kT}\right)} \simeq \exp\left(\frac{E_f}{kT}\right) \exp\left(-\frac{E}{kT}\right) = $$

$$= \text{const} \cdot \exp\left(-\frac{E}{kT}\right).$$

In this manner, the electron states in metals may be described with the degenerate Fermi gas model, while the electron states in semiconductors are similar to the molecule states of a gas. When the temperature increases, the degenerate Fermi gas in metals turns also into an ordinary gas, but the metal electrons can be described with the ordinary gas model only at too high temperatures to be of any practical interest. For example for silver, the condition $E_f = kT$ becomes true at about 70 000°K.

2.8. EQUILIBRIUM DENSITY OF CHARGE CARRIERS IN SEMICONDUCTORS

The electron quantum states of a semiconductor crystal may bunch in three energy bands (section 2-6), the valency band, the conduction band and the energy gap. To begin with, we shall consider the pure semiconductor case, when the charge carriers appear only due to thermal agitation. At the temperature $T = 0$, the electron states of the valency band are thoroughly filled and the conduction band is unoccupied with electrons. At a certain temperature, part of the valency band electrons go to the conduction band, because of the thermal agitation, letting an equal number of holes in the valency band. Let E_c be the minimum energy in the conduction band and E_v the maximum energy in the valency band. And now let us calculate the electron density in the conduction band. The electron energy distribution is given by expression (2.97) where the energy is estimated as to E_c and the distribution function is of Maxwell-Boltzmann type

$$\nu_e(E)\,\mathrm{d}E = 4\pi \mathcal{V} \left(\frac{2m_e^*}{h^2}\right)^{3/2} \exp\left(\frac{E_f}{kT}\right)(E-E_c)^{1/2} \exp\left(-\frac{E}{kT}\right)\mathrm{d}E. \quad (2.103)$$

The electron number in the conduction band at temperature T may be estimated by integrating expression (2.103) with respect to energy

$$N_e = \int_{E_c}^{\infty} \nu_e(E)\mathrm{d}E = 4\pi \mathcal{V} \left(\frac{2m_e^*}{h^2}\right)^{3/2} \exp\left(\frac{E_f}{kT}\right)\int_{E_c}^{\infty} (E-E_c)^{1/2} \exp\left(-\frac{E}{kT}\right)\mathrm{d}E. \quad (2.104)$$

Inserting the new variable $\varepsilon = E - E_c$

$$N_e = 4\pi\mathcal{V} \left(\frac{2m_e^*}{h^2}\right)^{3/2} \exp\left(\frac{E_f - E_c}{k\,T}\right)\int_0^{\infty} \varepsilon^{1/2} \exp\left(-\frac{\varepsilon}{kT}\right)\mathrm{d}\varepsilon, \quad (2.105)$$

and noticing that

$$\int_0^{\infty} \varepsilon^{1/2} \exp\left(-\frac{\varepsilon}{k\,T}\right)\mathrm{d}\varepsilon = \frac{\sqrt{\pi}}{2}(kT)^{3/2}.$$

we obtain finally

$$N_e = 2\frac{(2\,\pi\,m_e^*\,k\,T)^{3/2}}{h^3}\,\mathcal{V}\exp\left(-\frac{E_c - E_f}{k\,T}\right). \tag{2.106}$$

The electron density in the conduction band is

$$n_e = \frac{N_e}{\mathcal{V}} = 2\,\frac{(2\,\pi\,m_e^*\,k\,T)^{3/2}}{h^3}\exp\left(-\frac{E_c - E_f}{k\,T}\right). \tag{2.107}$$

We observe that the density (2.107) is indeed zero at $T = 0$ and increases very abruptly at the same time with the crystal temperature.

The hole density in the valency band can be calculated in a similar way, provided the hole energy is estimated with respect to E_v, the distribution function is considered in opposite direction with respect to the Fermi level and the integral limits are $-\infty$ and E_v. Assuming the hole-energy distribution

$$\nu_h(E)\,\mathrm{d}E = 4\,\pi\,\mathcal{V}\left(\frac{2\,m_h^*}{h^2}\right)^{3/2}\exp\left(-\frac{E_f}{k\,T}\right)(E_v - E)^{1/2}\left(\frac{E}{k\,T}\right)\mathrm{d}E, \tag{2.108}$$

the hole density is

$$n_h = \frac{N_h}{\mathcal{V}} = 2\,\frac{(2\,\pi\,m_h^*\,k\,T)^{3/2}}{h^3}\exp\left(-\frac{E_f - E_v}{k\,T}\right). \tag{2.109}$$

As shown, the electron density in the conduction band and the hole density in the valency band must be equal, at a given temperature.

Making the respective expressions equal

$$2\,\frac{(2\,\pi m_e^*\,k\,T)^{3/2}}{h^3}\exp\left(-\frac{E_c - E_f}{k\,T}\right) = 2\,\frac{(2\,\pi\,m_h^*\,k\,T)^{3/2}}{h^3}\exp\left(-\frac{E_f - E_v}{k\,T}\right),$$

the Fermi energy can be estimated as

$$E_f = \frac{E_v + E_c}{2} - \frac{3\,k\,T}{4}\ln\left(\frac{m_e^*}{m_h^*}\right). \tag{2.110}$$

It is to be noticed that, generally, the Fermi energy depends on the crystal temperature. At the temperature $T = 0$, the Fermi level lies symmetrically with respect to the conduction and valency bands at the middle of the energy gap.

When the electron and hole effective masses are equal, the Fermi level of the semiconductor crystal keeps lying at the middle of the energy gap at any temperature. Generally, as the crystal temperature increases, the Fermi level shifts towards the conduction or the valency

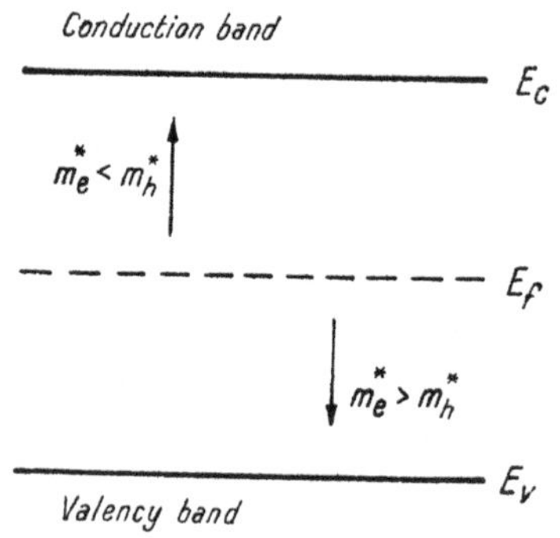

Fig. 2.17 — The Fermi level in intrinsic semiconductors.

band according as the electron effective mass is smaller or larger than the hole effective mass.

The electron and hole density given by expressions (2.107) and (2.109), respectively, explain the mechanism of the intrinsic electric conductivity, a characteristic of semiconductor crystals (section 2.6). Usually the semiconductor crystals hold impurities playing a determining part in the electric conductivity mechanism.

Let us examine now how the Fermi level is placed and how the charge carrier density is estimated for a semiconductor holding impurities. We assume that most of the impurities are of a donor type, the crystal having thus N-type conductivity. Let N_d be the total number of donor atoms, all supposed to be ionized and E_d the energy difference between the donor levels and the lower limit of the conduction band (Fig. 2.18). As origin of the energy we choose E_c, the lower limit of the conduction band.

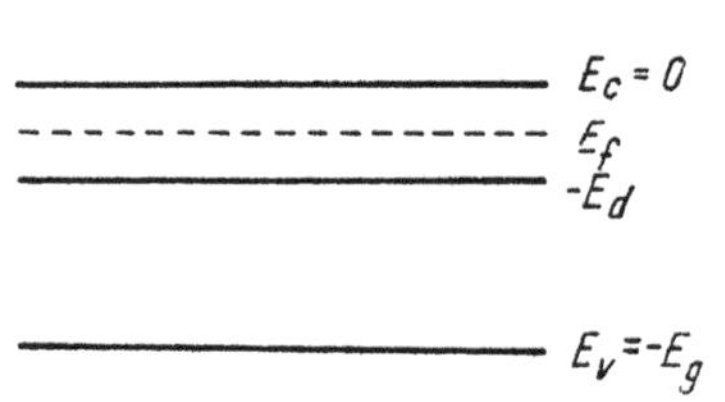

Fig. 2.18 — The Fermi level in an N-type semiconductor.

Denoting by $E_g = E_c - E_v$ the energy gap width, the state density is given by the expression [E.9]

$$\rho(E) = \begin{cases} 4\pi\mathcal{V}\left(\dfrac{2m_e^*}{h^2}\right)^{3/2} E^{1/2} \;.\;.\;.\;.\;.\;.\;.\; E > 0, \\ 4\pi\mathcal{V}\left(\dfrac{2m_e^*}{h^2}\right)^{3/2} (-E_g - E)^{1/2} \ldots E < -E_g \\ N_d\,\delta(E+E_d) \;.\;.\;.\;.\;.\;.\;.\;.\; -E_g < E < 0. \end{cases} \tag{2.111}$$

The hole density of the valency band is usually negligible as compared to the donor atom density and the condition that the crystal be neutral is

$$N_d = \int_{-\infty}^{+\infty} \rho(E)\, w_e\, \mathrm{d}E =$$

$$= \int_{-E_g}^{0} N_d \delta(E + E_d) \frac{\mathrm{d}E}{1 + \exp\left(\dfrac{E - E_f}{k\,T}\right)} + 4\pi\mathcal{V}\left(\frac{2m_e^*}{h^2}\right)^{3/2} \exp\left(\frac{E_f}{k\,T}\right) \times$$

$$\times \int_0^{\infty} E^{1/2} \exp\left(-\frac{E}{k\,T}\right)\mathrm{d}E = \frac{N_d}{1 + \exp\left(-\dfrac{E_d + E_f}{k\,T}\right)} + \tag{2.112}$$

$$+ 2\,\frac{(2\,\pi\, m_e^*\, k\,T)^{3/2}}{h^2}\,\mathcal{V} \exp\left(\frac{E_f}{k\,T}\right).$$

whence

$$N_d \frac{1}{1 + \exp\dfrac{E_d + E_f}{k\,T}} = 2\,\frac{(2\,\pi\, m_e^*\, k\,T)^{3/2}}{h^3}\,\mathcal{V} \exp\left(\frac{E_f}{k\,T}\right). \tag{2.113}$$

The donor level energies are usually about the order of tenths of eV [S 23], such that

$$\exp\left(\frac{E_d + E_f}{k\,T}\right) \gg 1\,,$$

and writing the donor atom density as n_d, we obtain the relation

$$n_d \exp\left(-\frac{E_d + E_f}{k\,T}\right) = 2\,\frac{(2\,\pi\,m_e^*\,k\,T)^{3/2}}{h^3}\exp\left(\frac{E_f}{k\,T}\right), \tag{2.114}$$

where the Fermi energy is

$$E_f = -\frac{E_d}{2} + \frac{k\,T}{2}\ln\frac{h^3 n_d}{2\,(2\,\pi\,m_e^*\,k\,T)^{3/2}}. \tag{2.115}$$

At the temperature $T = 0$, the Fermi level lies at the middle of the interval between the lower limit of the conduction band and the donor level energy. As the temperature increases, the Fermi level shifts towards the donor levels.

Using the Fermi energy estimate given by (2.115), the electron density in the valency band can be deduced

$$n_e = 2\,\frac{(2\,\pi\,m_e^*\,k\,T)^{3/2}}{h^3}\exp\left(\frac{E_f}{k\,T}\right) = \sqrt{2}\,\sqrt{n_d}\,\frac{(2\,\pi\,m_e^*\,k\,T)^{3/4}}{h^{3/2}}\exp\left(-\frac{E_d}{2k\,T}\right). \tag{2.115'}$$

The electron density increases with the crystal temperature, the donor impurity density and the donor level energy.

If acceptor impurities are predominant (P-type semiconductor), the position of the Fermi level (in this case, lying in the neighbourhood of the valency band) and the hole density in the valency band are calculated in a similar way.

The estimation of the hole density in the valency band of an N-type semiconductor gives

$$n_h = 2\,\frac{(2\pi\,m_h^*\,k\,T)^{3/2}}{h^3}\exp\left(-\frac{E_f + E_g}{k\,T}\right). \tag{2.116}$$

Notice that the product between the electron and hole density

$$n_e\,n_h = 4\,\frac{(2\pi\,k\,T)^3}{h^6}\,(m_e^* m_h^*)^{3/2}\exp\left(-\frac{E_g}{k\,T}\right), \tag{2.117}$$

is no longer dependent on the Fermi level position and equals the product of the same quantities for the pure semiconductor. Thus, the impurity inclusion, which shifts the Fermi level towards the limit of the energy gap leads to the increase of the hole density in the conduction band, to the detriment of the hole density in the valency band, or inversely, the product of the two densities remaining unchanged.

Expressions (2.115) and (2.116) of the electron and hole density, respectively, can be written under a more symmetrical form by inserting the intrinsic level "E_I" and the chemical potential $\mathcal{X}$ [N5]. The quantity E_I coincides with the Fermi energy in a pure semiconductor, given by relation (2.110)

$$E_I = \frac{E_v + E_c}{2} - \frac{3kT}{4} \ln\left(\frac{m_e^*}{m_h^*}\right) = -\frac{E_g}{2} + \frac{kT}{2} \ln\left(\frac{m_h^*}{m_e^*}\right)^{3/2}. \qquad (2.118)$$

The chemical potential is equal to the difference between the Fermi energies of the impure and pure semiconductor. From relations (2.115) and (2.118) it follows

$$\mathcal{X} = E_f - E_I = -\frac{E_d}{2} + \frac{E_g}{2} + \frac{kT}{2}\left[\ln \frac{h^3 n_d}{2(2\pi m_e^* kT)^{3/2}} - \ln\left(\frac{m_h^*}{m_e^*}\right)^{3/2}\right]. \qquad (2.119)$$

By setting

$$N_e = 2\frac{(2\pi m_e^* kT)^{3/2}}{h^3}, \qquad N_h = 2\frac{(2\pi m_h^* kT)^{3/2}}{h^3} \qquad (2.120)$$

relations (2.118) and (2.119) become

$$E_I = -\frac{E_g}{2} + \frac{kT}{2} \ln \frac{N_h}{N_e}, \qquad (2.121)$$

$$\mathcal{X} = -\frac{E_d}{2} + \frac{E_g}{2} + \frac{kT}{2} \ln \frac{n_d}{N_h}. \qquad (2.122)$$

The electron density in the conduction band is given by relation (2.115)

$$n_e = \sqrt{2}\sqrt{n_d}\frac{(2\pi m_e^* kT)^{3/4}}{h^{3/2}} \exp\left(-\frac{E_d}{2kT}\right) =$$

$$= \sqrt{n_a N_e}\,\exp\left(-\frac{E_a}{2kT}\right) = \sqrt{n_a N_e}\sqrt{\frac{n_a}{N_h}}\exp\frac{2\mathcal{K}-E_g}{2kT} =$$

$$= \sqrt{N_e N_h}\,\exp\left(-\frac{E_g}{2kT}\right)\exp\left(\frac{\mathcal{K}}{kT}\right). \tag{2.123}$$

The hole density in the valency band is given by relation (2.116)

$$n_h = 2\,\frac{(2\pi\, m_h^*\, kT)^{3/2}}{h^3}\exp\left(-\frac{E_f + E_g}{kT}\right) =$$

$$= N_h \exp\left(-\frac{\mathcal{K}}{kT}\right)\exp\left(-\frac{E_g}{2kT}\right)\sqrt{\frac{N_e}{n_a}}\sqrt{\frac{n_a}{N_h}} = \tag{2.124}$$

$$= \sqrt{N_e N_h}\,\exp\left(-\frac{E_g}{2kT}\right)\exp\left(-\frac{\mathcal{K}}{kT}\right).$$

Let n be the charge-carrier density defined as

$$n = \sqrt{N_e N_h}\,\exp\left(-\frac{E_g}{2kT}\right), \tag{2.125}$$

and having the following physical meaning : n^2 is the product between the electron and hole density, given by the relation (2.117), independent of the nature and quantity of impurities.

The insertion of (2.125) into (2.123) and (2.124) simplifies them

$$n_e = n\,\exp\frac{\mathcal{K}}{kT}, \tag{2.126}$$

$$n_h = n\,\exp\left(-\frac{\mathcal{K}}{kT}\right). \tag{2.127}$$

The examination of relations (2.126) and (2.127) shows that the electron and hole density are equal when the chemical potential becomes zero. When $\mathcal{K} \neq 0$, the more different become the densities, the higher is the chemical potential (its modulus). The chemical potential will be considered again in connection with transport phenomena.

2.9. TRANSPORT PHENOMENA

The electrons in the conduction band of a semiconductor crystal form a thermodynamic system, the total energy of which generally is a function of four variables i.e., the system enthropy S, the volume $\mathcal{V}$, the number of particles N_e and the total electric charge Q. The variation of the system energy, produced by that of the above variables is given by the differential relation [L. 8].

$$\begin{aligned} \mathrm{d}E &= \frac{\partial E}{\partial S}\mathrm{d}S + \frac{\partial E}{\partial \mathcal{V}}\mathrm{d}\mathcal{V} + \frac{\partial E}{\partial N_e}\mathrm{d}N_e + \frac{\partial E}{\partial Q}\mathrm{d}Q = \\ &= T\mathrm{d}S - p\mathrm{d}\mathcal{V} + \mathcal{X}\mathrm{d}N_e + V\mathrm{d}Q, \end{aligned} \tag{2.128}$$

where T is the temperature and p the pressure of the system, $\mathcal{X}$ the chemical potential and V the electric potential. The four terms represent the heat transfer, the mechanical, the chemical and the electric field work, respectively.

The last two terms expressing the matter and the electric charge transfer are particularly important in the semiconductor thermodynamic study.

It is to be observed firstly that for the conduction electron system, the number of variables actually becomes three due to the relationship

$$Q = -e\,N_e \tag{2.129}$$

and then

$$\mathrm{d}Q = -e\mathrm{d}N_e, \tag{2.130}$$

"e" being the electron charge. Inserting the expression of the differential $\mathrm{d}N_e$, resulting from (2.130) into (2.129) we obtain

$$\mathrm{d}E = T\mathrm{d}S - p\mathrm{d}\mathcal{V} + \left(-\frac{\mathcal{X}}{e} + V\right)\mathrm{d}Q. \tag{2.131}$$

The form of expression (2.131) suggests the introduction of the "effective" electrical work concept, taking as a force variable the "effective" electric potential

$$V_{eff} = -\frac{\mathcal{X}}{e} + V, \tag{2.132}$$

the algebraic sum of the chemical and electrical potential.

In a similar way, inserting the differential dQ given by (2.130) into relation (2.128)

$$\mathrm{d}E = T\mathrm{d}S - p d\mathcal{V} + (\mathcal{X} - eV)\ \mathrm{d}N_e, \tag{2.131'}$$

the electrochemical potential

$$\mu = \mathcal{X} - eV, \tag{2.133}$$

can be introduced as a force variable in the "effective" chemical work.

As it has been demonstrated in [N 5], in metals and semiconductors the electrochemical potential identifies itself with the Fermi energy.

$$\mu = E_f\,. \tag{2.134}$$

The total current produced in shifting the conduction electrons can be evaluated by means of the effective electric potential, given by relation (2.132). For this purpose, let us consider Ohm's law for a point

$$\vec{J} = \sigma\vec{\mathcal{E}} = -\sigma\nabla V, \tag{2.135}$$

with $\vec{J}$ the current density vector, σ the electric conductivity, $\vec{\mathcal{E}}$ the electric field intensity, and V the electric potential. To simplify the calculation, we project the vectorial relation (2.135) on the x-axis

$$J_x = \sigma\,\mathcal{E}_x = -\sigma\frac{\partial V}{\partial x}. \tag{2.136}$$

Replacing the electric potential in (2.136) by the effective electric potential we obtain the total current

$$J'_x = -\sigma\frac{\partial V_{ef}}{\partial x} = \frac{\sigma}{e}\frac{\partial\mathcal{X}}{\partial x} - \sigma\frac{\partial V}{\partial x}. \tag{2.137}$$

Notice in the electron-current expression the appearance of the component due to conduction

$$J_x^{(c)} = -\sigma\frac{\partial V}{\partial x} = \sigma\,\mathcal{E}_x \tag{2.138}$$

and the component due to the diffusion process

$$J_x^{(d)} = \frac{\sigma}{e}\frac{\partial \mathcal{X}}{\partial x}. \tag{2.139}$$

The electron diffusion is a result of the chemical potential non-uniformity along the direction we have considered, and takes place to the effect of making uniform the chemical potential. As soon as the condition

$$\mathcal{X}(x) = \text{const},$$

is satisfied, the diffusion process stops.

The diffusion current is usually expressed in terms of the diffusion constant and the electron density gradient. Therefore, the Fermy energy (2.134) is introduced in the electron density expression (2.115′) for the conduction band

$$n_e = 2\frac{(2\pi m_e^* kT)^{3/2}}{h^3}\exp\left(\frac{\mathcal{X} - eV}{kT}\right) = \text{const. } \exp\left(\frac{\mathcal{X}}{kT}\right) \tag{2.140}$$

and the chemical potential is deduced as

$$\mathcal{X} = k\,T\,(\ln n_e + \text{const}), \tag{2.141}$$

Hence

$$\frac{\partial \mathcal{X}}{\partial x} = \frac{kT}{n_e}\cdot\frac{\partial n_c}{\partial x}. \tag{2.142}$$

The component due to diffusion has the form

$$J_x^{(d)} = \frac{\sigma}{e}\cdot\frac{kT}{n_e}\cdot\frac{\partial n_e}{\partial x}. \tag{2.143}$$

Substituting the conductivity σ for the expression given by the electron theory

$$\sigma = e\,n_e\,\mu_e, \tag{2.144}$$

where μ_e represents the electron mobility in crystal, we obtain

$$J_x^{(d)} = \mu_e kT\frac{\partial n_e}{\partial x} = eD_e\frac{\partial n_e}{\partial x}, \tag{2.145}$$

where $D_e = \frac{\mu_e kT}{e}$ is the electron diffusion-constant. Relation (2.145) shows that the diffusion process is actually a result of the electron-density non-uniformity inside the crystal, similarly to diffusion in gases. The density of the resulting electron current is obtained from (2.137)

$$J'_x = e\, D_e \frac{\partial n_e}{\partial x} + e n_e\, \mu_e\, \mathcal{E}_x. \tag{2.146}$$

In the same way, the expression of the hole current density can be deduced as

$$J''_x = -\, e D_h \frac{\partial n_h}{\partial x} + e\, n_h\, \mu_h\, \mathcal{E}_x. \tag{2.147}$$

The thermodynamical-equilibrium state of the electrons inside a semiconductor material is perturbed when an external electric field and a temperature gradient are applied.

Let $f(\vec{r}, \vec{k}, t)$ be the fraction of the total number of electrons the position coordinates of which are within $\vec{r}$ and $\vec{r} + \mathrm{d}\vec{r}$ and the wave vectors within $(\vec{k}, \vec{k} + \mathrm{d}\vec{k})$ at the moment t. The mean number of electrons, in the volume element taken into consideration, having wave vectors within $\vec{k}$ and $\vec{k} + \mathrm{d}\vec{k}$ will be

$$\frac{1}{4\pi^3} \cdot f(\vec{k}, \vec{r})\ \mathrm{d}\vec{k}\ \mathrm{d}\vec{r}.$$

if the distribution function f is normalized to unit volume. In the absence of some external fields and of a temperature gradient, $f(\vec{k}, \vec{r}) = f_0(\vec{k}) = f_0\,(E)$ is the Fermi-Dirac distribution function. If the electrons do not undergo a discontinuous process, as, e.g. the collision with a lattice atom, at the moment $t + \mathrm{d}t$ their coordinates and wave vectors change, becoming $\vec{r} + \dot{\vec{r}}\,\mathrm{d}t$ and $\vec{k} + \dot{\vec{k}}\,\mathrm{d}t$, respectively. Keeping only the first terms of the Taylor series expansion, we obtain

$$f\,(\vec{r} + \dot{\vec{r}}\,\mathrm{d}t,\ \vec{k} + \dot{\vec{k}}\,\mathrm{d}t,\ t + \mathrm{d}t) - f\,(\vec{r}, \vec{k}, t) = \left(\dot{\vec{r}}\,\nabla_r f + \dot{\vec{k}}\,\nabla_k f + \frac{\partial f}{\partial t}\right)\mathrm{d}t. \tag{2.148}$$

Because of the collisions, part of the electrons will have wave vector values outside the specified volume, and thus a variation with opposite sign of the distribution function takes place which we write as $-\left.\frac{\partial f}{\partial t}\right|_{\text{collisions}}$.

Inserting this term into (2.148) and considering $\dot{\vec{r}} = \vec{v}$ and $\dot{\vec{k}} = \frac{\vec{F}}{\hbar}$, the Boltzmann transport equation is

$$\frac{\partial f}{\partial t} + \vec{v}\nabla_r f + \frac{\vec{F}}{\hbar}\nabla_k f = \left.\frac{\partial f}{\partial t}\right|_{\text{collisions}} \tag{2.149}$$

whence there results that the distribution function variation in time is due to some diffusion phenomena ($\nabla_r f \neq 0$), to the application of an electric field or of an external temperature gradient ($\vec{F} \neq 0$) and to collisions. The Boltzmann equation given under this general form is difficult to interpret. Assuming a small perturbation, the collision term can be written under the form

$$\left.\frac{\partial f}{\partial t}\right|_{\text{collisions}} = \frac{f - f_0}{\tau}, \tag{2.150}$$

where τ is a relaxation time equal to the mean interval of time between two successive collisions. Then (2.149) becomes

$$\frac{\partial f}{\partial t} + \vec{v}\nabla_r f + \frac{\vec{F}}{\hbar}\nabla_k f = \frac{f - f_0}{\tau}, \tag{2.151}$$

where, for a zero magnetic field [B 33]

$$\vec{F} = e\vec{\mathcal{E}} - T\nabla\left(\frac{E_F}{T}\right) - \frac{E}{T}\nabla T. \tag{2.152}$$

In a steady state $\left(\frac{\partial f}{\partial t} = 0\right)$ a comparatively simple expression is deduced for the electric conductivity, if we consider a homogeneous material ($\nabla_r f = 0$) at constant temperature ($\vec{F} = e\,\vec{\mathcal{E}}$); in such a case (2.151) becomes

$$\frac{f - f_0}{\tau} = \frac{\vec{E}}{\hbar} \nabla_k f, \tag{2.153}$$

For small perturbations about the equilibrium one may consider

$$\nabla_k f = \nabla_k f_0 = \frac{\partial f_0}{\partial E} \nabla_k E, \tag{2.154}$$

and thus

$$f = f_0 + \frac{\tau e \vec{\mathcal{E}}}{\hbar} \frac{\partial f_0}{\partial E} \nabla_k E. \tag{2.155}$$

Here $\frac{\partial f_0}{\partial E} = -\frac{f_0 (1 - f_0)}{kT}$ for a Fermi-Dirac distribution and $E = \frac{\hbar^2 k^2}{2m}$ assuming spherical surfaces of constant energy in the wave vector space(k_x, k_y, k_z).

Knowing the function f, the electric current density can be calculated

$$\vec{J}_e = -\frac{e}{4\pi^3} \int \vec{v} f \,\mathrm{d} k. \tag{2.156}$$

The conductivity σ will be written under the form

$$\sigma_e = \frac{e^2 n_e}{m^*} < \tau > = n_e e \mu_e, \tag{2.157}$$

where

$$< \tau > = \frac{\int_0^\infty \tau (E)\, E^{3/2} f_0 \,\mathrm{d}E}{\int_0^\infty E^{3/2} f_0 \,\mathrm{d}E}. \tag{2.158}$$

The mobility of the carriers

$$\mu_e = \frac{e < \tau >}{m^*}, \tag{2.159}$$

is limited by their scattering on the lattice and on ionized impurity centers. Proceeding from (2.158) and (2.159) it is shown that when the scattering on the lattice prevails, the mobility variation with the temperature has the form $\mu \sim T^{3/2}$; when the scattering on ionized impurities prevails $\mu \sim T^{2/3}$.

2.10. RECOMBINATION PHENOMENA

The thermodynamical equilibrium conditions of a semiconductor are significantly disturbed when the charge carrier density changes under the action of some external agent (light, temperature, ionizing radiations, injection through rectifier contact).

Available mechanisms of restoring the equilibrium are: the recombination, the diffusion and the carrier drift in the electric field.

In fact, even in condition of thermodynamical equilibrium, the electrons and the holes will continuously exchange energy with the crystal lattice, a dynamical equilibrium between generation and recombination processes being thus achieved.

The principal kinetic models of an electron-hole pair recombination are assumed to be:

— the direct recombination (Fig. 2.19 *a*).

— the recombination through intermediate states (Fig. 2.19 *b*).

In each model, there are different mechanisms available to excess carriers for losing their exceeding energy under the form of photons (radiative recombination), phonon cascades (giant capture) or yielding of energy to other electrons (Auger process).

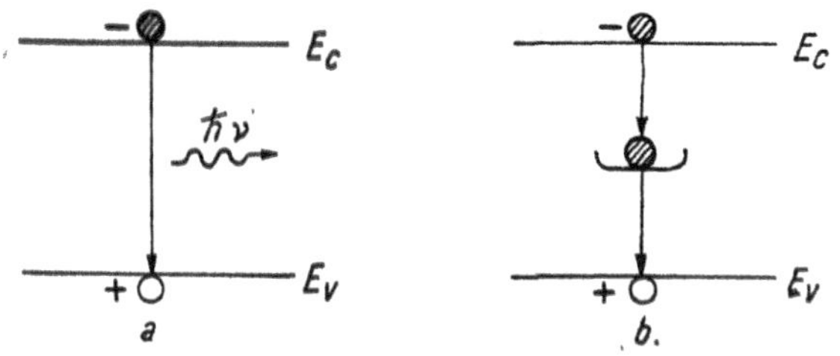

Fig. 2.19 — The recombination mechanism inside a semiconducting crystal: *a* — direct recombination, *b* — recombination through intermediate states.

In the direct recombination, the energy loss through multiple phonons is hardly probable, but a complex electron-hole (exciton) — an excited state of a valency electron — can arise, able to deexcite through radiation or Auger process.

Observe that in the semiconductor detector operation, besides the volume recombination, the surface recombination which plays a significant part in many cases, must be also taken into account.

In silicon and germanium, it is the recombination mechanism through intermediate states which determines the carrier lifetime, defined for electrons by the relation

$$\tau_e = \frac{\Delta n_e}{\frac{dn_e}{dt}}, \tag{2.160}$$

where Δn_e is the number of excess electrons and $\frac{dn_e}{dt}$ their recombination rate.

If at thermodynamical equilibrium the carrier densities are n_{e_0} and n_{h_0}, in altered conditions they become $n_{e_0} + \Delta n_e$ and $n_{h_0} + \Delta n_e$, because the carriers are assumed to be born in pairs (electron-hole).

The lifetime of the minority (or majority) carriers, inside a given material is a significant parameter for many applications. According to the definition (2.160) its estimation amounts to the calculation of the recombination rate $\frac{dn_e}{dt}$, in terms of excess carrier density Δn_e. In the following, a calculation outline in the case of Schockley-Read-Hall [M 46] recombination model for a single intermediate capture level will be given.

The fundamental processes involved in the generation and the recombination through capture are represented in Fig. 2.20.

If the trap is neutral at the beginning, an electron may occupy it when coming from the conduction band (electron capture) or from the valency band (hole emission).

When there is an electron inside the trap, it may go to the conduction band (electron emission) or to the valency band (hole capture). The recombination consists of process (*a*) followed by (*c*) and the generation of (*d*) followed by (*b*).

In a steady state the recombination centers placed on a single capture level are characterized by the electron capture cross-section $\sigma_e(\vec{k})$ and the energies involved by the corresponding transitions.

The recombination rate of electrons with wave vector $\vec{k}$ is

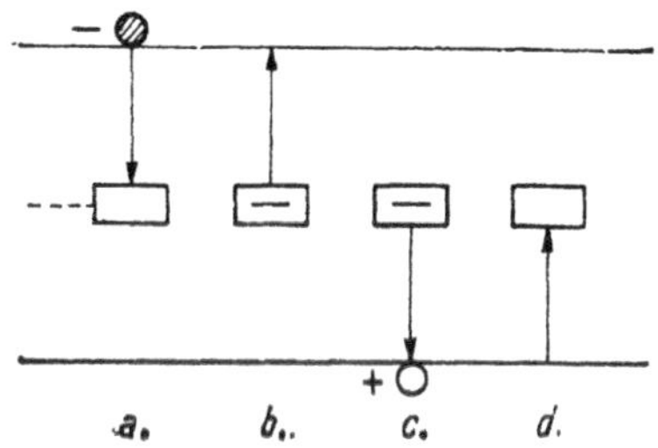

Fig. 2.20 — The main processes implied by the recombination through trapping centers: *a* — electron trapping, *b* — electron emission, *c* — hole trapping, *d* — hole emission.

$$R_e = \left< \frac{\hbar}{m} \vec{k} \sigma_e(\vec{k}) \right> N_T^0 n_e = \vec{v}_t \, \sigma_e N_T^0 n_e, \tag{2.161}$$

where

$$\left< \frac{\hbar}{m} \vec{k} \sigma_e(\vec{k}) \right> = \vec{v}_t \, \sigma_e,$$

and N_T^0 represents the density of the empty traps.

Similarly, the negatively charged traps give a hole recombination rate

$$R_h = v_t \, \sigma_h \, N_T^- \, n_h. \tag{2.162}$$

From these relations, the mean lifetimes of electrons and holes before capture result as

$$\tau_{1e} = \frac{n_e}{R_e} = \frac{1}{v_t \sigma_e N_T^0}; \qquad \tau_{1h} = \frac{1}{v_t \sigma_h N_T^-}. \tag{2.163}$$

It is shown [M 46] that the electron and hole mean lifetimes on capture levels are

$$\tau_{2e} = \frac{1}{v_t \sigma_e n_{e1}}; \qquad \tau_{2h} = \frac{1}{v_t \sigma_h n_{h1}}, \tag{2.164}$$

where n_{e_1} and n_{h_1} are the carrier densities corresponding to the Fermi level lying on the capture level ($E_f = E_T$).

The plain electron recombination rate will be

$$\frac{dn_e}{dt} = -\frac{n_e}{\tau_{1e}} + \frac{N_T^-}{\tau_{2e}} = -v_t\sigma_e\,(n_e N_T^0 - n_e N_T^-)\,, \tag{2.165}$$

and similarly, for holes

$$\frac{dn_h}{dt} = -\frac{n_h}{\tau_{1h}} + \frac{N_T^0}{\tau_{2h}} = -\,v_t\sigma_h\,(n_h N_T^- - n_{h1} N_T^0)\,. \tag{2.166}$$

It is evident that $N_T^- + N_T^0 = N_T$ = the total number of traps. At thermodynamical equilibrium

$$\frac{dn_e}{dt} = \frac{dn_h}{dt} = 0\,. \tag{2.167}$$

The final expression of the recombination rate in a steady state is [M 46]

$$\frac{dn_e}{dt} = \frac{dn_h}{dt} = -\frac{n_e\,n_h - n_i^2}{(n_e + n_{e_1})\,\tau_{h0} + (n_h + n_{e_1})\,\tau_{e_0}}\,, \tag{2.168}$$

where

$$\tau_{e_0} = \frac{1}{N_T v_t \sigma_e}\,, \tag{2.169}$$

is the minimum possible lifetime of the electrons for a recombination through capture, and in a similar manner, for holes we have

$$\tau_{h_0} = \frac{1}{N_T\,v_t\,\sigma_h}\,. \tag{2.170}$$

It is shown that in a steady state the lifetime of the majority carriers is longer than that of the minority carriers. For an N-type material

$$\tau_e = \tau_h\left(1 + \frac{N_T\,\sigma_h}{n_{e_0}\,\sigma_e}\right). \tag{2.171}$$

In the transient state, for the injection of small quantities $\Delta n_e \ll n_{e_0}$ and a sufficiently high concentration of the majority carriers with respect to the recombination center density $n_{e_0} \gg N_T$, the lifetime obtained for excess carriers [S 12] is

$$\tau = \tau_{h_0} \frac{n_{e_0} + n_{e_1}}{n_{e_0} + n_{h_0}} + \tau_{e_0} \frac{n_{h_0} + n_{h_1}}{n_{e_0} + n_{h_0}}. \quad (2.172)$$

Hence, we see that the lifetime depends on the doping type (Fig. 2.21). When the doping is heavy, the recombinations are numerous due to the great number of majority carriers available.

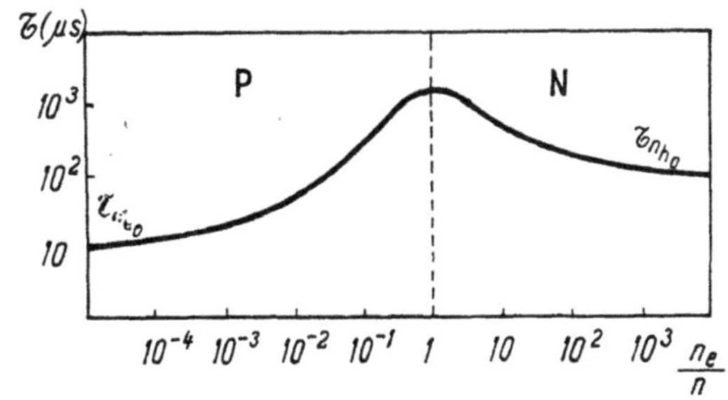

Fig. 2.21 — Dependence of minority carrier lifetime on impurity concentration.

When the impurity density decreases, the minority carrier lifetime increases, reaching a maximum value for the intrinsic material.

The lifetime τ_{eff} of the minority carriers inside a semiconductor material has two components: of volume (τ_v) and of surface (τ_s) with a composition law

$$\frac{1}{\tau_{eff}} = \frac{1}{\tau_v} + \frac{1}{\tau_s}. \quad (2.173)$$

The energy states of the energy gap lying at the surface can behave as donors, acceptors or recombination centers. Their statistical behaviour is similar to that of states lying inside the volume, having however a specific characteristic, i.e., the influence of the surface potential on the recombination.

The surface recombination rate in a P-type material is defined as the ratio of the recombination rate on surface unit $\left(\frac{dn_{es}}{dt}\right)_s$ to the excess density of the carriers at the surface

$$s_e = \frac{\left(\frac{dn_{es}}{dt}\right)_s}{n_{es} - n_{e_0}}. \qquad (2.174)$$

Its dimensions are cm/s and its value is strongly dependent on the mechanical working out and chemical treatment of the surface. In the case of recombination through a single capture level [M 46]

$$\left(\frac{dn_{es}}{dt}\right)_s = \frac{n_{hs} n_{es} - n_i^2}{\frac{n_{e_1} + n_{es}}{s_{h_0}} + \frac{n_{e_1} + n_{es}}{s_{e_0}}}, \qquad (2.175)$$

where $s_{h_0} = N_T R_2$ and $s_{e_0} = N_T R_1$. When $n_e \gg n_{e_0}$

$$s_{max} = \sqrt{\frac{kT}{2\pi m}} = \frac{\mathcal{V}_t}{\sqrt{6\pi}}. \qquad (2.176)$$

For semiconductor filaments with square cross-sections of side a

$$\frac{1}{\tau_{eff}} = \frac{1}{\tau} + \begin{cases} \frac{4s}{a}, & \text{for low } s \\ \frac{2\pi^2 D}{a^2}, & \text{for high } s \end{cases} \qquad (2.177)$$

D being here the diffusion constant of excess carriers.

The diffusion and the drift in an electric field, already presented in the above sections, are also processes leading to the restoring of thermodynamical equilibrium conditions inside a region of the semiconductor.

The continuity equations for electrons and holes are

$$\frac{dn_e}{dt} = G_e - \frac{n_e - n_{e_0}}{\tau_e} + \frac{\nabla \vec{J_e}}{e}, \qquad (2.178)$$

$$\frac{\mathrm{d}n_h}{\mathrm{d}t} = G_h \frac{n_h - n_{h_0}}{\tau_h} - \frac{\nabla \vec{J}_h}{e} \tag{2.179}$$

with the current densities given by the so-called transport equations (relations (2.146) and (2.147))

$$\vec{J}_e = e\mu_e n_e \vec{\mathcal{E}} + eD_e \nabla n_e, \tag{2.180}$$

$$\vec{J}_h = e\mu_h n_h \vec{\mathcal{E}} - eD_h \Delta n_h. \tag{2.181}$$

Therefore, the continuity equations involve all the processes contributing to the variation of the carrier number inside a given volume: the generation, the recombination and the charge drift (from inside or from outside) due to diffusion or to an electric field.

In connection with the recombination processes we also introduce the concept of diffusion length for electrons (or holes). Assume a semiconductor in a thermodynamical equilibrium state in which a hole excess with n_h' density is injected, by a certain method. The continuity equation (2.179) is then

$$\frac{d\,(n_h + n_h')}{\mathrm{d}t} = G_h - \frac{n_h + n_h' - n_{h_0}}{\tau_h} - \frac{\nabla \vec{J}_h}{e}, \tag{2.182}$$

n_h being the equilibrium density. Subtracting Eq. (2.179) from (2.182), we obtain

$$\frac{\mathrm{d}n_{h_1}}{\mathrm{d}t} = -\frac{n_h'}{\tau_h}. \tag{2.183}$$

Integrating Eq. (2.183), it results

$$n_h' = n_{h_0}' \exp\left(-\frac{t}{\tau_h}\right), \tag{2.184}$$

where τ_h is the hole lifetime defined by (2.160).

Before recombination, a hole diffusion takes place over a certain distance. The diffusion current appearing as a result of the hole excess flow, is

$$J_h = -eD_h \frac{\mathrm{d}n_h'}{\mathrm{d}x} = -eD_h \frac{1}{v_h} \frac{\mathrm{d}n_h'}{\mathrm{d}t}, \tag{2.185}$$

where "v_h" is the hole diffusion velocity. But, from the electronic theory of conductivity we know that

$$J_h = en'_h v_h = -e v_h \tau_h \frac{dn'_h}{dt}. \tag{2.186}$$

By equating Eqs. (2.185) and (2.186), we obtain

$$v_h^2 = \frac{D_h}{\tau_h}. \tag{2.187}$$

Introducing the hole diffusion length, defined as the distance the hole travels along within the time τ_h

$$L_h = v_h \tau_h , \tag{2.188}$$

and taking into account Eq. (2.187) it is determined as

$$L_h = \tau_h \sqrt{\frac{D_h}{\tau_h}} = \sqrt{D_h \tau_h}, \tag{2.189}$$

a relationship enabling us to estimate the diffusion length when the diffusion constant and the lifetime are known. The diffusion length has also a more direct meaning. Integrating the equation

$$\frac{dn'_h}{dx} = \frac{1}{v_h}\frac{dn'_h}{dt} = -\frac{n'_h}{v_h \tau_h} = -\frac{n'_h}{L_h}, \tag{2.190}$$

we obtain

$$n'_h = n'_{h0} \exp\left(-\frac{x}{L_h}\right). \tag{2.191}$$

A comparison between Eqs. (2.183) and (2.191) shows that the diffusion length means on the distance scale the same thing as the lifetime on the time scale, namely the mean of an exponential distribution. Very similar quantities are also introduced for electrons.

2.11. *P—N* JUNCTION

A *P—N* junction is considered to be the boundary region between two joined semiconductor crystals, with *P*-type and *N*-type conductivity, respectively (Fig. 2.22). Without other technological details, we mention here that the joining must be achieved so that it should not affect the structure continuity of both crystals. The processes occurring at the junction are examined, assuming the two semiconductors to be two thermodynamic systems coming into touch. Since the N-type semiconductors have available electrons in the conduction band and the P—type ones have available holes in the valency band, an electron diffusion towards the P-region and a hole diffusion towards the N-region will take place.

The electrons leaving the N-region and the holes leaving the P-region, let them positively and negatively charged, respectively. The electric field produced by these charges opposes to electron and hole flow and an equilibrium comes to be set up, when the electron and hole motion ceases. From a thermodynamic point of view, the equilibrium sets up in the whole system at the moment the chemical

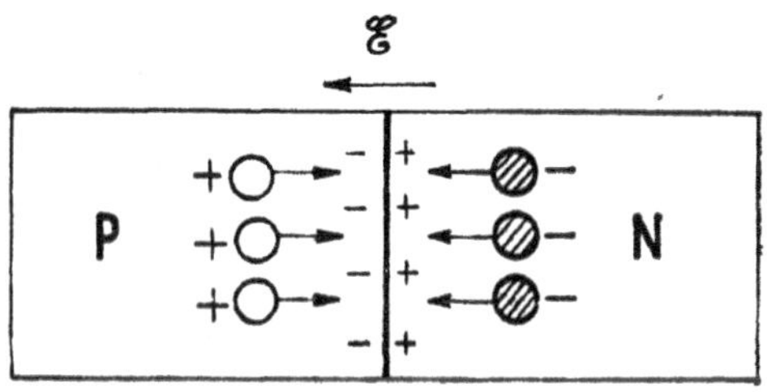

Fig. 2.22 – Formation of the $P-N$ junction.

potentials (the Fermi energies) become equal. In this way, three regions can be defined—of N-type conductivity, of P-type conductivity and of transition, or the junction itself. In the transition region the density of the free charge carriers is much lower than in the other two regions, whence the name of depletion layer.

The potential distribution at the $P-N$ junction can be obtained from Poisson's equation

$$\nabla^2 V = -\frac{\rho_v}{\varepsilon}, \tag{2.192}$$

combined with the thermodynamical equilibrium condition — the equality of the chemical potentials — already mentioned. The potential distribution along a perpendicular direction to the junction plane is examined, supposing a symmetrical junction, i.e., the Fermi level lies equally spaced from the conduction and the valency band in the

N and P-region, respectively [N 5] (Fig. 2.23). In the case of a symmetrical junction the calculations are simplified, but the generality of the conclusions is not restricted. Supposing all donor and acceptor atoms being ionized, the charge density is

$$\rho_v = e[N_d - N_a + n_h - n_e]. \tag{2.193}$$

As concerns the potential V in the Poisson equation, we could choose any of the potentials corresponding to the lower limit E_c of the conduction band, to the upper limit E_v, of the valency band or to the chemical potential itself, all of them being equivalent (they differ by a constant). For the sake of simplicity we shall choose the chemical potential*); for expressions (2.126) and (2.127) of the electron and hole density may then be used. The Poisson equation becomes

$$\nabla^2 \mathcal{X} = e \nabla^2 V = - \frac{e^2}{\varepsilon} [N_d - N_a + n_h - n_e]. \tag{2.194}$$

Substituting the electron and hole density expressions (2.126) and (2.127) we obtain

$$\frac{d^2\mathcal{X}}{dx^2} = -\frac{e^2}{\varepsilon}\left[N_d - N_a + n \exp\left(-\frac{\mathcal{X}}{kT}\right) - n \exp\left(\frac{\mathcal{X}}{kT}\right)\right] =$$

$$= \frac{2\, ne^2}{\varepsilon}\left[\operatorname{sh}\left(\frac{\mathcal{X}}{kT}\right) - q\right], \tag{2.195}$$

where

$$q = \frac{N_d - N_a}{2n},$$

is the relative density of the impurity ions. Equation (2.195) takes a simpler form when inserting the new variables

$$r = 2e \sqrt{\frac{nq}{\varepsilon k T}}\, x, \quad u = \frac{\mathcal{X}}{kT},$$

---

*) As resulting from its definition (section 2.8), the chemical potential is in fact a potential energy. A potential may be the quantity $\frac{1}{e}\mathcal{X}$.

namely,

$$2\frac{d^2u}{dr^2} = \frac{\mathrm{sh}u}{q} - 1. \tag{2.196}$$

Observe, that outside the depletion layer the potential is constant, and then

$$\left(\frac{du}{dr}\right)_{r=\pm\infty} = 0,$$

$$\left(\frac{d^2u}{dr^2}\right)_{r=\pm\infty} = 0,$$

whence

$$q = \mathrm{sh}\, u_m \qquad u_m = \frac{\mathcal{K}_N}{kT} = \frac{\mathcal{K}_P}{kT}, \tag{2.197}$$

where $\mathcal{K}_N = \mathcal{K}_P$ are the absolute values of the chemical potentials in N and P type semiconductors.

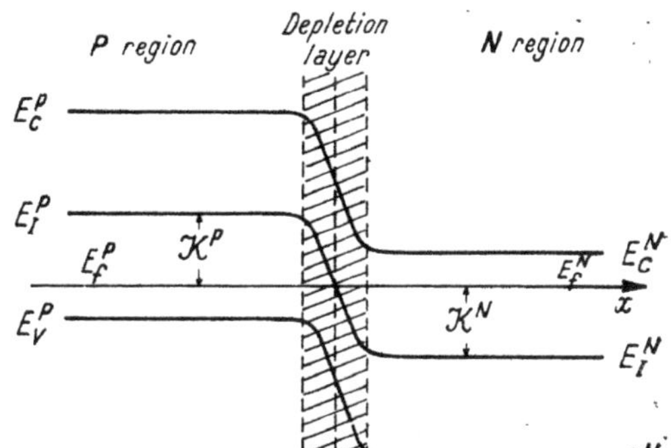

Fig. 2.23 — Energy diagram of a $P-N$ junction.

In the case of an abrupt symmetrical junction (Fig. 2.23), the bound charge distribution has the form shown in Fig. 2.24 and equation (2.196) can be simply integrated

$$\left(\frac{du}{dr}\right)^2 = \frac{\mathrm{ch}u}{\mathrm{sh}u_m} - u + C_1. \tag{2.198}$$

The constant of integration results from the boundary condition

$$r = \pm \infty, \quad u = u_m, \frac{du}{dr} = 0, \tag{2.199}$$

and equation (2.198) becomes

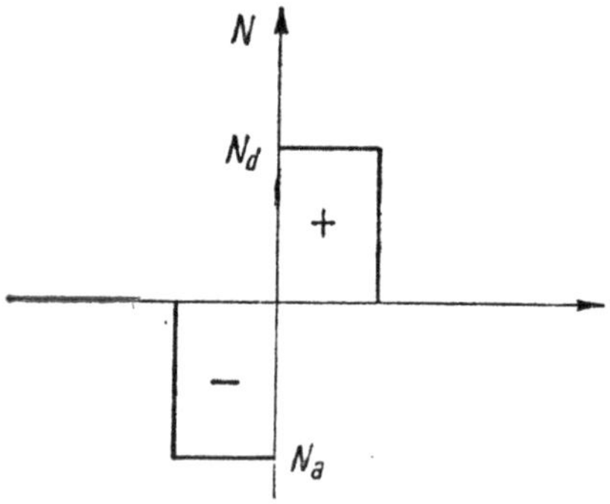

Fig. 2.24 — Distribution of bound electric charges in an abrupt symmetrical $P-N$ junction.

$$\left(\frac{du}{dr}\right)^2 = \frac{\text{ch}u - \text{ch}u_m}{\text{sh}u_m} - (u - u_m). \tag{2.200}$$

The integration of equation (2.200) leads to a solution $u(r)$ under implicit form

$$r = \int_0^u \frac{du}{\pm \sqrt{\frac{\text{ch}u - \text{ch}u_m}{\text{sh}u_m} - (u - u_m)}}. \tag{2.201}$$

the plus sign corresponding to the domain $r>0$ (towards the N-region), and the minus sign to the domain $r<0$ (towards the P-region). It has been found that, when two semiconductor crystals are joined, a potential difference appears between them, behaving like a barrier for the majority charge carriers (electrons from N-type semiconductors and holes from P-type semiconductors)

$$V_0 = \frac{1}{e}(\mathcal{X}^P + \mathcal{X}^N) = \frac{1}{e}(E_f^N - E_f^P). \tag{2.202}$$

However, the barrier V_0 allows the electron and the hole (minority charge carriers) circulation from the P-region towards the N-region and in opposite direction, respectively. The height of the potential barrier (also called junction built-in potential), can be easily estimated, supposing the impurity levels very close to the forbidden band limits $E_f^N \simeq E_c$, $E_f^P \simeq E_v$ and then

$$V_0 = \frac{1}{e}(E_c - E_v) = \frac{1}{e} E_g . \tag{2.203}$$

The junction built-in potential value is equal to the width of the forbidden band (in eV units), therefore is about one volt in germanium and silicon.

In the case of an abrupt asymmetrical junction, the problem is solved in the same way, with the difference that the lower limit of the integral (2.201) is no longer zero, because at $r = 0$, $u = u_0 \neq 0$. The value u_0 is deduced from the continuity condition of the potential derivatives at the point $r = 0$

$$\left(\frac{\mathrm{d}u}{\mathrm{d}r}\right)_{r=0-\varepsilon} = \left(\frac{\mathrm{d}u}{\mathrm{d}r}\right)_{r=0+\varepsilon}$$

In practice, because of reasons we shall discuss later on, the P—N junction works with an external voltage applied. Let us estimate the current resulting at the junction when applying this external voltage V_{ext} (Fig. 2.25).

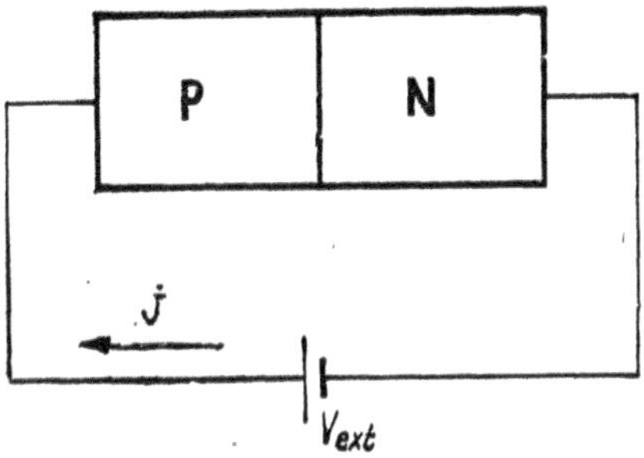

Fig. 2.25 — P—N junction with external voltage applied on.

Obviously, the electric field disturbs the thermodynamical equilibrium state and the two chemical potentials are no longer equal.

Since the free charge density is very low at the junction, the disturbing electric field actually appears only in the junction region. It follows that the difference between the chemical potentials in the

N and P-region, from outside the junction, is equal to the Coulombian energy

$$E_f^N - E_f^P = eV_{ext}\,. \tag{2.204}$$

In the depletion layer, both chemical potentials, E_f^N for electrons and E_f^P for holes, will appear, as shown in Fig. 2.26.

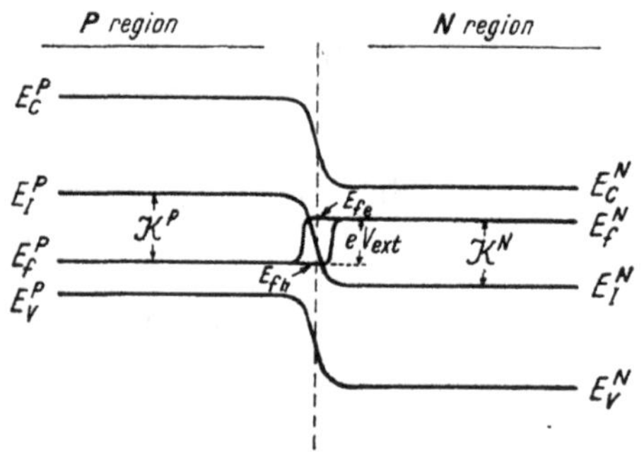

Fig. 2.26 — Energy diagram in a $P-N$ junction with external voltage applied on.

Because the chemical potentials (the Fermi levels) are defined only for equilibrium states, E_f^N and E_f^P are usually called "Fermi quasilevels".

Consider at the beginning the P-region ($r<0$). In the zero electric field region, the chemical potentials for electrons and holes are identical and the electron and hole densities are similar to those in the equilibrium state (Eqs. (2.126) and (2.127))

$$\begin{aligned} n_e^P &= n \exp\left(-\frac{\mathcal{K}^P}{kT}\right), \\ n_h^P &= n \exp\left(\frac{\mathcal{K}^P}{kT}\right). \end{aligned} \tag{2.205}$$

The appearance of different chemical potentials for electrons and holes

$$\begin{cases} \mathcal{K}_e^P = E_I^P - E_f^N, \\ \mathcal{K}_h^P = E_I^P - E_f^P, \end{cases} \tag{2.206}$$

leads to a change of the equilibrium densities (2.205).

Considering Eq. (2.204), the chemical potentials can be written under the form

$$\mathcal{X}_e^P = E_I^P - (E_f^P + eV_{ext}) = \mathcal{X}^P - eV_{ext},$$
$$\mathcal{X}_h^P = E_I^P - E_f^P = \mathcal{X}^P, \tag{2.207}$$

and then

$$(n_e^P)' = n \exp\left(-\frac{\mathcal{X}_e^P}{kT}\right) = n_e^P \exp\left(\frac{eV_{ext}}{kT}\right), \tag{2.208}$$
$$(n_h^P)' = n \exp\left(\frac{\mathcal{X}_h^P}{kT}\right) = n_h^P.$$

We observe that by applying an external electric field, an electron (minority carrier) excess is created while the hole (majority carrier) density remains unchanged. We call this an "electron injection". A diffusion of this electron excess

$$\Delta n_e^P = (n_e^P)' - n_e^P = n_e^P \left[\exp\left(\frac{eV_{ext}}{kT}\right) - 1\right], \tag{2.209}$$

takes place towards the N-region, giving rise to an electric current. The current density is given by the relationship (2.146), the conduction component

$$J_e = eD \frac{\partial n_e}{\partial x} \tag{2.210}$$

being neglected for not very strong fields.

Before recombining, every electron injected inside the P-region travels a mean distance L_e, called "electron diffusion length". In this way, only the electrons injected at distances shorter than the diffusion length L_e (from the junction) will contribute to the formation of the diffusion current (2.210).

The ratio of the electron excess to "effective" thickness of the P-region gives the density gradient in (2.210) and hence

$$J_e' = eD_e \frac{\Delta n_e^P}{L_e} = \frac{eD_e n_e^P}{L_e}\left[\exp\left(\frac{eV_{ext}}{kT}\right) - 1\right], \tag{2.211}$$

under the assumption that we may neglect the electron recombination inside the depletion layer (this last hypothesis limits the calculation validity, in the case of not very thick junction).

A hole injection in the N-region, leading to a hole diffusion current generation

$$J_h = \frac{eD_h n_h^N}{L_h}\left[\exp\left(\frac{eV_{ext}}{kT}\right) - 1\right] \tag{2.212}$$

takes place in a similar way.

The density of the total current across the junction is

$$J = J_e + J_h = J_s\left[\exp\left(\frac{eV_{ext}}{kT}\right) - 1\right], \tag{2.213}$$

where

$$J_s = e\left[\frac{D_h n_h^N}{L_h} + \frac{D_e n_e^P}{L_e}\right]. \tag{2.214}$$

Notice that when negative voltages are applied ($V_{ext} < 0$) (in the case of semiconductor detectors), the current reaches soon the saturation value $-J_s$. In fact, since at room temperature $kT \sim 0.03$ eV, a negative voltage higher than 0.1 V is sufficient to make the exponential term negligible in comparison with unity. The current

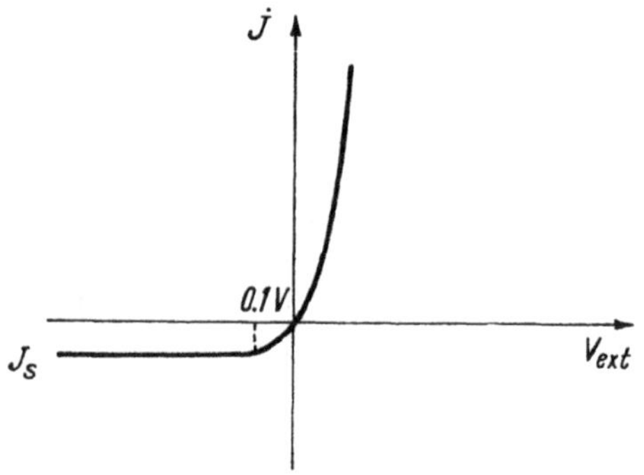

Fig. 2.27 — The volt-ampere characteristic of an ideal abrupt $P-N$ junction.

increases exponentially as the temperature increases, when $V_{ext} > 0.1$ V. At voltages still higher ($V_{ext} \sim 1$ V), the potential barrier no longer exists and the majority carriers provide the electric current, but this is of no practical interest for radiation detectors. The volt-ampere characteristic $J-V_{ext}$ of an abrupt ideal $P-N$ junction is plotted in Fig. 2.27.

The $P-N$ junction may be considered as a capacitor with the depletion layer, poor in free charge carriers, as dielectric and the P and N-regions as metallic plates. For an abrupt, possibly asymmetrical, junction with a bound charge distribution as shown in Fig. 2.28, the Poisson equation takes the form

$$\frac{d^2V}{dx^2} = -\frac{\rho_v}{\varepsilon} \qquad \rho_v = \begin{cases} -eN_a \ldots & -x_a \leqslant x \leqslant 0, \\ eN_d \ldots & 0 \leqslant x \leqslant x_d \end{cases} \tag{2.215}$$

when the free charges are neglected.

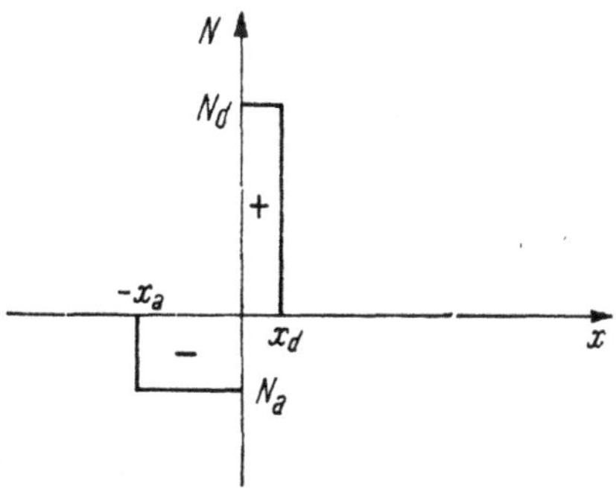

Fig. 2.28—Bound electric charge distribution at an abrupt, asymmetrical $P-N$ junction.

Integrating equation (2.215) over the range $-x_a \leqslant x \leqslant 0$, we obtain

$$\frac{dV_1}{dx} = \frac{eN_a}{\varepsilon} x + C_1, \tag{2.216}$$

$$V_1(x) = \frac{eN_a}{\varepsilon} \cdot \frac{x^2}{2} + C_1 x + C_1'. \tag{2.217}$$

From the condition $x = 0$, $V_1 = 0$, there results $C_1' = 0$.

From the boundary condition of the derivatives at the point $x = -x_a$, there results

$$\left(\frac{dV_1}{dx}\right)_{x=-x_a} = -\frac{eN_a}{\varepsilon} x_a + C_1 = 0, \tag{2.218}$$

$$C_1 = \frac{eN_a}{\varepsilon} x_a. \tag{2.219}$$

The electric potential inside the region $-x_a \leqslant x \leqslant 0$ has the expression

$$V_1(x) = \frac{eN_a}{\varepsilon} x\left(x_a + \frac{x}{2}\right). \tag{2.220}$$

In the same way, inside the region $0 \leqslant x \leqslant x_d$ it is

$$V_2(x) = \frac{eN_d}{\varepsilon} x\left(x_d - \frac{x}{2}\right). \tag{2.221}$$

The abscissae delimiting the bound charge distribution (and implicitly, the depletion layer) can be obtained writing the neutrality condition for the system

$$N_a x_a = N_d x_d, \tag{2.222}$$

and the obvious relationship

$$x_a + x_d = d, \tag{2.223}$$

where d is the total thickness of the depletion layer, equal to the distance between the plates of the condenser represented by the junction, Solving the system of equations (2.222) and (2.223) for x_a and x_d, we obtain

$$\begin{cases} x_a = \dfrac{N_d}{N_a + N_d} d, \\ x_d = \dfrac{N_a}{N_a + N_d} d. \end{cases} \tag{2.224}$$

Imposing the condition that the difference of the potentials at the depletion layer boundaries should be equal to the total voltage

$$V_2(x_d) - V_1(-x_a) = V_0 + V_{ext}, \tag{2.225}$$

where V_0 is the junction built-in potential in the equilibrium state and V_{ext} the reverse external voltage applied, it results

$$\frac{eN_d}{\varepsilon} x_d\left(x_d - \frac{x_d}{2}\right) + \frac{eN_a}{\varepsilon} x_a\left(x_a - \frac{x_a}{2}\right) = V_0 + V_{ext}, \tag{2.226}$$

and considering expressions (2.224)

$$V_{ext} + V_0 = \frac{eN_d}{\varepsilon}\frac{x_d^2}{2} + \frac{eN_a}{\varepsilon}\frac{x_a^2}{2} = \frac{ed^2}{2\varepsilon}\frac{N_a N_d}{N_a + N_d}. \qquad (2.227)$$

The depletion layer thickness results

$$d = \sqrt{\frac{2\varepsilon}{e}\left[\frac{1}{N_a} + \frac{1}{N_d}\right][V_0 + V_{ext}]}, \qquad (2.228)$$

and also the $P - N$ junction capacity (on surface unit)

$$C_s = \frac{\varepsilon}{4\pi d} = \frac{1}{4\pi\sqrt{2}}\frac{\sqrt{\varepsilon e}}{\sqrt{\left(\frac{1}{N_a} + \frac{1}{N_d}\right)(V_0 + V_{ext})}}. \qquad (2.229)$$

Examining the relationship (2.228), we observe that the depletion layer thickness control is possible through the external reverse applied voltage, which is essential in the radiation detector working.

2.12. PHENOMENA AT THE METAL-SEMICONDUCTOR INTERFACE

Processes similar to those met with at a $P-N$ junction also occur in the system obtained by joining a metal to a semiconductor. The investigation of such a system is particularly important for the radiation detection problem, since the working of a whole class of semiconductor detectors — the surface barrier detectors — relies on the metal-semiconductor system properties. Unfortunately, the understanding of the processes occurring at the metal-semiconductor boundary is at its beginning [W 8] and much behind that of the processes at the $P-N$ junction, for example. This is a fact objectively due to the complexity of the processes involving the surface state excitation. In the following, we shall use the analogy with the $P-N$ junction in the qualitative analysis of the phenomena occurring at the metal-semiconductor frontier.

Assume a metal joined to an N-type semiconductor. Due to the difference between the metal and semiconductor work function, an electric field appears at the interface [I 3]. If the metal has a work function higher than the semiconductor, this field tends to orientate the semiconductor conduction electrons inwards, depriving of free charges the region between the metal and the semiconductor. Thus, the electric potential increases in this region of the semiconductor and the conduction band shifts from the semiconductor Fermi level, coincident at thermodynamical equilibrium with the electron Fermi level of the metal (Fig. 2.29 *a*).

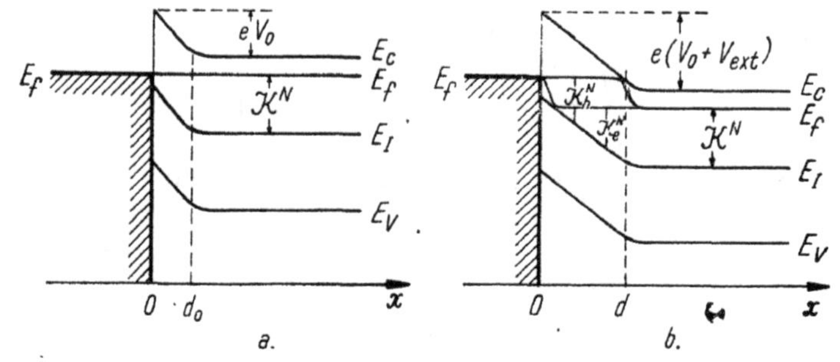

Fig. 2.29 — Energy diagram of a metal — N-type semiconductor system : a — without voltage applied, b — with external voltage applied on.

Similarly, it is shown that depletion layers with very low free charge-carrier density can be obtained when a P-type semiconductor is joined to a metal. But, in this case the metal must have a lower work function than the semiconductor.

The field and the potential of the metal — semiconductor system are estimated according to Schottky's model [G 19] which assumes that :

— the junction built-in potential satisfies the condition : $eV_0 \gg kT$,

— the depletion layer width is much larger than the Debye length

$$L_D = \sqrt{\frac{\varepsilon k T}{2 n_e n_h e^2}},$$

— the bound charge distribution has the form

$$\begin{cases} N_a \ldots 0 \leqslant x \leqslant d, \\ 0 \ldots d < x, \end{cases}$$

— the hole density at the metal — semiconductor interface is not higher than the bound charge density

$$n_h(0) \leqslant N_a .$$

When a reverse external voltage is applied (Fig. 2.30), as in the case of a $P-N$ junction, the depletion layer thickness increases and two different Fermi quasilevels and chemical potentials appear for electrons and holes (Fig. 2.29 *b*).

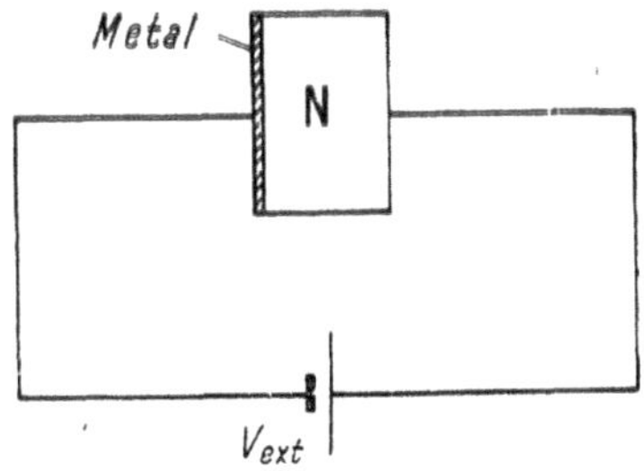

Fig. 2.30 — Metal — N-type semiconductor system with reverse voltage applied.

The integration of the Poisson equation

$$\frac{d^2V}{dx^2} = -\frac{eN_d}{\varepsilon}, \tag{2.230}$$

gives

$$V(x) = -\frac{eN_d}{\varepsilon}\frac{x^2}{2} + C_1 x + C_2. \tag{2.231}$$

Choosing the potential origin so that $V(0) = 0$, we obtain : $C_2 = 0$. The limit condition of the potential derivatives at $x = d$, gives the value of the other constant

$$C_1 = \frac{eN_d d}{\varepsilon}, \tag{2.232}$$

and Eq. (2.231) becomes

$$V(x) = \frac{eN_d}{\varepsilon} x\left(d - \frac{x}{2}\right). \tag{2.233}$$

The electric field behaves within the domain $0 \leqslant x \leqslant d$ as

$$\mathcal{E} = -\frac{dV}{dx} = \frac{eN_d}{\varepsilon}(d - x). \tag{2.234}$$

The potential and electric field variation inside the depletion layer are drawn in Fig. 2.31. Obviously, inside the remaining region of the semiconductor

$$V = \text{const.}, \ \mathcal{E} = 0.$$

The depletion layer thickness may be estimated depending on the external voltage applied, from the potential continuity condition at $x = d$

$$V(d) = \frac{eN_d}{\varepsilon}\frac{d^2}{2} = V_0 + V_{ext}. \tag{2.235}$$

It follows that

$$d = \sqrt{\frac{2\varepsilon(V_0 + V_{ext})}{eN_d}}. \tag{2.236}$$

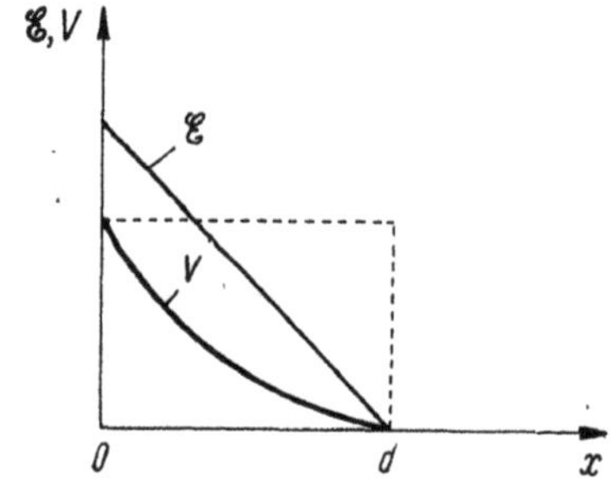

Fig. 2.31 — Electric field and potential distribution inside the depletion region.

The metal-semiconductor system may also be considered as a condenser, the unit surface capacity resulting from Eq. (2.236)

$$C_s = \frac{\varepsilon}{4\pi d} = \frac{1}{4\pi\sqrt{2}}\frac{\sqrt{\varepsilon e}}{\sqrt{\frac{1}{N_d}(V_0 + V_{ext})}}. \tag{2.237}$$

Comparing this expression to (2.229), we find that in this case, as far as the depletion layer width is concerned, the metal is equivalent to a P-type semiconductor, with $N_a \gg N_d$.

In spite of its qualitative character, Schottky's model allows a good description of the electrostatic phenomena i. e., potential, electric field, capacity, inside the depletion layer.

Another model, proposed by Bethe [G 19], though similar, enables us to consider the effects produced by semiconducting oxide layers appearing at the metal — semiconductor interface. The Bethe model, that we shall not present in detail, gives for the depletion layer thickness

$$d = \sqrt{\frac{2\varepsilon(V_0 + V_{ext})}{eN_d} + \delta^2} \tag{2.238}$$

an expression, different from that of the Schottky model (2.236) by a δ^2 term. The quantity δ represents the thickness of the insulating region between metal and semiconductor and has been introduced in order to take into account the effects of the oxide layer.

No one of these models provides an accurate description of the electrodynamic phenomena and particularly with the deduction of the current-voltage characteristic for a metal—semiconductor system.

CHAPTER III

WORKING PRINCIPLES OF NUCLEAR RADIATION SEMICONDUCTOR DETECTORS

The collection of the charges yielded by ionization in diamond, NaI, CsI, AgCl monocrystals has been comprehensively studied in the decade following the Second World War, without obtaining however remarkable results for detection. Only since 1949, very pure germanium and silicon (used in the transistor technology) have been available. McKay [M 28] has undertaken the study of the germanium as a charged particle detector. In a reversely biased junction, he could obtain a rather strong electric field for the charge collection with a reasonable reverse current and thus with sufficiently low noise.

Between 1949 and 1958, besides McKay from the Bell Telephone Laboratory, another group (Mayer, Gossik *et al.*) from Purdue University, led by Lark Horowitz, has studied this problem, being the first in using thin junctions with wide areas [G 18]. In the same period of time, good results have also been obtained and published in 1957 by Airapetiantz and Ryvkin [A 2] from the Institute of Technical Physics in Leningrad.

The detection geometry was not the most adequate at the beginning [M 14]. Thick *PN* junctions, of rectifier type, were used at that time. The detector sensitive region front surface, used by McKay was the narrow edge of an $N-P$ junction depletion layer (Fig. 3.1.a).

In this arrangement, the effective available surface was small and thus poor efficiency and energy resolution was obtained. The particles, falling on the neutral N and P region, generate electron-hole pairs which diffuse towards the junction, and there are collected by the strong field of the depletion layer. During the diffusion, the pair velocity is comparatively low and therefore, many charges recombine before being collected. It is evident that particles with equal energies falling on the N or P region will give rise to electric charge pulses with different amplitudes depending on the distance between their tracks and the junction.

In the geometry suggested by Airapetiantz and Ryvkin (Fig. 3.1 b) all electron-hole pairs were collected by diffusion and travelled identical paths, no matter the place the alpha particles were falling on. Because of important and non-uniform recombination losses within the detector volume, the resolution was not significantly improved.

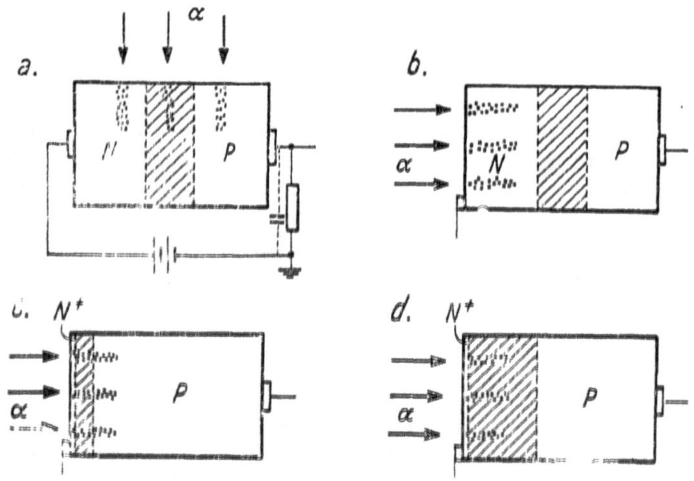

Fig. 3.1 — **Detection geometries : *a* — thick diffused junction sidewise irradiated, *b* — thick diffused junction under frontal irradiation, *c* — thin surface barrier layer, *d* — thick surface barrier layer.**

For the same reason, Gossik and Mayer could not obtain in 1956 a good energy resolution with the first barrier layer detectors (Au—Ge) (Fig. 3.1.c).

It was only when using very pure semiconductor materials that surface barrier layers with thickness equal to the range of detected alpha particles (Fig. 3.1.d) could be obtained and the solid state advantages emphasized.

The successful work of the Purdue group and the use in 1958 of Ge junctions at low temperature in nuclear physics experiments by Dabbs, Walter, Roberts, determined a rapid advance of the studies pertaining to the technology of such detectors. In 1960 at the VI-th IRE Scintillation and Semiconductor Detector Symposium [4], 25 papers concerning radiation semiconductor detectors were already presented.

Besides Au-Si surface barrier devices, N^+P diffused silicon junction and lithium drifted *PIN* diodes starting from a *P*-type basic material of comparatively low resistivity are also used. The silicon is used in many applications because it can operate at room temperature. The devices made from germanium need cooling at liquid nitrogen or helium temperature.

The fundamental processes occurring in semiconductor devices used for radiation detection and inducing the desired signal in the external circuit represent the object of this chapter.

3.1. CHARGE-CARRIER INJECTION. THE MEAN ENERGY FOR AN ELECTRON-HOLE PAIR PRODUCTION

The heavy charged-particles of moderate energies penetrating in matter are quickly slowed down and produce ionizations by interactions with the outer electrons.

The maximum energy E_M, a heavy particle with mass M and energy E is able to transfer to an electron (with mass m) is [B 45]

$$E_M \simeq 4 \frac{m}{M} E, \tag{3.1}$$

e. g., it is about 2 keV for alpha particles of 4 MeV energy.

Due to this collision process, the semiconductor electrons can be raised from the valency to the conduction band or from the lower occupied to the upper empty bands (Fig. 3.2) with electron-hole pair yield.

The semiconductor remains in this excited state for about 10^{-12} s, afterwards the electrons falling to the bottom of the conduction band and the holes rising to the top of the valency band, due to interactions among electrons and holes.

During this de-excitation process many holes and electrons appear. This many step process is, on an average, followed by an electron-hole pair creation for an $\in = 3.6$ eV energy loss of the heavy particle in Si. Therefore, for a total energy loss ΔE in Si, an electron-hole pair number N_p appear, no matter the nature and the initial energy of the particles

$$N_p = \frac{\Delta E}{\in}, \tag{3.2}$$

We may say that the charged particles inject the charge-carriere into the semiconductor.

The mean energy ($\in$) required for the production of an electron-hole pair is a constant of the material, very important for the nuclear radiation spectrometry with semiconductor devices.

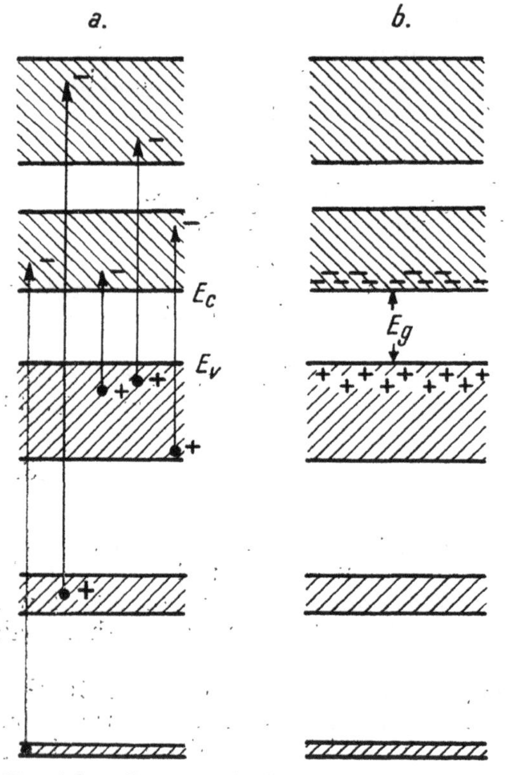

Fig. 3.2 — Charge carrier injection : *a* — initial excitation of electrons ; *b* — remanent excitation of electrons and holes after 10^{-12} s [B 45].

Shockley [S15] has given a phenomenological model of the secondary ionization effects in the solid state, explaining rather satisfactorily the experimental results for silicon.

Each ionization requires an energy E_i from the kinetic energy of the carrier producing it. In addition, before the ionization takes place, the particle transfers an energy $r.E_R$ to the phonon (lattice vibrations). The final kinetic energy of the carrier produced by ionization, energy which is also transferred to phonons, is denoted by E_f. For parabolic surfaces of constant energy in the momentum space $E_f = 0.6\ E_i$. Summing up all the contributions, we obtain

$$\in = 2E_f + E_i + rE_R, \quad (3.3)$$

or

$$\in = 2{,}2E_i + rE_R. \quad (3.4)$$

The significant phonon-losses of charge-carriers in strong fields involve phonons with high energies (frequencies). Phonons with high energies (Raman phonons) represent vibrations in opposite directions of both face-centered cubic subsystems of the diamond-type crystalline structures met with in Si and Ge.

The energy of a phonon with Raman frequency is $E_R = 0.063$ eV for Si and 0.037 eV for Ge. E_i is the carrier threshold energy required for an ionization and is taken $E_i \equiv E_g$. Here r is the ratio of the mean free path between two ionizations of a carrier with energy higher than E_i (L_i) to the mean free path between two scatterings on Raman phonons (L_R)

$$r = \frac{L_i}{L_R}. \tag{3.5}$$

For silicon, $r = 18$ and for germanium $r = 57$ [G 11]. For these values we obtain $\epsilon = 3.5$ eV in Si and $\epsilon = 3.6$ eV in Ge. The experimental result of McKay and Fabri $\epsilon = 3.6$ eV was checked up for Si, but for Ge, the measurements of McKay, Zareba, McKenzie and Tavendale gave a lower value (2.8—2.95 eV). $\epsilon = 2.9$ eV for Ge and 3.6 eV for Si [G 11] are the values usually adopted.

The measurements of Fabri, Gati and Svelto [F 2] show for silicon a dependence on temperature of the material constant ϵ of the form plotted in Fig. 3.3. Interesting results were also obtained by Emery and Rabson [E 7].

In solids, the range of low energy electrons yielded by collisions is short (~ 0.1 μm), hence, the primary heavy particle track is surrounded by a thin cylindrical plasma of holes and electrons. The plasma pair density is very high ($\sim 10^{16}$ pairs/cm^3) even for primary particles of relativistic velocity, with a stopping power in Si about 4 MeV/cm, (i. e. $\sim 1.2 \times 10^6$ pairs/cm are yielded).

The circumstances change for incident electrons with high energies. By collisions with the atomic electrons, a greater quantity of energy is transferred, so that the secondary electrons have longer ranges and the electron-hole pairs yielded by the subsequent cascade are much more spread over and their density is correspondingly lower.

The slowed down particle reaches a limit energy below which the probability of pair yield becomes very low and the coulombian collisions with the solid nuclei are the prevailing interactions. This energy value is considered [B 45] to be

$$E_i = \frac{1}{8} \cdot \frac{M}{m_0} I, \tag{3.6}$$

where I is the minimum excitation energy of the electron system, equal to E_g in the case of semiconductors.

In Si, the alpha particles give $E_i \simeq 1$ keV, an energy lower than the best resolution obtained up to now in solid state detectors. E_i is higher for heavy ions and may contribute to a significant increase of $\in$.

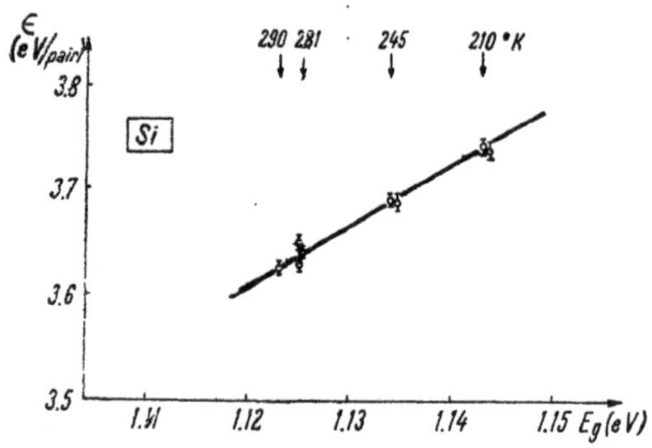

Fig. 3.3 — The mean energy necessary for an electron-hole pair production as a function of the energy gap variation vs temperature in Si [F 2].

The particle stopping time can be determined [T 12] from the relationship

$$t_s = \int_0^R \frac{dx}{v}, \tag{3.7}$$

where v is the particle velocity and R its range.

In the case of a uniform slowing down, the upper limit of the stopping time is

$$t_{sM} = \frac{2F}{v_0}, \tag{3.8}$$

when v_0 is the initial velocity of the particle.

In semiconductor devices, a typical case is that of $R = 0.5$ mm when $t_{sM} = 4 \times 10^{-12}$ s for electrons and 2.6×10^{-11} s for protons and alpha particles.

Taking into account the variation of dE/dx with the energy, the slowing down can be considered as uniform and since $\frac{dE}{dx}$ increases with x, the stopping time obtained is still lower. For electrons, for instance the stopping time is 2.8×10^{-12} s instead of 4×10^{-12} s.

In most cases, the stopping time is negligible in comparison with the carrier collection time.

3.2. THE DRIFT OF CHARGE-CARRIERS IN THE ELECTRIC FIELD. THE SHAPE OF THE CURRENT AND VOLTAGE PULSE GIVEN BY THE COLLECTION OF A SINGLE PAIR

The drift in an electric field $\mathcal{E}$ of charge carrier pairs (electrons and holes) injected by ionization is similar to electron and ion pair drift inside a gas-filled ionization chamber (Fig. 3.4) and produces an electric current in the external circuit. The collected charge (the current

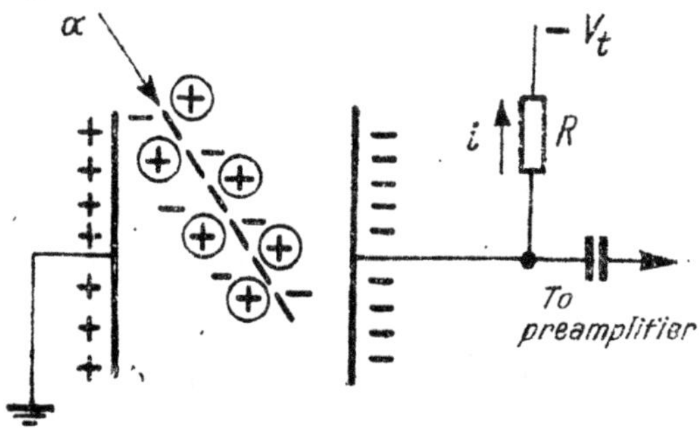

Fig. 3.4 — Solid state ionization chamber.

integral) can be used as a measure of the primary particle energy loss because ϵ in a given material does not depend on the particle energy and type.

Drifting in opposite directions in the electric field (due to the opposite signs of their electric charges), the electrons and the holes gain energy from the field at a rate

$$\frac{\mathrm{d}E}{\mathrm{d}t} = ev\mathcal{E}\cdot \tag{3.9}$$

The energy E is extracted from the energy accumulated by the capacitor C, having a charge Q due to the applied voltage V_t. Hence

$$\frac{\mathrm{d}E}{\mathrm{d}t} = \frac{\mathrm{d}}{\mathrm{d}t}\left(\frac{Q^2}{2C}\right) = \frac{Q}{C}i = V_t i. \tag{3.10}$$

The current i is given by the condenser charge variation $\mathrm{d}Q'$ due to the carrier drift for a time $\mathrm{d}t$.

From Egs (3.10) and (3.11) it follows that

$$i = \frac{ev\mathcal{E}}{V_t}, \tag{3.11}$$

or

$$i = e\frac{\mathrm{d}x}{\mathrm{d}t}\cdot\frac{\mathrm{d}V}{\mathrm{d}x}\cdot\frac{1}{V_t} = \frac{e}{V_t}\cdot\frac{\mathrm{d}V}{\mathrm{d}t}, \tag{3.12}$$

where $\frac{\mathrm{d}V}{\mathrm{d}t}$ represents the time derivative of the potential at the point where the carrier is due to its drift in the field. Integrating this relation over time, we obtain the known expression of the effective variation of the electrode charge, due to a carrier travelling inside the fraction $\frac{\Delta V}{V_t}$ of the total voltage on the condenser

$$Q_{\mathrm{ef}} = \int i\mathrm{d}t = \frac{e}{V_t}\,\Delta V. \tag{3.13}$$

This result is valid for a uniform electric field between the condenser plates (it does not change with the distance). i. e. when the potential distribution is linear.

For devices like $P-N$ junctions, whithin the sensitive region of which the electric field has a linear dependence on the distance we obtain a more general relationship [C 3], from which Eq. (3.13) results as a particular case. In the case of a fixed space charge, the potential energy variation must be taken into account in the energy balance of this process. Instead of Eq. (3.13) using Ramo's theorem, generalized by Jen [J 2], one obtains [C 3], [T 13] the relation

$$Q_{eff} = e\frac{\Delta x}{d}, \tag{3.14}$$

where Q_{eff} is the charge induced on the condenser plates when the charge carrier travels the distance Δx. The time differential of this expression gives the current induced in the external circuit

$$i = \frac{e}{d}\cdot\frac{\mathrm{d}x}{\mathrm{d}t}. \tag{3.15}$$

The same expression can be also obtained [T 13] from the continuity equation of the electric current

$$\operatorname{div}\left(\vec{I} + \frac{\mathrm{d}\vec{D}}{\mathrm{d}t}\right) = 0, \tag{3.16}$$

by integrating it over a volume, the boundaries of which are a plane passing through the moving particle and one of the electrodes. It may be also derived from the general expression found by Gunn [G 32] and written for semiconductor devices.

3.2.1. *Uniform electric field.* With the plane-parallel geometry shown in Fig. 3.4, and with a homogeneous material between the condenser plates, the potential gradient is constant and the electric field is uniform all over the thickness d of the device (except the fringe effects, not taken into account here). This is the case of *SD* with homogeneous semiconducting material and of *NIP*-type *SD*.

If V_t is the voltage applied on the condenser, the value of the electric field $\mathcal{E}$ at any point between its plates is

$$\mathcal{E} = \frac{V_t}{d}. \tag{3.17}$$

In not very strong fields inside the semiconductor, the carrier drift velocity is proportional to the field intensity

$$v = \mu\mathcal{E} \tag{3.18}$$

and then

$$i = \frac{ev\mathcal{E}}{V_t} = \frac{e\mu\mathcal{E}^2}{V_t} = \frac{e\mu}{d^2} V_t. \tag{3.19}$$

As distinguished from gases where the electron mobility is two orders of magnitude higher than the ionic one, in Si $\mu_e = 1{,}200$ and $\mu_h \simeq 500$ cm²/V. s. (at room temperature).

The total current produced by a particle is equal to the sum of the contributions of the form (3.19) of all created pairs.

The current pulse shapes resulting from the movement of an electron-hole pair created at different points inside the condenser (Fig. 3.5*a*) are shown in Fig. 3.5 *c*. In the case denoted by (*1*), the electron goes through all the field (Fig. 3.5.*b*); in the case (*2*) the hole is the active carrier, and in (*3*) both carriers move from the middle of the condenser.

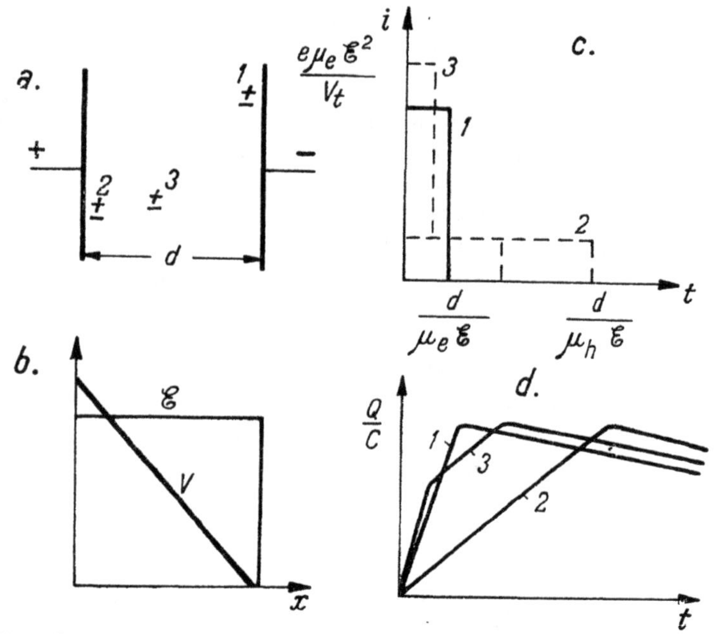

Fig. 3.5 — Uniform electric field : *a* — three characteristic electron-hole pairs, *b* — electric field and potential distribution, *c* — total current given by every pair, *d* — amplifier output voltage due to the currents from (*c*) [B 45].

Each carrier contribution to the current is determined with the aid of relation (3.19), and the current pulse duration is given by the carrier collection time

$$t_i = \frac{x}{v} = \frac{x}{\mu \mathcal{E}}. \tag{3.20}$$

The current time integral in every case is the same and equal to the electron charge. It should be mentioned that when both the electron and the hole are drifting, the total charge transferred is equal to e and not to $2e$.

The voltage pulse shape given by the equivalent *RC* integrating circuit is drawn in Fig. **3.5** *d*. The differences between the rise times show the differences between the transit times of the carriers

$$t_{tM} = \left(\frac{x}{v}\right)_M = \frac{d}{\mu_h \mathcal{E}}$$

and for $d = 1$ mm and $\mathcal{E} = 10^3$V/cm one obtains $t_{tM} \sim 10^{-7}$ s; therefore the pulse is much faster than in a gas-filled ionization chamber,

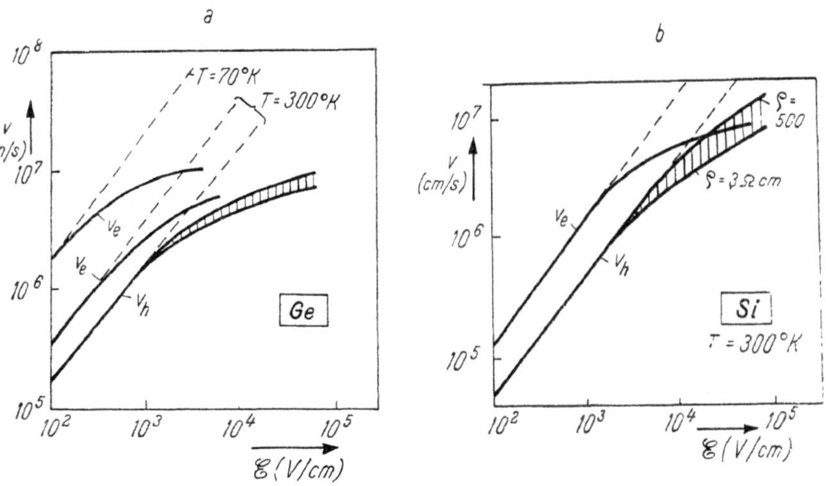

Fig. 3.6 — Electron and hole velocity in Si and Ge at the room temperature, versus the electric field intensity [M 46].

of comparable stopping power, due to a shorter transit path inside a solid-state detector. The pair space distribution yielded by a particle inside the detector gives an output voltage between the limiting cases shown in Fig. 3.5. *d*

The field intensity $\mathcal{E}$ must be increased in order to decrease t_t, but in strong fields (beginning with $\sim 10^3$ V/cm) μ is no longer a constant (Fig. 3.5 *b*) and the electron maximum drift velocity of about 10^7 cm/s (in Si) is obtained only with high field intensities (about 10^5 V/cm.)

The decrease of the electron mobility in strong fields follows from the carrier mean free-path decrease when the carrier mean energy is increased by the field.

3.2.2. *Linearly variable electric field.* In an abrupt $P-N$ junction, the electric field is a linear function (see section 2.11) of distance

$$\mathcal{E} = \mathcal{E}_M \frac{d-x}{d}. \tag{3.21}$$

The electron drift velocity is

$$v = \frac{\mathrm{d}x}{\mathrm{d}t} = \mu\mathcal{E} = \frac{\mu\mathcal{E}_M}{d}(d-x), \tag{3.22}$$

and therefore the time used for travelling the distance between 0 and x is

$$t = \int_0^x \frac{\mathrm{d}x}{v} = \frac{d}{\mu\mathcal{E}_M}\int_0^x \frac{\mathrm{d}x}{d-x} = \frac{d}{\mu\mathcal{E}_M}\ln\frac{d}{d-x}, \tag{3.23}$$

whence

$$x = d\left[1-\exp\left(-\frac{\mu\mathcal{E}_M}{d}t\right)\right], \tag{3.24}$$

and from Eq. (3.15) there results the current pulse expression

$$i = \frac{e}{d}\cdot\frac{\mathrm{d}x}{\mathrm{d}t} = \frac{e}{d}d\cdot\frac{\mu\mathcal{E}_M}{d}\exp\left(-\frac{\mu\mathcal{E}_M}{d}t\right),$$

where substituting the time constant $\dfrac{d}{\mu\mathcal{E}_m} = \tau$,

$$i = \frac{e}{\tau}\exp\left(-\frac{t}{\tau}\right), \tag{3.25}$$

and from Eqs. (3.14) and (3.24) we obtain the charge pulse

$$q = e\left[1-\exp\left(-\frac{t}{\tau}\right)\right]. \tag{3.26}$$

The voltage pulse on the detector capacity C will be q/C, i. e.,

$$V_s = \frac{e}{C}\left[1-\exp\left(-\frac{t}{\tau}\right)\right]. \tag{3.27}$$

The relations (3.25), (3.26), (3.27) were obtained by assuming that the pair was created at a junction boundary and thus the charge collection involved only one type of carrier. In the following section those cases in which the current pulse is obtained by the summation of contributions given by both types of carriers will be also analysed.

3.3. COLLECTION TIME OF ELECTRON-HOLE PAIRS IN AN ABRUPT P—N JUNCTION

Assume an abrupt N^+P junction with an electric field potential and intensity distribution inside the depletion layer shown in Fig. 3.7 [T 12].

From the condition that the total charge of the depletion layer is zero, it results that

$$\frac{d_p}{d_n} = \frac{N_d}{N_a} \gg 1, \tag{3.28}$$

and therefore $d_n \ll d_p$, so that the contribution of the carriers from the region $x < 0$, may be neglected.

In the region $0 < x < d_p \simeq d$, the hole drift (towards the right) and the electron drift (towards the left) are considered separately. Inside this region, the electric field intensity and the potential V vary linearly and parabolically, respectively, and have the expressions

$$\mathcal{E} = \frac{eN_a}{\varepsilon}(d - x), \tag{3.29}$$

$$V = \frac{eN_a}{2\varepsilon}(d - x)^2. \tag{3.30}$$

3.3.1. *The hole drift.* The drift velocity of a hole in an electric field $\mathcal{E}$ is

$$v_h = \frac{dx}{dt} = \mu_h \mathcal{E} = \mu_h \frac{eN_a}{\varepsilon}(d - x),$$

$$\frac{dx}{dt} = \mu_h k_a (d - x), \tag{3.31}$$

where k_a denotes the quantity

$$\frac{eN_a}{\varepsilon} = k_a.$$

From Eq. (3.31) it results that the transit time of a hole from the point $x = x_0$, where it has been created, to another point x, will be

$$t = \frac{1}{\mu_h k_a} \int_{x_0}^{x} \frac{dx}{d - x} = \frac{1}{\mu_h k_a} \ln \frac{d - x_0}{d - x},$$

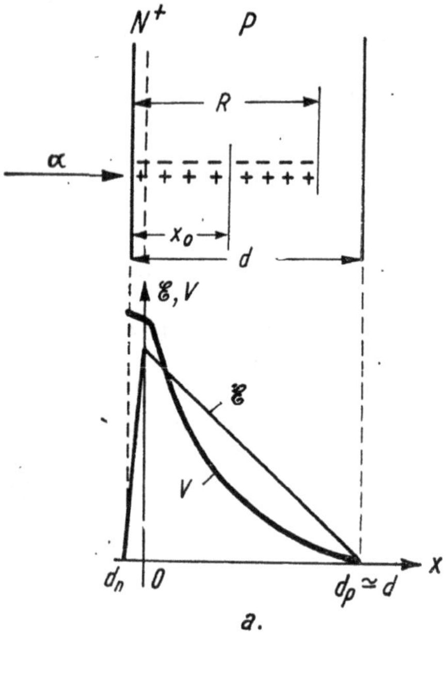

Fig. 3.7 — Depletion region of the abrupt $N-P$ junction, potential (V) and field (E) distribution.

which means that the hole coordinate at the moment t is

$$x = d\left[1 - \frac{d - x_0}{d} \exp(-\mu_h k_a t)\right]. \tag{3.32}$$

The hole reaches the negative electrode ($x = d$) after an infinitely long time. This hole drift gives rise in the external circuit to a current which according to Eq. (3.15) has the expression

$$i_h = \frac{e}{d} \cdot \frac{dx}{dt} = e \cdot \frac{d - x_0}{d} \mu_h \cdot k_a \cdot \exp(-\mu_h k_a t),$$

or

$$i_h = \frac{e}{\tau_h} \cdot \frac{d - x_0}{d} \exp\left(-\frac{t}{\tau_h}\right), \tag{3.33}$$

where

$$\tau_h = \frac{1}{\mu_h k_a} = \frac{\varepsilon}{eN_a \mu_h}, \tag{3.34}$$

and to a corresponding electric charge variation

$$q_h = e\frac{\Delta x}{d} = e\frac{x - x_0}{d} = e\frac{d - x_0}{d}\left[1 - \exp\left(-\frac{t}{\tau_h}\right)\right], \tag{3.35}$$

i. e., a contribution to the voltage pulse on the capacity C

$$V_{sh} = \frac{q_h}{C} = \frac{e}{C} \cdot \frac{d - x_0}{d}\left[1 - \exp\left(-\frac{t}{\tau_h}\right)\right]. \tag{3.36}$$

Because the stopping time is short enough, all the carriers may be considered as simultaneously created, so that, assuming a common time origin, the total current, charge, or voltage pulse can be calculated by summing up all the contributions of the form (3.33), (3.35) and (3.36), respectively, of all carriers created along the whole particle range R, *i. e.*, for $0 < x_0 < R$.

The number of holes created along the path dx_0 by a nuclear particle with a stopping power $\frac{dE}{dx_0}$ is

$$\frac{dE}{\epsilon} = \frac{1}{\epsilon} \cdot \frac{dE}{dx_0} \cdot dx_0 \tag{3.37}$$

and therefore the hole contribution to the total current pulse given by a particle is

$$I_h = \frac{e}{d\tau_h \epsilon} \exp\left(-\frac{t}{\tau_h}\right) \cdot \int_0^R (d - x_0) \cdot \frac{dE}{dx_0} \cdot dx_0, \tag{3.38}$$

or assuming $\frac{dE}{dx_0} = \text{const.}$, we obtain

$$I_h = \frac{e}{\tau_h \epsilon} \cdot \frac{dE}{dx_0} \cdot \frac{R}{d}\left(d - \frac{R}{2}\right) \cdot \exp\left(-\frac{t}{\tau_h}\right). \tag{3.39}$$

The corresponding charge pulse is

$$Q_h = \frac{e}{d\epsilon}\left[1 - \exp\left(-\frac{t}{\tau_h}\right)\right] \cdot \int_0^R (d - x_0) \cdot \frac{dE}{dx_0} \cdot dx_0, \tag{3.40}$$

where, for $\frac{dE}{dx_0} = \text{const.}$, the integral may be easily solved, and gives

$$Q_h = \frac{e}{\epsilon} \cdot \frac{dE}{dx_0} \cdot \frac{R}{d}\left(d - \frac{R}{2}\right) \left[1 - \exp\left(-\frac{t}{\tau_h}\right)\right]. \tag{3.41}$$

The voltage pulse is given by the relation

$$V_{sh} = \frac{e}{C\epsilon} \cdot \frac{dE}{dx_0} \frac{R}{d}\left(d - \frac{R}{2}\right)\left[1 - \exp\left(-\frac{t}{\tau_h}\right)\right]. \tag{3.42}$$

3.3.2. *The electron drift.* Using a method similar to that for holes, we obtain the variation in time of the velocity of an electron created at the point $x = x_0$.

$$v_e = \frac{dx}{dt} = -\mu_e \mathcal{E} = -\mu_e k_a (d - x), \tag{3.43}$$

i. e., the electron transit time from x_0 to x is

$$t = -\frac{1}{\mu_e k_a} \int_{x_0}^{x} \frac{dx}{d-x} = \frac{1}{\mu_e k_a} \ln \frac{d-x}{d-x_0}, \tag{3.44}$$

whence

$$x = d\left[1 - \frac{d-x_0}{d} \exp\ (\mu_e k_a t)\right]. \tag{3.45}$$

Comparing this expression to (3.32) obtained in the case of a hole drift, we observe that unlike the first case, when the hole reaches the electrode in an infinitely long time and the expression is valid for any time value t, in the case of an electron, the limit condition $x > 0$

restricts the validity of Eq. (3.45) to a time interval between $t = 0$ and $t = t'$, the time required by the electron to arrive at $x = 0$.

Substituting $x = 0$ and $t = t'$ in Eq. (3.45) it results

$$t' = \frac{1}{\mu_e k_a} = \ln \frac{d}{d - x_0} \cdot \tag{3.46}$$

For the time interval $0 < t < t'$ we have

$$i_e = \frac{e}{\tau_e} \cdot \frac{d - x_0}{d} \cdot \exp\left(\frac{t}{\tau_e}\right). \tag{3.47}$$

where

$$\tau_e = \frac{1}{\mu_e k_a} = \frac{\varepsilon}{e N_a \mu_e}. \tag{3.48}$$

The charge variation given by the electron drift between x_0 and x is

$$q_e = (-e) \frac{d - x_0}{d}\left[1 - \exp\left(\frac{t}{\tau_e}\right)\right] = e \frac{d - x_0}{d}\left[\exp\left(\frac{t}{\tau_e}\right) - 1\right]. \tag{3.49}$$

At $t \geqslant t'$, this expression is replaced by

$$q_e = q_{eM} = e \frac{d - x_0}{d}\left(\frac{d}{d - x_0} - 1\right) = e \cdot \frac{x_0}{d}. \tag{3.50}$$

The corresponding voltage pulse may be written as

$$V_{se} = \frac{e}{C} \cdot \frac{x_0}{d}\left[\exp\left(\frac{t}{\tau_e}\right) - 1\right]. \tag{3.51}$$

The electron contribution to the total charge pulse collected till the moment t' is [T 13]

$$Q_e = \frac{e}{\epsilon} \cdot \frac{dE}{dx_0} \cdot \frac{d}{2}\left\{1 - \exp\left(-\frac{t}{\tau_e}\right) - \left(1 - \frac{R}{d}\right)^2 \left[\exp\left(\frac{t}{\tau_e}\right) - 1\right]\right\}. \tag{3.52}$$

3.3.3. *The total charge pulse due to electron and hole collection.* In the case of a single electron-hole pair drift, the current, voltage and charge pulse are obtained by summation

$$i = i_h + i_e, \tag{3.53}$$

$$q = q_h + q_e, \tag{3.54}$$

$$V_s = V_{sh} + V_{se}. \tag{3.55}$$

The total charge collected after a sufficiently long time is according to Eqs. (3.35), (3.50) and (3.54)

$$q = q_{h\infty} + q_{e\infty} = e\left(1 - \frac{d - x_0}{d}\right) + e\frac{d - x_0}{d} = e \tag{3.56}$$

and not $2e$, as it could seem to be; hence, in the case of $N = \dfrac{\Delta E}{\in}$ pairs, the total charge collected is

$$Q = Ne = \frac{\Delta E}{\in} e. \tag{3.57}$$

The total charge pulse is a sum of all electron and hole contributions, a sum of exponentials. The rise time (0.1—0.9) of the charge pulse (or of the voltage pulse on the capacity C) equal with the falling time of the current pulse may be estimated from the nomograph in Fig. 3.8. [T 13].

When the particle range is short as compared to the depletion layer thickness ($R \ll d$), the signal waveform is determined by a carrier type (drifting the whole thicknes d).

In the case of an N^+P junction, only the hole collection is important and the rise time of the voltage pulse is $2.2.\tau_h$.

The results obtained for an N^+P junction can be also applied to a P^+N junction interchanging the rôle of holes and electrons. The quantities N_a, k_a, have also to be substituted by N_d and $k_d = \dfrac{eN_d}{\varepsilon}$. In this case

$$\tau_e = \frac{1}{\mu_e k_d} = \frac{\varepsilon}{eN_d \mu_e} = \varepsilon \rho_N, \tag{3.58}$$

where ρ_N is the resistivity of the junction material, and

$$\tau_h = \frac{1}{\mu_h k_h} = \frac{\varepsilon}{eN_d \mu_h}. \tag{3.59}$$

When the particle range is short ($R \ll d$) the electron contribution prevails, the voltage pulse form is exponential and its rise time (0.1—0.9) is 2.2 τ_e. For the same doping of the material ($N_d = N_a$) the rise time is shorter in a P^+N junction because $\mu_e > \mu_h$. In

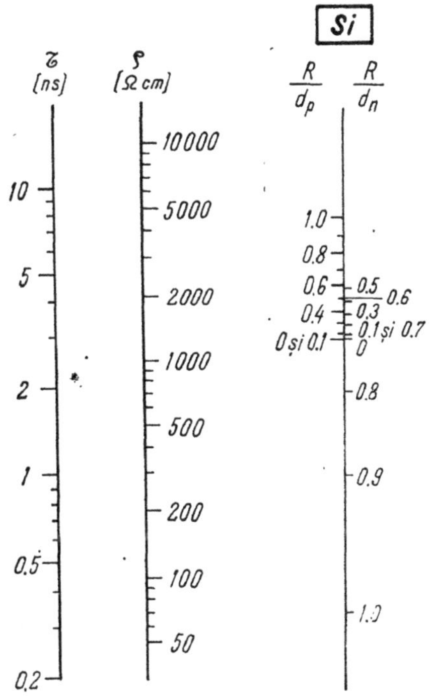

Fig. 3.8 — Nomograph used to determine the charge-carrier collection-time in a P—N silicon junction [T 13].

the general case, when the particle range is comparable to the depletion layer thickness, the expression for the rise time in the P^+N junction has the form [V 1]

$$t_r \simeq 2{,}2\tau_e\left\{2 - 2r(\beta+1) + r^2\left(1-\frac{r}{4}\right)(\beta+1)^2 - (1-r)\left[\frac{\beta}{r}\ln(1-r) + (\beta+1)\left(1-\frac{r}{2}\right)\right]^2\right\}^{1/2}, \tag{3.60}$$

where $r = \dfrac{R}{d}$ and $\beta = \dfrac{\mu_e}{\mu_h}$ is replaced by $\beta' = \dfrac{1}{\beta} = \dfrac{\mu_h}{\mu_e}$ in the case of an N^+P junction. For $r = 1$, Eq. (3.60) gets a simplified form

$$t_r = \tau_e \sqrt{3}\left((\beta^2 + 1 - \frac{2}{3}\beta\right)^{1/2}. \qquad (3.61)$$

For silicon ($\beta = 2.81$) relation (3.61) becomes

$$t_r = 4.6\,\tau_e. \qquad (3.61')$$

We may see, that in such a case ($R = d$) the rise time is double as against that for a short range ($R \ll d$). In any case the smaller the resistivity of the detector junction material, the shorter the rise time. Experimental checkings are described in [T. 15].

3.3.4. *Influence of the stopping power.* The variation of the stopping power $\frac{dE}{dx}$ may be approximated by a simple function. For instance the protons with a range $R_p \leqslant 1$ mm have an expression [T 12]

$$10 \log \left(\frac{dE}{dx_0}\right) = 8.51 - 0.442 \log (d - x_0) \qquad (3.62)$$

By considering the variation of $\frac{dE}{dx}$, and using the above expression in calculation, an increase of the charge pulse rise time has been found [T 12].

3.3.5. *Influence of the carrier mobility variation.* The electric field inside the junctions used in the nuclear radiation detection, is quite strong. From Fig. 3.6., it is obvious that, in strong electric fields, the electron and hole mobility decrease and the carrier velocities reach a limit value in Ge, for high enough field intensities. No clear saturation can be observed in Si. The maximum limit values of the field intensities for which one may consider that $\mu = \text{const.}$ are $\mathcal{E}_{lh} \simeq 7.5 \times 10^3$ V/cm for holes and $\mathcal{E}_{le} \simeq 2.5 \times 10^3$ V/cm for electrons in Si at 300° K [F 8]. It is considered [F. 8] that beyond these values, the mobility may be approximated by the expression

$$\mu = \mu_0 \left(\frac{\mathcal{E}_l}{\mathcal{E}}\right)^{1/2}. \qquad (3.63)$$

The calculations performed with the aid of this relationship [F 8] show an increase of the pulse rise time, due to the mobility decrease. Then the edge of the voltage pulses given by DS depends on the voltage value applied on the detector. For $\mu =$ const., the rise time does not depend on the applied voltage.

3.3.6. *Influence of the plasma.* An important problem is that of the electric field alteration in the region with high carrier density due to the thick plasma screening the inner carriers and modifying the potential distribution. It requires a correction of the simple theory of the particles moving in the undistorted field of the junction. Interesting results are reported in [M 34], [T 14], [K 17].

Within 10^{-12} s, the plasma width amounts to 10^{-8} m (the electron mean free path).

Due to the ambipolar diffusion, the ionization column quickly widens, reaching 5×10^{-7} m after 10^{-10} s and 1.5×10^{-6} m after 10^{-9} s. In this manner the microfield within the ionized column amounts to a small fraction of the junction field which becomes determinant.

The number of free charges represents only a negligible fraction of the number of charges due to ionized impurity atoms inside the detector. The junction field is distorted over a distance determined from the relationship between the ionized column and Si conductivity. The effect is less important in the case of a material with low resistivity. Due to the plasma, a longer carrier-collection time is observed during the fission fragment detection.

It may be concluded that there are many effects producing an increase of the carrier-collection time. An important one will be analysed in section(3.5). In this manner one can explain those cases [M 35] in which the experimental values of t_r are higher than those given by the nomograph in Fig. 3.8 obtained with simplifying assumptions.

3.4. COLLECTION TIME OF ELECTRON-HOLE PAIRS IN COAXIAL Ge (Li) DETECTORS

Large volume Ge (Li) detectors are required for obtaining high gamma-ray detection efficiencies. Practically attainable compensation depths are restricted to the value of some millimeters; thus coaxial geometry (Fig. 3.9) is a solution of this problem.

Measurements [M 4], [E 13], [G 30] have shown that double open-ended coaxial detectors (Fig. 3.9 *b*), which allow a better time resolution than single open-ended diodes (Fig. 3. 9. *a*), are well suited for time analysis.

Assuming that uniform compensation in the whole detector body (except the P^+ central core) was achieved, the coaxial detector will

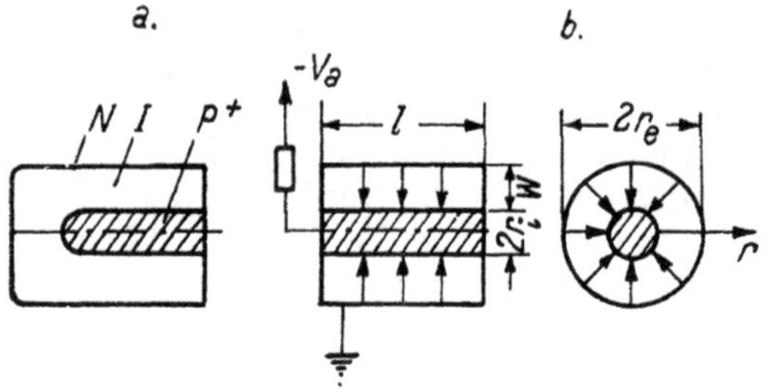

Fig. 3.9 — Coaxial detectors : *a* — single open-ended, *b*, — double open-ended.

be totally depleted for a low applied bias voltage (V_a). The potential ($V(r)$) and electric field intensity ($\mathcal{E}(r)$) distributions are given by

$$V(r) = -\frac{V_a}{\ln \frac{r_e}{r_i}} \ln \frac{r_e}{r} \tag{3.64}$$

and

$$\mathcal{E}(r) = \mathcal{E}_M \frac{r_i}{r} \tag{3.65}$$

where

$$\mathcal{E}_M = -\frac{V_a}{r_i \ln \frac{r_n}{r_i}} \tag{3.66}$$

is the maximum electric field intensity and r_i, r_e are the internal and external radii of the coaxial detector (see Fig. 3. 9 *b*). In the following the sign minus in the Eqs. (3.64) and (3.66) will not be taken into consideration because only the magnitude of the current pulse is of interest, its direction of flowing being easily determined. Eqs. (3.64) and (3.65) for the case $r_e = 2r_i$ are plotted in Fig. 3.10

Semiconductor detectors with such a distribution of the electric field will be called "H" detectors (abbreviation from *hyperbolic*) in contrast with "U" —and "L"-type detectors [P 14] (with uniform and linear electric field, respectively).

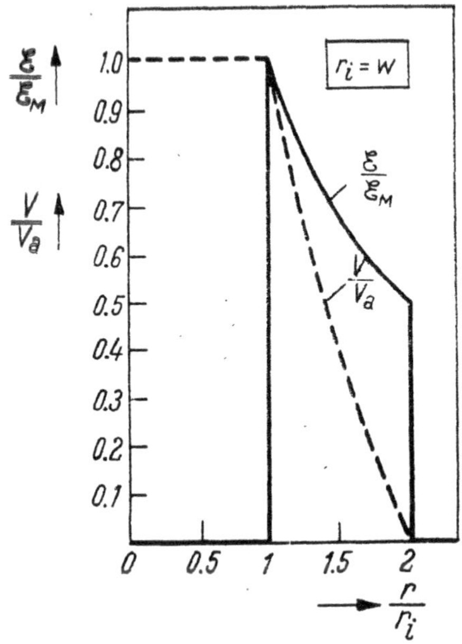

Fig. 3.10 — Potential and electric field intensity distribution in a coaxial detector.

Double open-ended coaxial Ge(Li) detector capacity may be found from the relationship

$$C = \frac{2\pi l \varepsilon}{\ln \frac{r_e}{r_i}} \tag{3.67}$$

(fringe effects being neglected).

The current pulse induced in the external circuit when an electron-hole pair is collected may be calculated by using a theorem due to S. Ramo and C. K. Jen [R 4], [J 2], [C 3], [T 13] according to which for the electron motion in a stationary electric field

$$i = e\,\frac{\mathcal{E}(r)}{V_a}\,\frac{dr}{dt}, \tag{3.68}$$

Thus, when a charge carrier travels the whole depletion layer width (w), the collected charge is $q = e =$ electron charge. Radial collection path was assumed.

The charge carrier motion law $r(t)$ may be found by solving the differential equation

$$v(r) \equiv \frac{\mathrm{d}r}{\mathrm{d}t} = \mu\,\mathcal{E}(r) \tag{3.69}$$

In Ge cooled at liquid nitrogen temperature, for low electric field intensities (up to 200—500 V/cm), the charge carrier mobility μ may be considered to be a constant with the same value for electrons and holes $\mu_h = \mu_e = \mu$ [G 28]. For $\mathcal{E} \gg \mathcal{E}_c \simeq 1\,000$ V/cm a constant carrier velocity is reached: $v = v_M = 1.5 \times 10^7$ cm/s, regardless of the electric field intensity value.

Electron-hole pairs are generated by gamma-radiation in the entire detector sensitive region. Three typical positions are shown in Fig. 3.11.

Only electron collection contributes to the electric pulse for position No 1, hole for position No 2, and both electron and hole for intermediate position No 3.

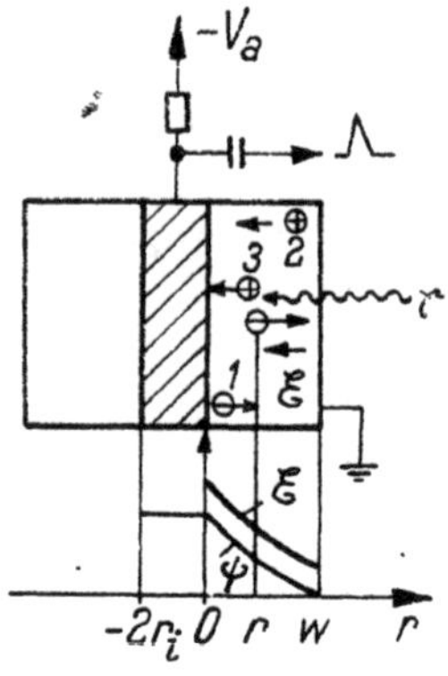

Fig. 3.11 — Coaxial detector section and electric field distribution.

Important conclusions can be drawn from the study of the collection of only one electron-hole pair created at various distances $r = r_0$ from the internal electrode, because the actual electric pulses may be found by adding such contributions of the whole range of electrons produced by gamma-rays.

It is convenient to make a translation of the origin for the r axis to the point $r = r_i$ (Fig. **3.11** b) because only the region between the two electrodes $r_i \leqslant r \leqslant r_e = r_i + W$ is of interest. In this system, equation (3.65) becomes

$$\mathcal{E} = \mathcal{E}_M \cdot \frac{r_i}{r + r_i} \tag{3.65'}$$

from which the maximum and minimum electric field intensity result

$$\mathcal{E}_{(0)} = \mathcal{E}_M ; \; \mathcal{E}_{(W)} = \mathcal{E}_m = \mathcal{E}_M \cdot \frac{r_i}{r_e}.$$

In the following, the collection of electrons and holes is considered separately.

3.4.1. *The collection of electrons.* The drift velocity of an electron generated at the distance $r = r_0$, which should travel a path length $W - r_0$, in the positive direction of the r axis, is given by Eqs. (3.69) and (3.65')

$$\frac{dr}{dt} = \mu_e \, \mathcal{E}_M \frac{r_i}{r + r_i}, \tag{3.70}$$

which can be integrated

$$\int_{r_0}^{r} (r + r_i) \, dr = \int_0^t \mu_e \, \mathcal{E}_M \, r_i \, dt, \tag{3.71}$$

giving an algebraic quadratic equation in r. By selecting only the positive solution, we get

$$r(t) = [(r_i + r_0)^2 + 2\mu_e \, \mathcal{E}_M r_i t)]^{1/2} - r_i . \tag{3.72}$$

In order to simplify this equation, a time constant τ_e is defined by the following relation

$$\tau_e \equiv \frac{r_i}{2\mu_e \, \mathcal{E}_M} = \frac{r_i^2}{2\mu_e V_a} \ln \frac{r_e}{r_i}, \tag{3.73}$$

and thus equation (3.72) can be written in the normalized form

$$\frac{r(t)}{r_i} = \left[\left(1 + \frac{r_0}{r_i}\right)^2 + \frac{t}{\tau_e}\right]^{1/2} - 1. \tag{3.72'}$$

For $t = 0$ it follows that $r = r_0$, and from equation $r(\tau_{ie}) = W$, the electron collection time (τ_{ie}) may be found as

$$\tau_{ie} = \tau_e \frac{r_e^2 - (r_i + r_0)^2}{r_i^2} = \frac{r_e^2 - (r_i + r_0)^2}{2\mu_e V_a} \ln \frac{r_e}{r_i}, \tag{3.74}$$

which has a maximum for $r_0 = 0$ (position no. 1 in Fig. 3.11) and is zero for $r_0 = W$ (position no. 2 in Fig. 3.11).

The equation which describes the current pulse shape is obtained with the aid of Eqs. (3.68) and (3.72′)

$$\frac{i_e(t)}{I} = \left[\left(1 + \frac{r_0}{r_i}\right)^2 + \frac{t}{\tau_e}\right]^{-1}, \tag{3.75}$$

valid only for $0 < t < \tau_{ie}$. The current diminishes in time due to the electron slowing down (it moves in a decreasing electric field). Hence, the current pulse amplitude is

$$i_e(0) = I_e \left(\frac{r_i}{r_i + r_0}\right)^2, \tag{3.76}$$

where the maximum amplitude equal to

$$I_e = \frac{e}{2\tau_e} \cdot \frac{r_i \mathscr{E}_M}{V_a} = \frac{e\mu_e \mathscr{E}_M^2}{V_a} = \frac{e\mu_e V_a}{r_i^2} \cdot \frac{1}{\left(\ln \frac{r_e}{r_i}\right)^2}, \quad (14) \tag{3.77}$$

is obtained for $r_0 = 0$. At the time $t = \tau_{ie}$

$$i_e(\tau_{ie}) = I_e \left(\frac{r_i}{r_e}\right)^2. \tag{3.78}$$

Thus the ultimate value does not depend on r_0.

Charge pulse $q_e(t)$ may be found by integration

$$q_e(t) = \int_0^t i_e(t)\, dt = \frac{e}{\ln\left(\frac{r_e}{r_i}\right)^2} \ln \frac{\left(1 + \frac{r_0}{r_i}\right)^2 + \frac{t}{\tau_e}}{\left(1 + \frac{r_0}{r_i}\right)^2}, \tag{3.79}$$

and when $t = \tau_{ie}$

$$q_e(\tau_{ie}) = e \frac{\ln \frac{r_e}{r_i + r_0}}{\ln \frac{r_e}{r_i}}. \tag{3.80}$$

The voltage pulse $u(t)$ due to the integration of the current pulse on the capacity C is

$$u_e(t) = \frac{q_e(t)}{C}, \tag{3.81}$$

hence

$$\frac{\mu_e(t)}{U_e} = \frac{q_e(t)}{e} \tag{3.82}$$

where

$$U_e = \frac{e}{C} \tag{3.83}$$

Equations (3.75) and (3.79) are plotted in Fig. 3.12 for the cases $r_0 = 0$ and $r_e = 2r_i = 2\ W$.

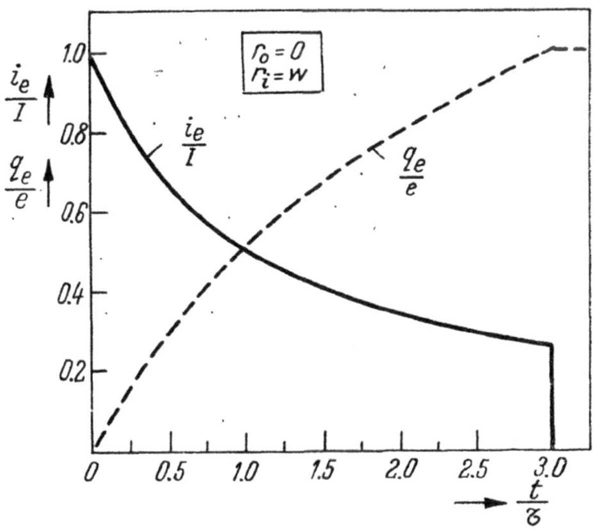

Fig. 3.12 — Current and electric charge waveforms obtained for electron collection in a coaxial detector.

3.4.2. *The collection of holes.* A hole created at the point $r = r_0$ travels the distance r_0 in the decreasing direction of the r axis; hence a formula similar to equation (3.70) can be written

$$-\frac{\mathrm{d}r}{\mathrm{d}t} = \mu_h \mathcal{E}_M r_i \frac{1}{r + r_i}. \tag{3.84}$$

By integration

$$-\int_{r_0}^{r} (r + r_i)\, \mathrm{d}r = \int_0^t \mu_h \mathcal{E}_M r_i\, \mathrm{d}t, \tag{3.85}$$

and after simple calculations, the following equation is obtained

$$\frac{r(t)}{r_i} = \left[\left(1 + \frac{r_0}{r_i}\right)^2 - \frac{t}{\tau_h}\right]^{1/2} - 1, \tag{3.86}$$

where

$$\tau_h \equiv \frac{r_i}{2\mu_h \mathcal{E}_M} = \frac{r_i^2}{2\mu_h V_a} \ln\frac{r_e}{r_i}, \tag{3.87}$$

and since

$$\mu_h = \mu_e = \mu,$$

$$\tau_h = \tau_e = \tau. \tag{3.88}$$

By making $t = 0$ in equation (3.85) it is verified that $r(0) = r_0$ and the condition $r(\tau_{ie}) = 0$ gives the hole pulse current duration

$$\tau_{ih} = \tau_h \frac{(r_i + r_0)^2 - r_i^2}{r_i^2} = \frac{(r_i + r_0)^2 - r_i^2}{2\mu_h V_a} \ln\frac{r_e}{r_i}, \tag{3.89}$$

which is maximum when $r_0 = W$ (position 2 in Fig. 3.11), and for $r_0 = 0$ (position 1 in Fig. 3.11) $\tau_{ih} = 0$. With equations (3.68) and (3.86) the hole current can be calculated as

$$\frac{i_h(t)}{I_h} = \left[\left(1 + \frac{r_0}{r_i}\right)^2 - \frac{t}{\tau_h}\right]^{-1}. \tag{3.90}$$

The hole travels the distance r_0 in the direction of the increasing electric field; it is accelerated and so the current pulse induced by its motion, increases from the value

$$i_h(0)/I_h = \left(\frac{r_i}{r_i + r_0}\right)^2, \tag{3.91}$$

to the amplitude

$$i_h(\tau_{ih}) = I_h = \frac{e\mu_h \mathcal{E}_M^2}{V_a} = \frac{e}{2\tau_h} \cdot \frac{r_i \mathcal{E}_M}{V_a} = \frac{e\mu_h V_a}{r_i^2} \cdot \frac{1}{\left(\ln \frac{r_e}{r_i}\right)^2} \tag{3.92}$$

and for $\mu_e = \mu_h = \mu$ we get $I_h = I_e = I$.

As in the case of electron collection, the ultimate value of the current does not depend on r_0.

The electric charge pulse is

$$q_h(t) = \int_0^t i_h(t)\, dt = \frac{e}{\ln\left(\frac{r_e}{r_i}\right)^2} \ln \frac{\left(1 + \frac{r_0}{r_i}\right)^2}{\left(1 + \frac{r_0}{r_i}\right)^2 - \frac{t}{\tau_e}} \cdot \tag{3.93}$$

At the time $t = \tau_{ih}$

$$q_h(\tau_{ih}) = e \cdot \frac{\ln \frac{r_i + r_0}{r_i}}{\ln \frac{r_e}{r_i}} \tag{3.94}$$

and by adding it to equation (3.80), the total amount of charge collected for an electron-hole-pair motion is

$$q_t = q_e(\tau_{ie}) + q_h(\tau_{ih}) = e. \tag{3.95}$$

Relations similar to Eqs. (3.81)—(3.83) may be found. In Fig. 3.12 current and charge wave forms are plotted for $r_0 = W$ and $r_e = 2r_i$.

3.4.3. *Collection of an electron-hole pair. Total current and charge pulses.* Pulse shapes resulting from the collection of only one charge carrier (electron or hole) generated at the particular distance from the internal electrode ($r = 0$ or $r_0 = W$), can be seen in Figs. 3.12 and 3.13. For these initial positions, minimum amplitude and maximum duration of the current pulses are obtained.

Both electron and hole contributions to the current pulse must be added for an electron-hole pair generated at any distance $r_0 \neq 0, W$.

By equating (3.89) and (3.74), the value of r_0 is obtained for

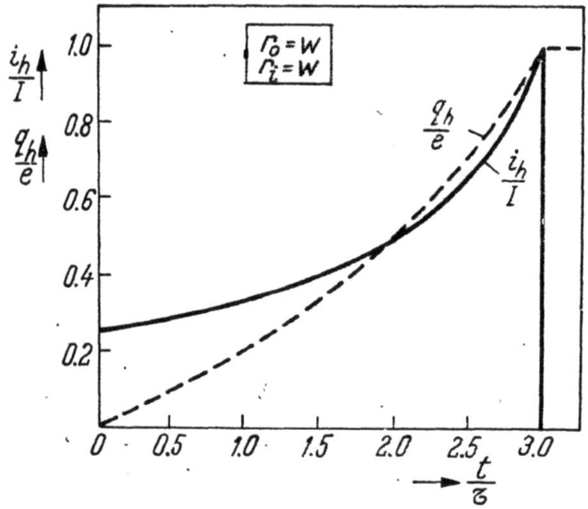

Fig. 3.13 – Current and charge pulse shapes obtained for hole collection in a coaxial detector.

which the electron and hole transit time are equal, i.e. $\tau_{ie} = \tau_{ih}$. Thus

$$r_0 = \left(\frac{r_i^2 + r_e^2}{2}\right)^{1/2} - r_i \tag{3.96}$$

The wave forms for $r_e = 2\, r_i$, i.e., $r_0 = 0.58\, r_i$, are plotted in Fig. 3.14 *a*. Current and charge pulse shapes for $r_0 = 0.25\ W$ and $r_0 = 0.75\ W$ are shown in Fig. 3.14 *b* and *c*, respectively.

Depending on r_0, the charge pulse rise time varies from a minimum value (at $r_0 = 0.58\ W$ for $W = r_i$) to the maximum one given by the limiting values $r_0 = 0$ and $r_0 = W$. This fact can be better seen in Fig. 3.15.

The total charge pulse supplied by the detector as a response to an ionizing particle of energy E may be found by adding the contributions of all electron-hole pairs generated along the charged particle path.

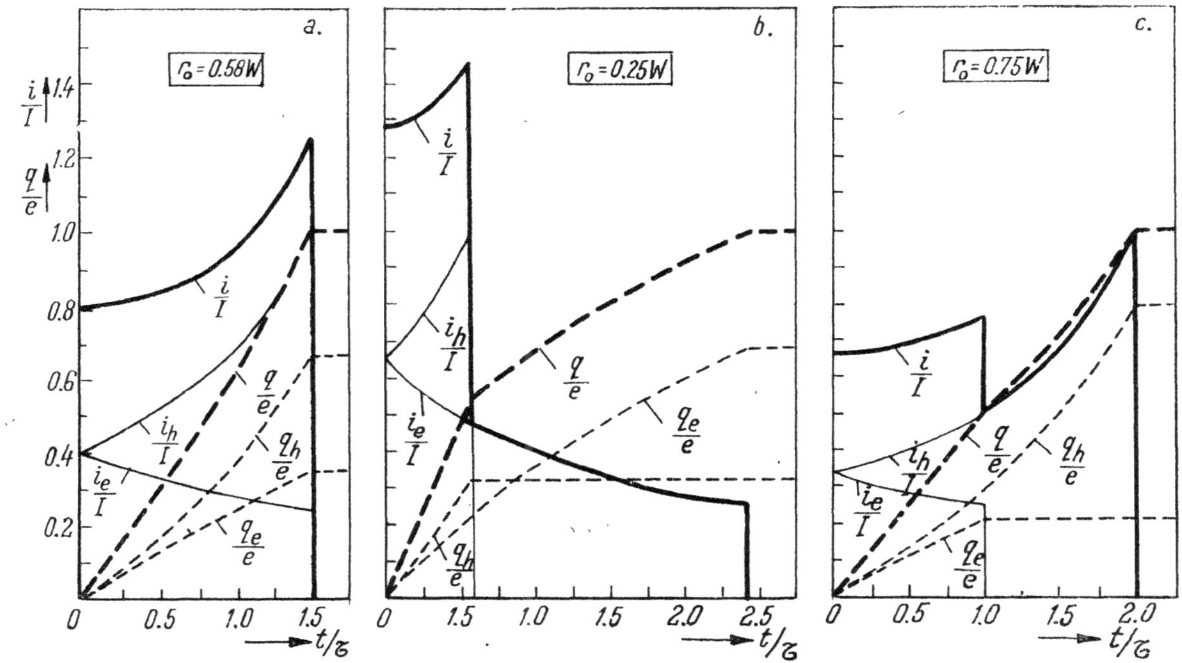

Fig. 3.14 — Current and charge waveforms induced during collection of an electron-hole pair generated at different distances r_0; (a) $r_0 = 0.58\ W$; (b) $r_0 = 0.25\ W$; (c) $r_0 = 0.75\ W$.

The rise time of this pulse will be determined by the collection times ranging from τ_{im} to τ_{iM}, and the amplitude will be

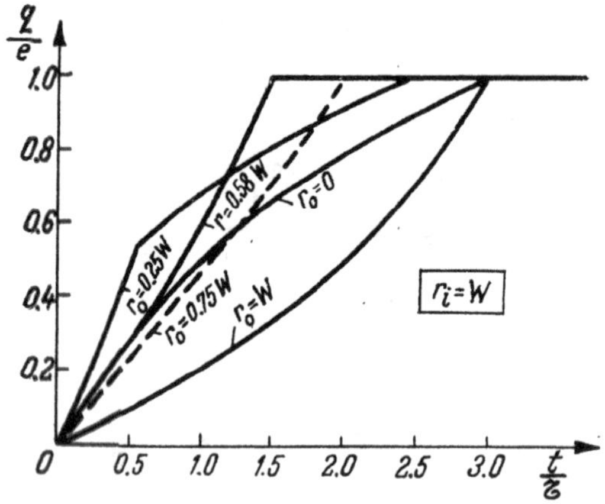

Fig. 3.15 – Ccaxial detector charge pulse shapes.

$$Q = \frac{E}{\epsilon} e, \qquad (3.97)$$

where ϵ is the mean energy required for an electron-hole pair generation (2.9 eV in Ge).

For $W = r_i$ the minimum duration of the current pulse (minimum collection time) is given by

$$\tau_{im} = 1.5\,\tau \qquad (3.98)$$

and the maximum

$$\tau_{iM} = 3\tau = 2\,\tau_{im} \qquad (3.99)$$

where

$$\tau = \frac{r_i}{2\mu\,\mathcal{E}_M} = \frac{r_i^2}{2\mu\,V_a} \ln \frac{r_e}{r_i}. \qquad (3.87')$$

The minimum amplitude of the current pulse can be determined from the formula

$$I_m \simeq \frac{E}{\epsilon} I = \frac{E}{\epsilon} \cdot \frac{e}{2\tau} \cdot \frac{r_i \mathcal{E}_m}{V_a} = \frac{E}{\epsilon} \cdot \frac{e\mu\,V_a}{r_i^2} \cdot \frac{1}{\left(\ln \frac{r_e}{r_i}\right)^2} \qquad (3.100)$$

So far, a low electric field intensity ($\mathcal{E} < \mathcal{E}_c$) was assumed: thus $\mu =$ const. The measured values [K 18] for Ge at liquid nitrogen temperature lie between $\mu = 8\,000$ cm²/V s and $\mu = 28000$ cm²/V s depending on the oxygen content of the basic material.

Taking as an example $W = r_i = 5$ mm and $\mu = 10\,000$ cm²/Vs at $V_a = 50$ V. we have $\mathcal{E}_M = 145$ V/cm; $\tau = 172$ ns; $\tau_{im} = 258$ ns; $\tau_{iM} = 516$ ns and $I_m E^{-1} = 0.23\ \mu$ A/MeV.

Usually a higher field intensity is applied. For $\mathcal{E}_m > \mathcal{E}_c = 1\,000$ V/cm, ($V_a > 690$ V in our example) a constant velocity of charge carriers is reached, $v = v_M = 1.5 \times 10^7$ cm/s. In this case

$$\tau_{im} = \frac{W}{2v_M}, \tag{3.101}$$

$$\tau_{iM} = \frac{W}{v_M} = 2\tau_{im}, \tag{3.101'}$$

and

$$I_m = \frac{E}{\epsilon} \cdot \frac{ev_M \mathcal{E}_M}{V_a} = \frac{E}{\epsilon} \cdot \frac{e v_M}{r_i} \cdot \frac{1}{\ln \frac{r_e}{r_i}}. \tag{3.102}$$

Consequently, in our example for $V_a \gg 700$ V we get

$$\tau_{im} = 16.7 \text{ ns}; \; \tau_{iM} = 33.4 \text{ ns}; \; I_m = 2.3 \; \mu\text{A/MeV}.$$

When designing coaxial detectors, with the purpose of using their current pulses [P 14], it is important to know the optimum geometry which ensures a minimum duration and maximum amplitude of the detector current pulses at a given sensitive width W.

From equation (3.102) it follows that for obtaining a maximum current pulse amplitude, a small value of the internal electrode radius (r_i) is preferable.

As far as the current pulse duration in a high electric field ($\mathcal{E} > \mathcal{E}_c$) is concerned, it is determined by the depletion layer width (W) as in the case of the planar devices (see Eqs. (3.101) and (3.101′)).

3.5. INFLUENCE OF SD EQUIVALENT CIRCUIT ELEMENTS ON THE VOLTAGE AND CURRENT PULSE SHAPE

The shape of the pulses given by the detector at the associated preamplifier input must be necessarily known in many applications of the semiconductor detectors. It is possible to calculate these waveforms on the basis of the *SD* junction equivalent circuit [W 5], shown in Fig. 3.16. Many works, such as for instance [F 4], [S 17], [C 18], [T 1], have experimentally proved the validity of this circuit. The circuit elements $R_d - C_d$ are associated with the junction depletion layer; $R_s - C_s$ take into account the undepleted bulk material, and R_c is a contact series resistance. Although each of these elements (except R_c) varies with the silicon resistivity ρ, the semiconductor wafer surface area S, the width d, and the applied reverse bias voltage V_a, the dielectric relaxation time constant

$$R_s C_s = \rho\varepsilon = \tau, \tag{3.103}$$

is not dependent on V_a. ε is the dielectric constant of silicon. However, when

$$V_a \geqq V_p = d^2/(2\varepsilon\rho\mu), \tag{3.104}$$

the detector is totally-depleted, $R_s = 0$ and C_d is no longer dependent on V_a; μ is the charge carrier mobility (electrons in the n-type basic material, and holes in the p-type silicon). Usually R_d is very high and may also be neglected for our purposes.

The current $i(t)$ is induced in the circuit due to the collection of the electrical charge Q generated by an ionizing event of energy E in the depletion layer.

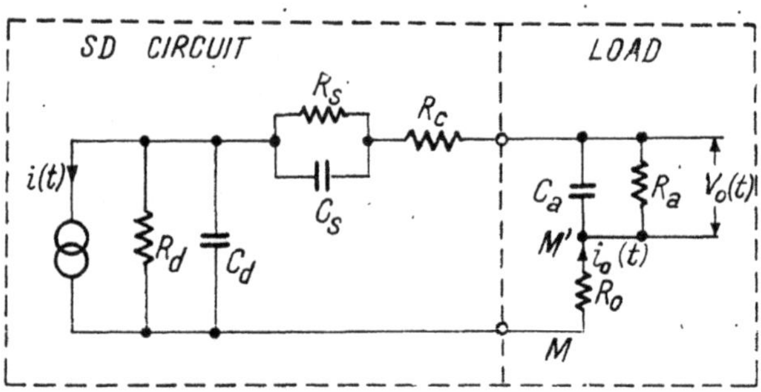

Fig. 3.16 — Semiconductor junction detector and preamplifier input equivalent circuit.

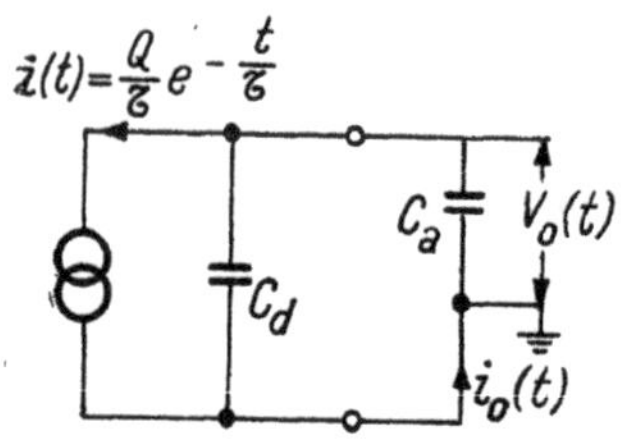

Fig. 3.17 — $L_a^0 Q I_0 P$ circuit configuration.

For an abrupt junction in which the electric field strength $\mathcal{E}$, is linearly distributed with the distance, it was found [C 18], [T 1], [C 3], [T 13], [G 24] that

$$i(t) = (Q/\tau)\,e^{-t/\tau} = I\,e^{-t/\tau}, \tag{3.105}$$

where

$$Q = (E/\epsilon)\,q, \tag{3.106}$$

in which ϵ represents the mean energy required for an electron-hole pair production in silicon and q is the electron charge.

The simple expression (3.105) is valid only if a number of restrictive assumptions are accomplished. However, when these assumptions are not satisfied, almost always the relation for $i(t)$ can be put in the equivalent form

$$i(t) = (Q/\tau_i)\,e^{-t/\tau_i} = I_i\,e^{-t/\tau_i}, \tag{3.107}$$

where the collection time constant $\tau_i > \tau$ and consequently $I_i < I$. The "plasma time" brings the major contribution to the increase of the collection time, when heavy charged particles are detected.

If a strong overdepletion ($V_a \gg V_p$) is obtained, the electric field may be considered uniform and the shape of the current pulse becomes rectangular, with amplitude

$$I_i = Q/\tau_i, \tag{3.108}$$

and length

$$\tau_i = d/(\mu\mathcal{E}) = d^2/(\mu V_a). \tag{3.109}$$

At a very strong field intensity, the maximum charge carrier velocity, v_M, is reached and equation (3.109) becomes

$$\tau_i = d/v_M. \tag{3.110}$$

Like in the case of overdepleted junction detectors, the equivalent circuit of *NIP* detectors with plane geometry (in which a uniform electric field is assumed), may be obtained making $R_s = 0$ in Fig. 3.16, and taking a rectangular shape for the "generator" current pulse, the relations (3.108) and (3.109) being assumed valid.

The load in Fig. 3.16, which represents the preamplifier input circuit impedance, may be taken for the cases of practical importance as follows: $R_0 = 0$ and $R_a = \infty$ for a charge-sensitive preamplifier with C_a as dynamic input capacity; $R_0 = 0$ for a voltage amplifier; $R_a = 0$ for a current amplifier [K 6], [B 38] and only $R_a = \infty$ when both charge-sensitive and current preamplifiers [P 1] [B 10] are used.

The purpose of this calculation is to give the relations for the current $i_0(t)$ or the voltage $V_0(t)$ supplied by various types of *SD* in different working conditions and with several load impedances, on the basis of equivalent circuit analysis, particularized for a given case. The expressions for the Laplace transforms $I_0(p)$ and $V_0(p)$ are also tabulated with their poles and zeros which may be used for instance when the pole-zero cancellation technique [N 4] is employed.

3.5.1. *Equivalent circuit specification.* Unified notations for various detector-preamplifier ensembles are introduced in this section, in order to obtain some shortened denominations of the equivalent-circuit particular configurations. Thus, indiferently of the detector technology, the shape of the current pulse is rectangular for a uniform electric field and exponential if a linear distribution of the collecting field is assumed. The SD will be called "U" in the former case and "L" in the latter. The current pulse Laplace transform will be

$$I(p) = I_i\{1 - e^{-\tau_i p}\}/p, \tag{3.111}$$

for the U detector, and

$$I(p) = I_i/\{p + (1/\tau_i)\}, \tag{3.112}$$

for the L-type detector.

U detectors have always $R_s = 0$. L detectors in which $R_s = 0$ (totally depleted) will be specified with the subscript d as L_d. When $R_c = 0$, L or U will be marked with the zero exponent : L^0 or U^0.

The load impedance will be specified by a capital letter following L or U. This letter will be "V" for the voltage preamplifier ($R_0 = 0$); "Q" for a charge sensitive preamplifier ($R_0 = 0$; $R_a = \infty$); "I" for a current preamplifier ($R_0 = 0$); "I" for a current preamplifier with both $R_0 = R_a = 0$; QI for a charge-sensitive and current preamplifier simultaneously connected at the SD output ($R_a = \infty$).

When the L detectors (with $R_s \neq 0$) have $\tau_i = \tau$, the relations for $i_0(t)$ and $V_0(t)$ become simpler than for $\tau_i > \tau$, because $R_sC_s = \tau$. These particular circumstances will be denoted by a third capital letter "P". For instance if the detector-input preamplifier circuit takes the simple form shown in Fig. 3.17 this will be called $L_d^0QI_0P$.

3.5.2. *Current and voltage pulse relations.* Laplace transforms $I_0(p)$ or $V_0(p)$ are easily obtained from the equivalent circuit transfer function and equations (3.111) or (3.112). From these expressions we obtain $i_0(t)$ or $V_0(t)$ in the time domain. Voltage pulses $V_0(t)$ are interesting for V and Q amplifiers; current pulses $i_0(t)$ — for I amplifiers; both $i_0(t)$ and $V_0(t)$ for QI amplifiers. These results are given in Tables 3.1—3.4.

The Laplace transform poles ($-p_k$) are located in the "p" complex plane on the negative real frequency semiaxis. Their moduli

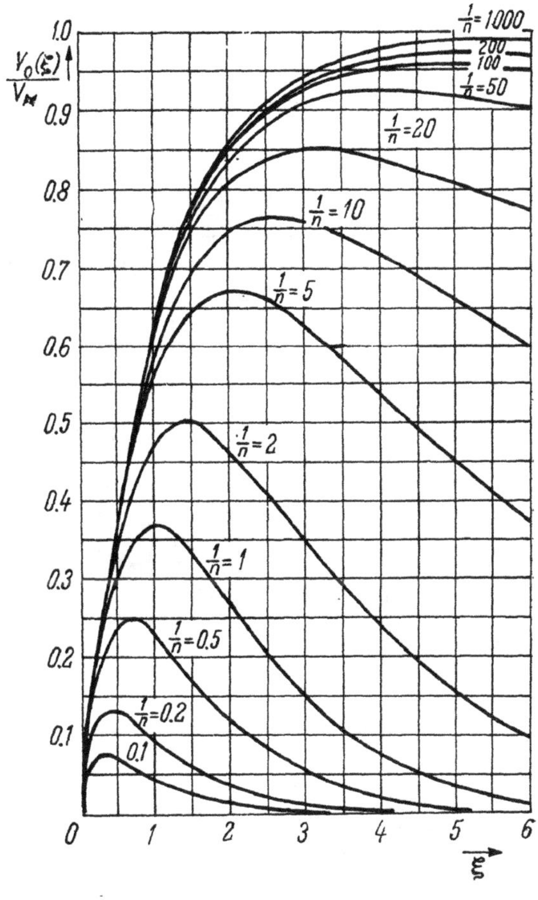

Fig. 3.18 — Voltage waveforms given by $L_d^0 V$ configuration.

depending on the equivalent circuit time constant $\tau_{ik} = R_i C_k$ are tabulated; zero values are not specified.

The analysis of the current waveform modification by the equivalent circuit elements is easily performed and universal curves may be calculated if the time is normalized

$$t/\tau_i = \xi. \tag{3.113}$$

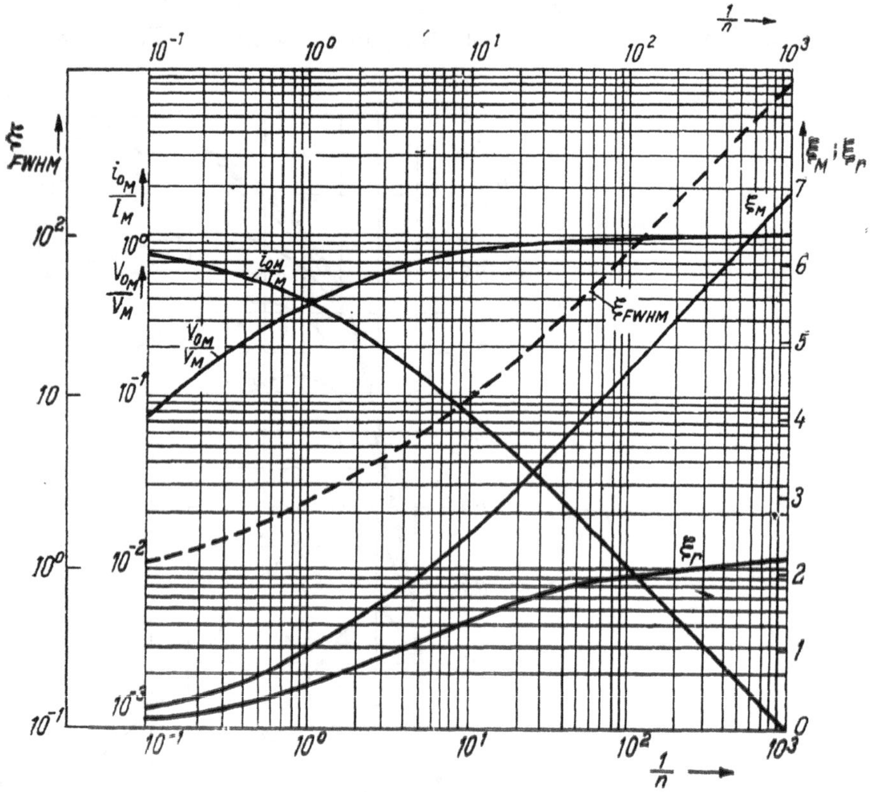

Fig. 3.19 — Voltage and current pulse, time and amplitude parameters for $L_a^0 V$; $L_a I$, $L_a I_0$ and $L_a^0 I$ configurations.

The time functions V_0 (ξ) and i_0 (ξ) for U detectors are valid only for a time length $0 < \xi \leqslant 1$; for $\xi > 1$ these become: $V_0(\xi) - V_0(\xi - 1)$ and $i_0(\xi) - i_0(\xi - 1)$ respectively.

The relations for the voltage pulses supplied by the *SD* at the voltage preamplifier input circuit $R_a - C_a$ are shown in Table 3.1. The quantity V_p in this table is not a voltage. An equivalent capacity C_a is introduced in some cases in order to simplify the equations.

The voltage waveform normalized analytical expressions V_0 (ξ) $< V_M$ include a sum of exponential functions with the A_i coefficients determined by zero and pole values in the normalized forms : *k*, *n*, *m*, 1.

Table 3

Voltage pulses supplied by various SD type

	LV	LVP	L⁰V	
$V_0(p)/V_p$	$\dfrac{p + p_{01}}{(p+p_1)(p+p_2)(p+p_3)(p+p_4)}$	$\dfrac{1}{(p+p_1)(p+p_2)(p+p_3)}$	$\dfrac{p + p_{01}}{(p+p_1)(p+p_2)(p+p_3)}$	$(p +$
V_p	$(Q/C_a)/(\tau_i\tau_{cd})$	$(Q/C_a)/(\tau\tau_{cd})$	$(Q/C_e)\ [C_s/(C_s + C_d)]/\tau_i$	(Q/C_e)
C_e	—	—	$C_a + [(C_dC_s)/(C_d + C_s)]$	$C_a + [($
$\|p_{01}\|$	$1/\tau$	—	$1/\tau$	
$\|p_k\|$	$p_{1.2.3}$ are the roots of the equation $p^3 + ap^2 + bp + c = 0.$ $a = (1/\tau) + (1/\tau_{aa}) + (1/\tau_{cd}) + (1/\tau_{cs}) + (1/\tau_{ca}),$ $b = \{1/(\tau\tau_{aa})\} + \{1/(\tau_{aa}\tau_{cd})\} + \{1/(\tau_{cd}\tau)\} + \{1/(\tau\tau_{ca})\} +$ $+ \{1/(\tau_{cs}\tau_{ad})\},$ $c = 1/(\tau\tau_{aa}\tau_{cd})\ ;\ \ p_4 = 1/\tau_i$		$p_{1.2} = \frac{1}{2}\{(\tau + \tau_{aa} + \tau_{ad} + \tau_{sd})$ $\tau_{ad} + \tau_{sd})^2 + 4\tau_{ad}\tau_{sd}]^{\frac{1}{2}}\}/(\tau_{aa}\tau$ $p_3 = 1/\tau_i$	
$V_0(\xi)/V_M$	$A_1 \exp(-n\xi) + A_2 \exp(-m\xi) +$ $+ A_3 \exp(-l\xi) + A_4 \exp(-\xi)$	$A_1 \exp(-n\xi) +$ $A_2 \exp(-m\xi) +$ $+ A_3 \exp(-l\xi)$	$A_1 \exp(-n\xi) +$ $+ A_2 \exp(-m\xi) +$ $+ A_3 \exp(-\xi)$	$A_1 \exp$ $+ A$
V_M	$(Q/C_a)\ (\tau_i/\tau_{cd})$	$(Q/C_a)(\tau/\tau_{cd})$	$(Q/C_e)\ \{C_s/(C_s + C_d)\}$	(Q/C_e)
A_k	$A_1 = \dfrac{k-n}{(m-n)\ (l-n)\ (1-n)}$ $A_2 = \dfrac{k-m}{(n-m)\ (l-m)\ (1-m)}$ $A_3 = \dfrac{k-l}{(n-l)\ (m-l)\ (1-l)}$ $A_4 = \dfrac{k-1}{(n-1)\ (m-1)\ (l-1)}$	$A_1 = \dfrac{1}{(m-n)\ (l-n)}$ $A_2 = \dfrac{1}{(n-m)\ (l-m)}$ $A_3 = \dfrac{1}{(n-l)\ (m-l)}$	$A_1 = \dfrac{k-n}{(m-n)\ (1-n)}$ $A_2 = \dfrac{k-m}{(n-m)(1-m)}$ $A_3 = \dfrac{k-1}{(n-1)\ (m-1)}$	A_1 A_2
Normalized zero and poles	$k = p_{01}\tau_i\ ;$ $n = p_1\tau_i\ ;\ m = p_2\tau_i\ ;$ $l = p_3\tau_i\ ;\ nml/k = \tau_i^2/(\tau_{aa}\tau_{cd})$	$n = p_1\tau\ ;\ m = p_2\tau\ ;$ $l = p_3\tau\ ;$ $nml = \tau^2/(\tau_{aa}\tau_{ca})$	$k = p_{01}\tau_i\ ;$ $n = p_1\tau_i\ ;\ m = p_2\tau_i\ ;$ $nm/k = \tau_i\tau/(\tau\tau_{aa} +$ $+ \tau_{sd}\tau_{aa} + \tau\tau_{ad})$	$n =$ nm $+ \tau_a$

Table 3.1

SD types at the voltage preamplifier input.

L^0VP	L_dV	L_d^0V	UV	U^0V
$\dfrac{1}{(p+p_1)\,(p+p_2)}$	$\dfrac{1}{(p+p_1)\,(p+p_2)\,(p+p_3)}$	$\dfrac{1}{(p+p_1)\,(p+p_2)}$	$\dfrac{1-\exp(-\tau_i p)}{p(p+p_1)\,(p+p_2)}$	$\dfrac{1-\exp(-\tau_i p)}{p(p+p_1)}$
$(Q/C_e)\ [C_s/(C_s+C_d)]/\tau$	$(Q/C_a)/(\tau_i\tau_{cd})$	$(Q/C_e)/\tau_i$	$(Q/C_a)/(\tau_i\tau_{cd})$	$(Q/C_e)/\tau_i$
$C_a+[(C_dC_s)/(C_d+C_s)]$	—	C_a+C_d	—	C_a+C_d
—	—	—	—	—
$\ldots+\tau_{sd})\mp[(\tau-\tau_{aa}-\ldots$ $\ldots\}/(\tau_{aa}\tau+\tau_{aa}\tau_{sd}+\tau_{ad}\tau)$	$p_{1,2}=\{(\tau_{cd}+\tau_{aa}+\tau_{ad})\mp$ $\mp[(\tau_{cd}-\tau_{aa}+\tau_{ad})^2+$ $+4\tau_{aa}\tau_{ad}]^{\frac{1}{2}}\}/(2\tau_{cd}\tau_{aa})$ $p_3=1/\tau_i$	$p_1=1/\tau_{ae}$ $p_2=1/\tau_i$	$p_{1,2}$ are the same as in the L_d V case	$p_1=1/\tau_{ae}$
$A_1\exp(-n\xi)+$ $+A_2\exp(-m\xi)$	$A_1\exp(-n\xi)+$ $+A_2\exp(-m\xi)+$ $+A_3\exp(-\xi)$	$A_1\exp(-n\xi)+$ $+A_2\exp(-\xi)$	$A_1\exp(-n\xi)+$ $+A_2\exp(-m\xi)+$ $+A_3$	$A_1\exp(-n\xi)+A_2$
$(Q/C_e)\{C_s/(C_s+C_d)\}$	$(Q/C_a)\ (\tau_i/\tau_{cd})$	Q/C_e	$(Q/C_a)\ (\tau_i/\tau_{cd})$	Q/C_e
$A_1=1/(m-n)$ $A_2=1/(n-m)$	$A_1=\dfrac{1}{(m-n)\ (1-n)}$ $A_2=\dfrac{1}{(n-m)\ (1-m)}$ $A_3=\dfrac{1}{(n-1)\ (m-1)}$	$A_1=1/(1-n)$ $A_2=1/(n-1)$	$A_1=1/\{n(n-m)\}$ $A_2=1/\{m(m-n)\}$ $A_3=1/(mn)$	$A_1=-1/n$ $A_2=1/n$
$n=p_1\tau;\ m=p_2\tau;$ $nm=\tau^2/(\tau_{aa}\tau+$ $+\tau_{aa}\tau_{sd}+\tau_{ad}\tau)$	$n=p_1\tau_i;\ m=p_2\tau_i$ $nm=\tau_i^2/(\tau_{aa}\tau_{cd})$	$n=p_1\tau_i=\tau_i/\tau_{ae}$	$n=p_1\tau_i;\ m=p_2\tau_i$ $nm=\tau_i^2/(\tau_{aa}\tau_{cd})$	$n=p_1\tau_i=\tau_i/\tau_{ae}$

Tabl

Voltage pulses supplied by SD at th

	LQ	LQP	L^0Q
$V_0(p)/V_p$	$\dfrac{p+p_{01}}{p\,(p+p_1)\,(p+p_2)\,(p+p_3)}$	$\dfrac{1}{p\,(p+p_1)\,(p+p_2)}$	$\dfrac{p+p_{01}}{p(p+p_1)\,(p+p_2)}$
V_p	$[Q/(C_a+C_d)]/(\tau_i\tau_{ce})$	$[Q/(C_a+C_d)]/(\tau\tau_{ce})$	$[Q/(C_a+C_d)]\,\tau/(\tau_{se}\tau_i)$
C_e	$C_aC_d/(C_a+C_d)$	$C_aC_d/(C_a+C_d)$	$C_s+[C_aC_d/(C_a+C_d)]$
$\lvert p_{01}\rvert$	$1/\tau$	—	$1/\tau$
$\lvert p_k\rvert$	$p_{1,2}=\{(\tau+\tau_{ce}+\tau_{se})\mp[(\tau-\tau_{ce}+\tau_{se})^2+4\tau_{ce}\tau_{se}]^{\frac{1}{2}}\}/(2\tau\tau_{ce})$ $p_3=1/\tau_i$		$p_1=1/\tau_{se}$ $p_2=1/\tau_i$
$V_0(\xi)/V_M$	$A_1\exp(-n\xi)+A_2\exp(-m\xi)+$ $+A_3\exp(-\xi)+A_4$	$A_1\exp(-n\xi)+$ $+A_2\exp(-m\xi)+A_3$	$A_1\exp(-n\xi)+$ $+A_2\exp(-\xi)+A_3$
V_M	$[Q/(C_a+C_d)](\tau_i/\tau_{ce})$	$[Q/(C_a+C_d)](\tau/\tau_{ce})$	$[Q/(C_a+C_d)]\,(\tau/\tau_{se})$
A_k	$A_1=\dfrac{n-k}{n(m-n)\,(1-n)}$ $A_2=\dfrac{m-k}{m(n-m)\,(1-m)}$ $A_3=\dfrac{1-k}{(n-1)\,(m-1)}$ $A_4=k/(nm)$	$A_1=1/\{n\,(n-m)\}$ $A_2=1/\{m(m-n)\}$ $A_3=1/(nm)$	$A_1=(n-k)/\{n(1-n)\}$ $A_2=(1-k)\,/(n-1)$ $A_3=k/n$
Normalized zero and poles	$k=p_{01}\tau_i$; $n=p_1\tau_i$; $m=p_2\tau_i$; $k/(nm)=\tau_{ce}/\tau_i$	$n=p_1\tau$; $m=p_2\tau$; $nm=\tau/\tau_{ce}$	$k=p_{01}\tau_i$ $n=p_1\tau_i$; $k/n=\tau_{se}/\tau$

3.2

charge-sensitive preamplifier input.

L^0QP	L_dQ	L^0_dQ	UQ	U^0Q
$\frac{1}{p(p+p_1)}$	$\frac{1}{p(p+p_1)(p+p_2)}$	$\frac{1}{p(p+p_1)}$	$\frac{1-\exp(-\tau_i p)}{p^2(p+p_1)}$	$\frac{1-\exp(-\tau_i p)}{p^2}$
$Q/(C_a+C_d)]/\tau_{se}$	$[Q/(C_a+C_d)]/(\tau_i\tau_{ce})$	$[Q/(C_a+C_d)]/\tau_i$	$[Q/(C_a+C_d)]/(\tau_i\tau_{ce})$	$[Q/(C_a+C_d)]/\tau_i$
$+[C_aC_d/(C_a+C_d)]$	$C_a C_d/(C_a+C_d)$	—	$C_aC_d/(C_a+C_d)$	—
—	—	—	—	—
$p_1=1/\tau_{se}$	$p_1=1/\tau_{ce}$ $p_2=1/\tau_i$	$p_1=1/\tau_i$	$p_1=1/\tau_{ce}$	
$A_1\exp(-n\xi)+A_2$	$A_1\exp(-n\xi)+$ $+A_2\exp(-\xi)+A_3$	$A_1\exp(-\xi)+A_2$	$A_1\exp(-n\xi)+$ $+n\xi A_2+A_3$	ξ
$Q/(C_a+C_d)$	$[Q/(C_a+C_d)](\tau_i/\tau_{ce})$	$Q/(C_a+C_d)$	$[Q/(C_a+C_d)](\tau_i/\tau_{ce})$	$Q/(C_a+C_d)$
$A_1=-1$ $A_2=1$	$A_1=1/\{n(n-1)\}$ $A_2=1/(1-n)$ $A_3=1/n$	$A_1=-1$ $A_2=1$	$A_1=1/n^2$ $A_2=1/n^2$ $A_3=-1/n^2$	—
$n=p_1\tau=\tau/\tau_{se}$	$n=p_1\tau_i=\tau_i/\tau_{ce}$	—	$n=p_1\tau_i=\tau_i/\tau_{ce}$	—

Table

Current pulses given by SD at the c

	LI	LIP	L^0I_0
$I_0(p)/I_p$	$\dfrac{p+p_{01}}{(p+p_1)(p+p_2)(p+p_3)}$	$\dfrac{1}{(p+p_1)(p+p_2)}$	$\dfrac{p+p_{01}}{(p+p_1)(p+p_2)}$
I_p	I_1/τ_{ed}	I/τ_{ed}	$I_i\tau/\tau_{se}$
R_e	R_e+R_0	R_e+R_0	—
C_e	—	—	C_s+C_d
$\|p_{01}\|$	$1/\tau$	—	$1/\tau$
$\|p_k\|$	$p_{1,2}=\{(\tau+\tau_{ed}+\tau_{sd})\mp[(\tau-\tau_{ed}+\tau_{sd})^2+4\tau_{ed}\tau_{sd}]^{\frac{1}{2}}\}/(2\tau\tau_{ed})$ $p_3=1/\tau_i$		$p_1=1/\tau_{se}$ $p_2=1/\tau_i$
$i_0(\xi)/I_M$	$A_1\exp(-n\xi)+$ $+A_2\exp(-m\xi)+A_3\exp(-\xi)$	$A_1\exp(-n\xi)+$ $+A_2\exp(-m\xi)$	$A_1\exp(-n\xi)+$ $+A_2\exp(-\xi)$
I_M	$I_i\tau_i/\tau_{ed}$	$I\tau/\tau_{ed}$	$I_i\tau/\tau_{se}$
A_k	$A_1=\dfrac{k-n}{(m-n)(1-n)}$ $A_2=\dfrac{k-m}{(n-m)(1-m)}$ $A_3=\dfrac{k-1}{(n-1)(m-1)}$	$A_1=1/(m-n)$ $A_2=1/(n-m)$	$A_1=(k-n)/(1-n)$ $A_2=(1-k)/(1-n)$
Normalized zero and poles	$k=p_{01}\tau_i$; $n=p_1\tau_i$; $m=p_2\tau_i$; $k/(nm)=\tau_{ed}/\tau_i$	$n=p_1\tau$; $m=p_2\tau$; $nm=\tau/\tau_{ed}$	$k=p_{01}\tau_i$; $n=p_1\tau_i=\tau_i/\tau_{se}$ $k/n=\tau_{se}/\tau_i$

le 3.3

urrent-sensitive preamplifier input.

$L^0 I_0 P$	$L_d I$	$L_d^0 I_0$	UI	$U^0 I_0$
$\frac{1}{p+p_1}$	$\frac{1}{(p+p_1)(p+p_2)}$	$\frac{1}{p+p_1}$	$\frac{1-\exp(-\tau_i p)}{p(p+p_1)}$	$\frac{1-\exp(-\tau_i p)}{p}$
$I\tau/\tau_{se}$	I_i/τ_{ed}	I_i	I_i/τ_{ed}	I_i
—	R_c+R_0	—	R_c+R_0	—
C_s+C_d	—	—	—	—
—	—	—	—	—
$p_1=1/\tau_{se}$	$p_1=1/\tau_{ed}$ $p_2=1/\tau_i$	$p_1=1/\tau_i$	$p_1=1/\tau_{ed}$	—
$A_1\exp(-n\xi)$	$A_1\exp(-n\xi)+$ $+A_2\exp(-\xi)$	$A_1\exp(-\xi)$	$A_1\exp(-n\xi)+A_2$	1
$I\tau/\tau_{se}$	$I_i\tau_i/\tau_{ed}$	I_i	$I_i\tau_i/\tau_{ed}$	I_i
$A_1=1$	$A_1=1/(1-n)$ $A_2=1/(n-1)$	$A_1=1$	$A_1=-1/n$ $A_2=1/n$	—
$n=p_1\tau=\tau/\tau_{se}$	$n=p_1\tau_i=\tau_i/\tau_{ed}$	$n=p_1\tau_i=1$	$n=p_1\tau_i=\tau_i/\tau_{ed}$	—

Table 3.4

Current and voltage pulses for the current and charge-sensitive pr

	LQI	LQIP	L^0QI_0	L^0QI_0P
$I_0(p)/I_p$	$\frac{p+p_{01}}{(p+p_1)(p+p_2)(p+p_3)}$	$\frac{1}{(p+p_1)(p+p_2)}$	$\frac{p+p_{01}}{(p+p_1)(p+p_2)}$	$\frac{1}{(p+p_1)}$
$V_0(p)/V_p$	$\frac{p+p_{01}}{p(p+p_1)(p+p_2)(p+p_3)}$	$\frac{1}{p(p+p_1)(p+p_2)}$	$\frac{p+p_{01}}{p(p+p_1)(p+p_2)}$	$\frac{1}{p(p+p_1)}$
I_p	I_i/τ_{ed}	I/τ_{ed}	$[I_i C_a/(C_a+C_d)]\tau/\tau_{se}$	$[IC_a/(C_a+C_d)]$
V_p	$[Q/(C_a+C_d)]/(\tau_i\tau_{ee})$	$[Q/(C_a+C_d)]/(\tau\tau_{ee})$	$[Q/(C_a+C_d)]\tau/(\tau_{se}\tau_i)$	$[Q/(C_a+C_d)]/$
R_e	R_c+R_0	R_c+R_0	—	—
C_e	$C_aC_d/(C_a+C_d)$	$C_aC_d/(C_a+C_d)$	C_s+C_d	C_s+C_d
$\lvert p_{01}\rvert$	$1/\tau$	—	$1/\tau$	—
$\lvert p_k\rvert$	$p_{12}=\{(\tau+\tau_{ee}+\tau_{se})\mp[(\tau-\tau_{ee}+\tau_{se})^2+4\tau_{ee}\tau_{se}]^{\frac{1}{2}}\}/(2\tau\tau_{ee})$ $p_3=1/\tau_i$		$p_1=1/\tau_{se}$ $p_2=1/\tau_i$	$p_1=1/\tau_{se}$
$i_0(\xi)/I_M$	$A_{i1}\exp(-n\xi)+$ $+A_{i2}\exp(-m\xi)+$ $+A_{i3}\exp(-\xi)$	$A_{i1}\exp(-n\xi)+$ $+A_{i2}\exp(-m\xi)$	$A_{i1}\exp(-n\xi)+$ $+A_{i2}\exp(-m\xi)$	$A_{i1}\exp(-n$
$V_0\xi/V_M$	$A_{v1}\exp(-n\xi)+$ $+A_{v2}\exp(-m\xi)+$ $+A_{v3}\exp(-\xi)+A_{v4}$	$A_{v1}\exp(-n\xi)+$ $+A_{v2}\exp(-m\xi)+A_{v3}$	$A_{v1}\exp(-n\xi)+$ $+A_{v2}\exp(-m\xi)+A_{v3}$	$A_{v1}\exp(-n\xi)+$
I_M	$I_i\tau_i/\tau_{ee}$	$I\tau/\tau_{ee}$	$[I_iC_a/(C_a+C_d)]\tau/\tau_{se}$	$[IC_a/(C_a+C_d)]$
V_M	$[Q/(C_a+C_d)]\tau_i/\tau_{ee}$	$[Q/(C_a+C_d)]\tau/\tau_{ee}$	$[Q/(C_a+C_d)]\tau/\tau_{se}$	$[Q/(C_a+C_d)]\tau$
A_k	$A_{i1}=\frac{k-n}{(m-n)(1-n)}=-nA_{v1}$ $A_{i2}=\frac{k-m}{(n-m)(1-m)}=-mA_{v2}$ $A_{i3}=\frac{k-1}{(n-1)(m-1)}=-A_{v3}$ $k/(nm)=A_{v4}$	$A_{i1}=1/(m-n)=-nA_{v1}$ $A_{i2}=1/(n-m)=-mA_{v2}$ $1/(nm)=A_{v3}$	$A_{i1}=(k-n)/(1-n)=-nA_{v1}$ $A_{i2}=(1-k)/(1-n)=-A_{v2}$ $k/n=A_{v3}$	$A_{i1}=1=-n$ $1/n=A_{v2}$
Normalized zero and poles	$k=p_{01}\tau_i$; $n=p_1\tau_i$; $m=p_2\tau_i$; $k/(nm)=\tau_{ee}/\tau_i$	$n=p_1\tau$; $m=p_2\tau$; $nm=\tau/\tau_{ee}$	$k=p_{01}\tau_i$; $n=p_1\tau_i$; $k/n=\tau_{se}/\tau$	$n=p_1\tau=\tau/$

preamplifiers simultaneously connected at the SD output.

	$L_d QI$	$L_d^0 QI_0$	UQI	$U^0 QI_0$
	$\frac{1}{(p+p_1)(p+p_2)}$	$\frac{1}{(p+p_1)}$	$\frac{1-\exp(-\tau_i p)}{p(p+p_1)}$	$\frac{1-\exp(-\tau_i p)}{p}$
	$\frac{1}{p(p+p_1)(p+p_2)}$	$\frac{1}{p(p+p_1)}$	$\frac{1-\exp(-\tau_i p)}{p^2(p+p_1)}$	$\frac{1-\exp(-\tau_i p)}{p^2}$
	I_i/τ_{ed}	$I_i C_a/(C_a+C_d)$	I_i/τ_{ed}	$I_i C_a/(C_a+C_d)$
	$[Q/(C_a+C_d)]/(\tau_i \tau_{ee})$	$[Q/(C_a+C_d)]/\tau_i$	$[Q/(C_a+C_d)]/(\tau_i \tau_{ee})$	$[Q/(C_a+C_d)]/\tau_i$
	R_0+R_c	—	R_0+R_c	—
	$C_a C_d/(C_a+C_d)$	—	$C_a C_d/(C_a+C_d)$	—
	—	—	—	—
	$p_1 = 1/\tau_{ee}$ $p_2 = 1/\tau_i$	$p_1 = 1/\tau_i$	$p_1 = 1/\tau_{ee}$	
	$A_{i1} \exp(-n\xi) +$ $+ A_{i2} \exp(-\xi)$	$A_{i1} \exp(-\xi)$	$A_{i1} \exp(-n\xi) + A_{i2}$	1
	$A_{v1} \exp(-n\xi) +$ $+ A_{v2} \exp(-\xi) + A_{v3}$	$A_{v1} \exp(-\xi) + A_{v2}$	$A_{v1} \exp(-n\xi) +$ $+ n\xi A_{i2} + A_{v3}$	
	$I_i \tau_i/\tau_{ed}$	$I_i C_a/(C_a+C_d)$	$I_i \tau_i/\tau_{ed}$	$I_i C_a/(C_a+C_d)$
	$[Q/(C_a+C_d)]\tau_i/\tau_{ee}$	$Q/(C_a+C_d)$	$[Q/(C_a+C_d)]\tau_i/\tau_{ee}$	$Q/(C_a+C_d)$
	$A_{i1} = 1/(1-n) = -nA_{v1}$ $A_{i2} = 1/(n-1) = -A_{v2}$ $1/n = A_{v3}$	$A_{i1} = 1 = -A_{v1} = A_{v2}$	$A_{i1} = -1/n = -nA_{v1}$ $A_{i2} = 1/n = nA_{v2}$ $-1/n^2 = A_{v3}$	—
	$n = p_1 \tau_i = \tau_i/\tau_{ee}$	—	$n = p_1 \tau_i = \tau_i/\tau_{ee}$	—

Similarly, the equations found for a charge sensitive preamplifier are presented in Table 3.2.

The current preamplifier circuit impedance may be either an R_0 resistance (usually the characteristic impedance of a cable) or a short-circuit ($R_0 = 0$). In this way the number of interesting cases is doubled. However, Table 3.3 contains as much columns as Table 3.1 or Table 3.2. The cases which are not shown in Table 3.3, may be obtained from Table 3.3 by replacing the time constant τ_{ed}. In this way LI_0 and L^0I follow from LI by substituting τ_{ed} by τ_{cd} and τ_{od}, respectively. With the same substitutions are obtained LI_0P and L^0IP from LIP, L_aI_0 and L_a^0I from L_aI, UI_0 and U_0I from UI.

The QI systems [P 1] may be employed for fast-slow arrangements with both SD leads isolated from the case.

The corresponding grounded point in the equivalent circuit (Fig. 3.16) is denoted by M'. As distinguished from the preceding cases (V, Q, I) in which M was the grounded point and so negative pulses were obtained at the preamplifier input, the positive current pulses at the I-input were applied for a QI configuration. Both the current and voltage pulse relations are given in Table 3.4. With the substitutions $\tau_{ee} = \tau_{ce}$, $\tau_{ed} = \tau_{cd}$ and $\tau_{ee} = \tau_{oe}$, $\tau_{ed} = \tau_{od}$, LQI_0 and L^0QI may be obtained from LQI, LQI_0P and L^0QIP from $LQIP$, L_aQI and L_a^0QI from L_aQI, UQI_0 and U^0QI from UQI.

3.5.3. *Current and voltage waveforms.* The calculated current and voltage waveforms may be obtained by plotting the relations given in Tables 3.1—3.4. Generally this is not easy because there are many parameters involved (k, n, m, l). However in four particular cases : L_a^0V, L_aI, L_aI_0, L_a^0I, only one parameter n is needed for a complete determination of the normalized curves which take the form

$$V_0(\xi)/V_M \text{ or } i_0(\xi)/I_M = A_1 e^{-n\xi} + A_2 e^{-\xi}, \tag{3.114}$$

where

$$A_1 = -A_2 = 1/(1-n). \tag{3.115}$$

Thus the voltage pulse shape for an L_a^0V ($R_c = R_s = R_o = 0$) configuration is given by

$$V_0(\xi)/V_M = \{1/(1-n)\}(e^{-n\xi} - e^{-\xi}), \tag{3.116}$$

in which

$$\xi = t/\tau_i; \quad n = \tau_i/\tau_{0e}; \quad V_M = Q/C_e; \quad C_e = C_a + C_d. \tag{3.117}$$

The waveform was plotted in Fig. 3.18 for $1/n = 0.1—1000$. It may be seen that, when $1/n < 100$ the pulse amplitudes are strongly dependent on the $1/n$ value. Indeed for the normalized time ξ_M at which equation (3.116) is maximized one can write

$$\xi_M = \{1/(1-n)\} \ln (1/n), \tag{3.118}$$

and the relative amplitude is given by

$$V_{0M}/V_M = V_0(\xi_M)/V_M = (1/n)\, e^{-\xi_M} \tag{3.119}$$

The relative pulse amplitude variation V_{0M}/V_M was plotted against $1/n$ in Fig. 3.19 from which one can see that $\tau_{ae} \gg 200\ \tau_i$ is required in

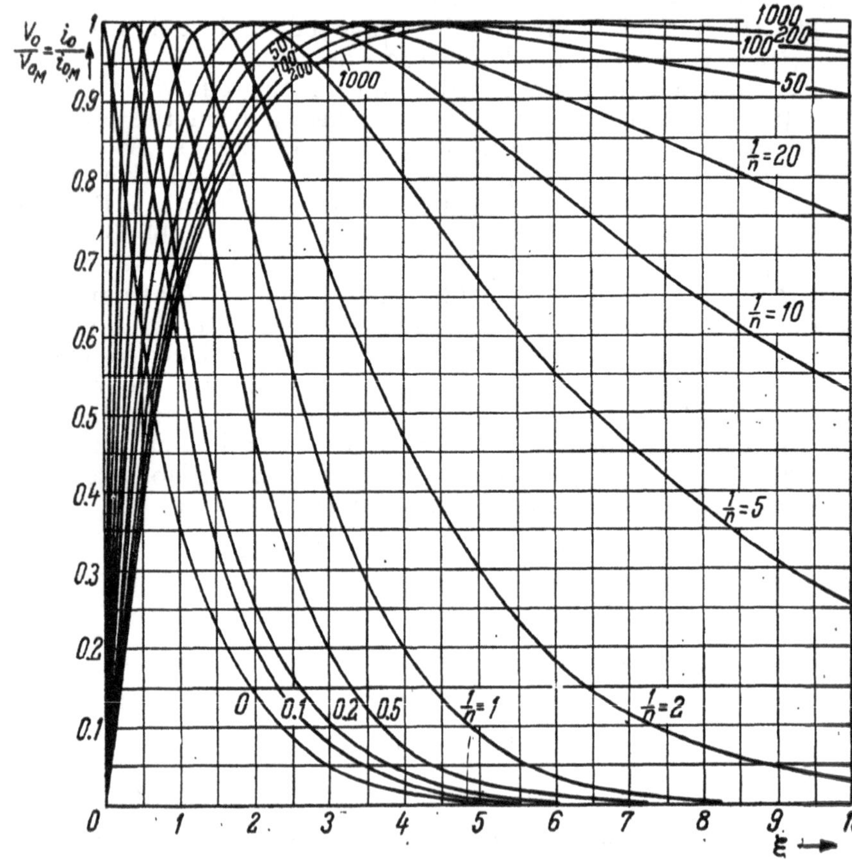

Fig. 3.20—Voltage and current waveforms for $L_a^0 V$; $L_a I$; $L_a I_0$ and $L_a^0 I$ configurations.

order to obtain a pulse amplitude V_{0M} which is lower than V_M at most by 1%. Only with a smaller value of the ratio τ_{ae}/τ_i, the rise time and the length of the voltage pulse become shorter, but in this case lower pulse amplitudes are obtained.

The pulse shape modification with the ratio $1/n$ may be better estimated from Fig. 3.20 in which $V_0(\xi)/V_{0M}$ was plotted, taking $1/n$ as parameter. When the time constant τ_{ae} is increased, the voltage pulse risetime ξ_r also increases up to the value $\xi_{rM} = 2.2$.

The pulse length specified by the full-width at halfmaximum ξ_{FWHM} also increases between 0.694 for $1/n = 0$ and $0.694/n$. The variations of ξ_M, ξ_r, ξ_{FWHM} with $1/n$ are plotted in Fig. 3.19. These results may be applied to the analysis of the current pulse waveforms obtained with an L_dI configuration and its equivalents (L_dI_0 and L_d^0I). Indeed, from Table 3.3 it follows that in this case

$$i_0(\xi) = I_i\,\{n/(1-n)\}(\mathrm{e}^{-n\xi} - \mathrm{e}^{-\xi}), \tag{3.120}$$

where

$$\frac{1}{n} = \begin{cases} \tau_{ed}/\tau_i \text{ for } L_dI, \\ \tau_{cd}/\tau_i \text{ for } L_dI_0,\ R_e = R_c + R_0;\ I_i = Q/\tau_i\,. \\ \tau_{0d}/\tau_i \text{ for } L_d^0I. \end{cases} \tag{3.121}$$

The normalized current pulse expression $i_0(\xi)/I_i$ was plotted in Fig. 3.21 with $1/n$ as parameter. Equation (3.118) is valid in this case, but the current pulse amplitude is given by

$$i_{0M}/I_i = \mathrm{e}^{-\xi_M}. \tag{3.122}$$

It follows that

$$i_0/i_{0M} = V_0/V_{0M}, \tag{3.123}$$

showing that the conclusions concerning current and voltage pulse-shape modification with $1/n$ are the same, but as far as the current pulse amplitude is concerned, an interesting conclusion can be drawn i.e. large values are obtained when the ratio $1/n$ is low. In this way, in order to obtain current pulses with maximum amplitude and minimum length from totally depleted SD, the parameters τ, C_d, R_c and R_0 should be minimized.

Now, if $1/n = 1$, equation (3.120) is no longer valid. In this case the response Laplace transform takes the form

$$I_0(p)/I_p = 1/\{p + (1/\tau_i)\}^n, \tag{3.124}$$

and in the time domain

$$i_0(\xi)/I_i = \{\xi^{m-1}/(m-1)!\}e^{-\xi}, \tag{3.125}$$

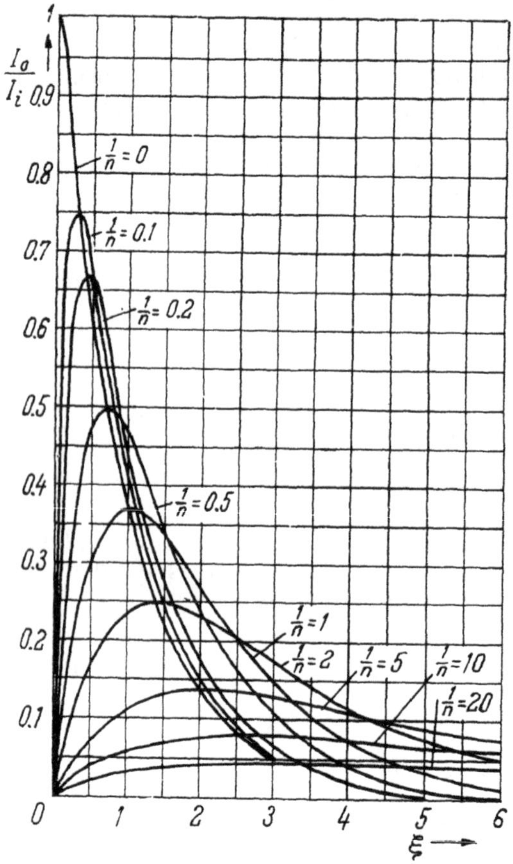

Fig. 3.21 — Current waveforms given by L_dI, $L_d I_0$ and $L_d^0 I$ configurations.

in which $m = 2$. If this current pulse passes through an amplification chain having a transfer impedance (at high frequencies) of the form

$$V_0(p)/I_{in}(p) = Z_{TO}/\{p + (1/\tau)\}^N, \tag{3.126}$$

the output voltage pulse shape will be given by

$$V_0(\xi)/V_{MN} = \{\xi^{N+1}/(N+1)!\}e^{-\xi}. \tag{3.127}$$

This function was plotted in Fig. 3.22*b* for a number of "integration" stages $N = 0;\ 1;\ 2$. The effect of "integration" on the waveform can be better seen from Fig. 3.22-a in which each pulse was normalized with respect to its amplitude V_{0M}. In comparison with the ideal case (no integration) the rise time ξ_r, the delay ξ_M, and the pulse length ξ_{FWHM} become longer. Obviously, for a longer integration time constant these effects will be more pronounced.

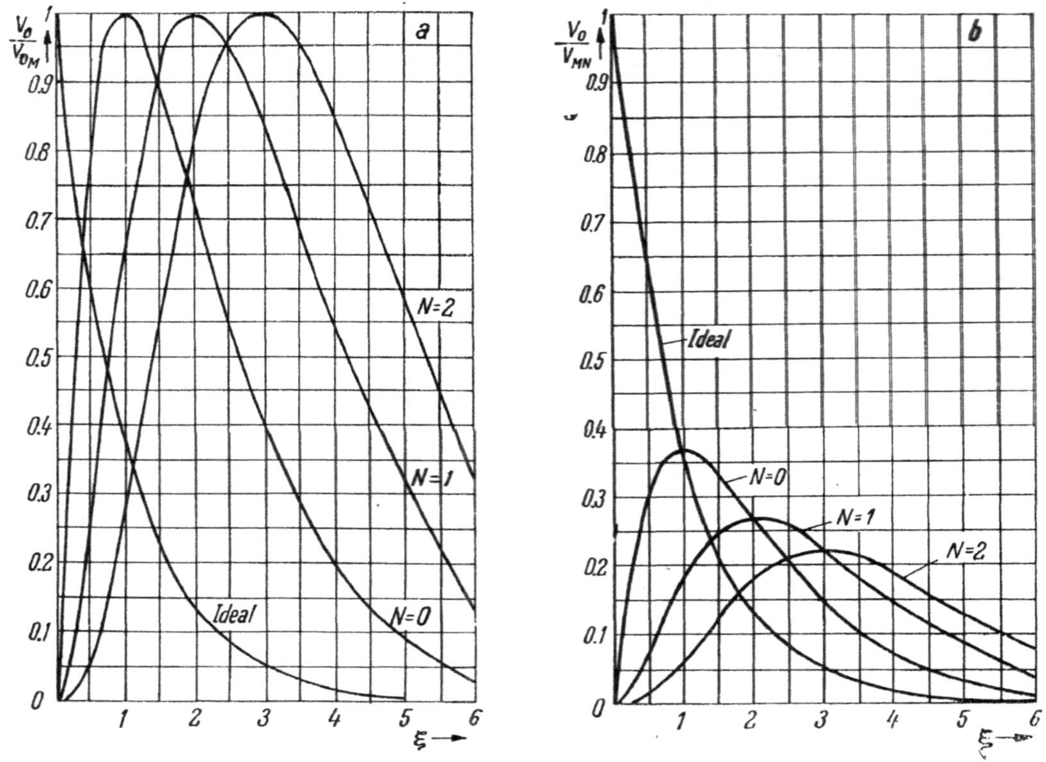

a *b*

Fig. 3.22 — Voltage waveforms after N-fold "integration" of the $L_d\, I$, $L_d\, I_0$, $\overset{0}{L_d} I$ current pulses with identical time constants ($n = 1$); (*a*) each curve is normalized at its amplitude; (*b*) normalization with respect to the ideal case.

The current or the voltage pulse waveforms supplied by an *SD* at the preamplifier input can be calculated using the tables (or the universal curves for some particular cases) given above, if the collection time constant, the elements of the *SD* equivalent circuit and the preamplifier input impedance are known.

The collection time constant τ_t can be estimated from the experimental results [M 34], [M 35] obtained with similar detectors operated in similar working conditions (the ionizing particle type and energy, the electric field strengths).

Capacity measurements give the value C_a. The total series resistance $R_s + R_c$ may be found from the *SD* forward voltage-current characterististics and if the resistivity q and the *SD* wafer geometry (S, d) are known, the series bulk resistance R_s may be calculated.

By using voltage preamplifiers one cannot obtain maximum amplitude and minimum rise time or pulse-length duration for a given ionizing particle energy. Only with a current sensitive preamplifier these two requirements, particularly useful in timing applications, can be achieved. In order to obtain the maximum pulse amplitude the *SD* works in open circuit conditions with voltage preamplifiers and in the short circuit with current sensitive preamplifiers.

3.6. COLLECTION OF CHARGE-CARRIERS IN REAL DEVICES

The simple processes of carrier drift in the electric field, studied with the idealized model in section 3.4, can be significantly altered inside real devices due either to unexpected phenomena or to an inadequate choice of devices and working conditions. In this section we shall analyse the influence of the recombination centers and traps (due to semiconductor crystal deffects) on the efficiency of the charge collection by diffusion from outside the space-charge region.

3.6.1. *Charge-collection efficiency. Influence of traps and recombination centers.* The ionizing particle yields inside the detector a charge

$$Q_L = e\frac{E}{\epsilon}, \tag{3.128}$$

and collects a charge Q_c. The collection efficiency [M 43] is

$$\eta = \frac{Q_c}{Q_L}. \tag{3.129}$$

It has been shown that

$$Q_c = Q_L \cdot \frac{\Delta x}{d}, \tag{3.130}$$

i.e., the collected charge is equal to the free charge multiplied by the path fraction travelled by the charge within the device.

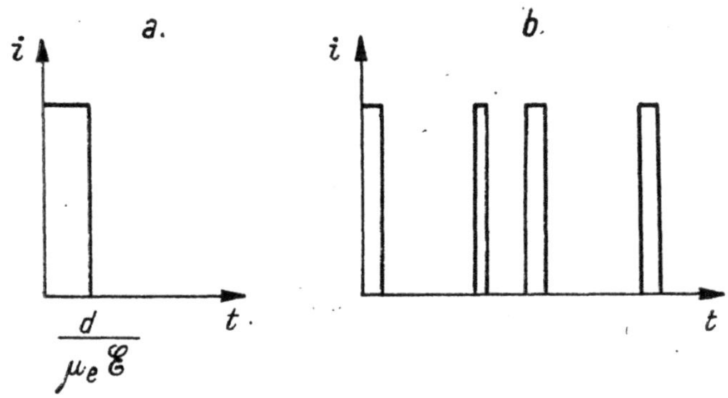

Fig. 3.23 — Current pulse shape corresponding to the electron drift with constant velocity in a uniform electric field : *a*) continuous drift ; *b*) intermittent drift due to trappings.

It follows that an inefficient collection of the charge is due to charge carriers being prevented to cross the whole distance up to the collecting electrode. That may take place in two ways : by trapping of carriers during their drift within the device and recombination along their initial track. In real crystals, there are traps, structure defects and chemical impurities with significant trapping cross-section for the drifting electron.

These trapping centers do not necessarily produce the electron recovering to the valency band, but immobilize it for a while and interrupt its contribution to the current. The electron can return in the field, drift and be trapped again, a.s.o. At last it ends its travel, reaching the positive electrode. In this case, the current waveform given by the electron drift with constant velocity within the field, uniformly changes (Fig. 3.23), and becomes a group of rectangles with random widths, spaced at random time intervals and with an integral equal to the electron charge.

In many cases when trappings take place, the recovery time from the trap is longer than 1 μs and thus the rise time of the pulse at the amplifier can be too long. Moreover, the recovery time depends much on temperature. The carrier transit may be also affected by the process of recombination.

In a trapping process, the carrier trapped by a center comes back to its former state before the center traps a carrier of opposite sign. If the carrier with opposite sign is trapped before the first carrier is set free, the defect behaves like a recombination center. The same impurity or defect can exhibit both types of behaviour (e.g. trapping or recombination depending on temperature).

Unlike trapping, by recombination both carrier contribution to the current ends and their current integral in time is smaller than the electron charge. Therefore, it is desirable that electrons and holes travel the distance d without being stopped. This requires a practical limit of d.

The mean time the electron remains free before being trapped by centers, characterized by a density N_t and a trapping cross-section σ_e, is

$$\tau_e = \frac{1}{N_t \cdot \sigma_e \cdot v_e}, \tag{3.131}$$

where v_e is the electron velocity.

For an electron transit without important trapping it is necessary that

$$t_t = \frac{d}{\mu_e \mathcal{E}} \ll \tau_e = \frac{1}{N_t \, \sigma_e v_e} \cdot \tag{3.132}$$

A similar situation exists for holes. Because τ_e can be longer than 100 μs, when reducing N_t inside the crystal and considering an electron velocity of about 10^7 cm/s, it is apparently possible that no capture happens over considerable distances of about 10^3 cm. A voltage of 10^8 V would be required to produce the necessary field of 10^5 V/cm. Actually, there exists another restriction in choosing the material which is not the best one as concerns τ_e. It seems that $\tau_e \sim 10^{-7} - 10^{-8}$ s and not 10^{-4} s. And then, the real thickness of the detector from the point of view of capture and recombination is about 1—10 mm. In the case of a $P - N$ junction characterized by a longer lifetime, there are other circumstances limiting the thickness d to the same range of values.

The most important regions of the detector are in the neighbourhood of $V(x)$ maxima. The potential distribution at the $P-N$ junction, the most used in particle detection, is a parabolic one; therefore, the most important region of th s detector, as concerns trappings and recombinations, is close to the frontal surface. Unfortunately, even this region has a maximum density of traps and recombination centers due to impurity diffusion from the crystal surface.

The effects of the incomplete collection of the charge appear as:

— Multiple peaks. In devices produced by diffusion at high temperature, the monoenergetic particles give a spectrum involving several close peaks. They may be also due to a nonuniform resistance of the surface of the detector basic material.

— Worsening of the detector resolution (multiple peaks unresolved).

There are cases when trapping and recombination effects can be separated. Thus, for instance, within the depletion layer of the junction there is practically no charge carrier and a trapped carrier meets no carrier with opposite sign to recombine with it. However, when the trapping occurs in the close vicinity of the initial, strongly ionized track, a carrier can be found to recombine with.

It is therefore expected that recombination effects should be important for fission fragments and trappings for weakly ionizing particles. We consider here only the charge collection within the depletion layer.

3.6.2. *Carrier losses by trapping.* We assume the particle trajectory to be normal to the detector surface. We denote by:

$N(y)$ — the initial density distribution of the electron-hole pairs yielded by a particle,
$V(x)$ — the potential distribution within the device,
$\mathcal{E}$ — the electric field,
R — the range of the incident particle and
d — the thickness of the depletion layer.

It can be shown [M 43] that the hole and electron charge, respectively, collected in the field is

$$Q_{eh} = \frac{e}{V(0)} \int_0^R N(y) \cdot \int_y^d \exp\left(-\int_y^x \frac{\mathrm{d}x}{\mu_g \tau_g \mathcal{E}}\right) \cdot \mathcal{E}\,\mathrm{d}x\,\mathrm{d}y, \qquad (3.133)$$

$$Q_{ee} = \frac{e}{V(0)} \int_0^R N(y) \cdot \int_0^y \exp\left(-\int_x^y \frac{\mathrm{d}x}{\mu_e \tau_e \mathcal{E}}\right) \mathcal{E}\,\mathrm{d}x\,\mathrm{d}y. \qquad (3.134)$$

The total collected charge is

$$Q_c = Q_{ch} + Q_{ce}.$$

3.6.3. *Charge collection in homogeneous semiconductor devices with uniform electric field*

In this case

$$N(y) = N \cdot \delta(y), \tag{3.135}$$

where $\delta(y)$ is the delta-function, because the device thickness is much larger than the alpha-particle range. The field intensity is also

$$\mathcal{E} = \text{const}, \tag{3.136}$$

because the electric field is assumed uniform. μ and τ may be also considered as constant quantities, because the material is supposed to be homogeneous.

Under this conditions

$$Q_a = Q_c = Q_L \frac{\mu_h \tau_h \mathcal{E}}{d}\left[1 - \exp\left(-\frac{d}{\mu_h \tau_h \mathcal{E}}\right)\right]. \tag{3.137}$$

The electron contribution is zero.

The quantity $\mu_h \tau_h \mathcal{E}$ has dimensions of length and is called the hole trapping length, a definition similar to that of the diffusion length. From (3.137) there results the condition of an efficient collection

$$\mu_h \tau_h \mathcal{E} \gg d. \tag{3.138}$$

The dependence of Q_c on $\mathcal{E}$ enables the use of (3.137) in estimating the product $\tau_h \mu_h$ and then by knowing μ_h we obtain τ_h.

In an electric field with reversed polarity, the hole contribution will be zero and a relationship similar to (3.137) is found for electrons. For minimum ionizing particles, we consider $N(y) = \text{const}$ in Eqs. (3.133) and (3.134).

3.6.4. *Charge collection in an abrupt P—N junction.* For an abrupt $P-N$ junction, the electric field distribution has the form

$$\mathcal{E}(x) = \frac{\mathcal{E}_M}{d}(d - x), \tag{3.139}$$

where $\mathcal{E}_M$ is the maximum electric field and d is the thickness of the depletion layer.

If $\mu\tau$ = const. the exponential factors in expressions (3.132) and (3.133) have the form

$$I = \exp\left(-\frac{1}{\mu\tau}\int_{u}^{v}\frac{dx}{\mathcal{E}(x)}\right). \tag{3.140}$$

From (3.132), (3.133), (3.138) and (3.139) it results

$$Q_{ce} = \frac{\mathcal{E}_M \cdot e}{d \cdot V(0) \cdot (2+\gamma_e)}\int_0^R N(y)\left[(d-y)^2 - d^{(2+\gamma_e)} \cdot (d-y)^{-\gamma_e}\right] dy, \tag{3.141}$$

$$Q_{ch} = \frac{\mathcal{E}_M \cdot e}{d \cdot V(0) \cdot (2+\gamma_h)}\int_0^R N(y)\,(d-y)^2\, dy, \tag{3.142}$$

where

$$\gamma = \frac{d}{\mu\tau\mathcal{E}_M}. \tag{3.143}$$

In the case of uniform ionizations

$$N(y) = \begin{cases} N(R), & 0 \leqslant y \leqslant R \leqslant d, \\ 0, & R \leqslant y. \end{cases} \tag{3.144}$$

There are two interesting cases :

$R \ll d$, short range ;

$R \gg d$, long range $\left(\frac{dE}{dx}\right.$ detectors and minimum ionizing particles$\left.\right)$.

Short range particles.

With (3.141), (3.143) and $\frac{R}{d} \to 0$, we obtain at the limit

$$\eta_h = \frac{2}{2+\gamma_h} \text{ şi } Q_{ce} = 0. \tag{3.145}$$

For a N^+P junction, $d \sim \sqrt{\rho\, V(0)}$.
We obtain

$$\mathcal{E}_M \sim \sqrt{\frac{V(0)}{\rho}}\ ;\ \gamma \sim \left(\frac{\rho}{\mu\tau}\right), \tag{3.146}$$

irrespective of V. The material with low ρ gives a better efficiency; this is obvious since when $\mathcal{E}$ increases, the carrier velocity increases and at the same time d decreases, such that the carriers do not travel long distances.

An estimation with reasonable values gives a low charge loss due to the mechanism decribed above (For $\rho = 10$ k Ω cm, $\tau = 1$ μs there results only a 0.5% charge loss). This mechanism prevails when τ is short (that is the case of a damaged region of the crystal or of the Si doped with Au, where $\tau \sim 1$ ns).

It should be mentioned that the assumption $\mu =$ const. made in deducing equation (3.144) imposes restrictions on the value $\mathcal{E}$ ($\mathcal{E} < 2.10^3$ V/cm) i.e.

$$\sqrt{\frac{V(0)}{\rho}} \leqslant \frac{1}{10}. \tag{3.147}$$

If this condition is not fulfilled η decreases below the value from (3.144) and for detectors with low ρ and very short τ, the collection efficiency at sufficiently high bias is expected to decrease monotonically with the increase of V. This analysis may be applied to devices with high ρ.

Long range particles

Substituting in Eqs. (3.140)—(3.143) $R = d$, we obtain

$$\eta = \frac{1}{3}\left(\frac{2}{2+\gamma_h}\right) + \frac{2}{3}\left(\frac{2}{2+\gamma_e}\right), \tag{3.148}$$

(in (3.144) an N^+P junction has been supposed). For a P^+N surface barrier the situation changes and

$$\eta_e = \frac{2}{2+\gamma_e}. \tag{3.145'}$$

2/3 of the signal are due to electrons and only 1/3 to holes, which results from the distribution asymetry of the diode potential.

From (3.147) it can be estimated that the signal given by the detector for minimum ionizing particles is proportional to its thickness, i.e. to $\sqrt{V(0)}$. This is true provided that the depletion layer spreads all over the device volume reaching the ohmic back contact.

In some materials the trapping length is so short that it controls the collection. For example, at junctions made by Zn diffusion in Ga As of 10^8 Ω cm when the device is illuminated, η increases by 30%. This fact can be explained by considering some of the traps occupied by optically generated carriers and thus inefficient, what leads to an increased trapping length. A detailed study of "trapping effects" has been made by Eisler [E 1], [E 2].

3.6.5. *Carrier losses by recombination.* This process takes place when both types of carriers are present and therefore depends on their density. The region of maximum electron and hole density is in the close vicinity of the initial track of the particle, just after its passing. If the ionization density is sufficiently high, the plasma is preserved by its own Coulombian forces for a time of the order of nanoseconds. This plasma is destroyed by diffusion of both types of carriers inside its volume where the field is zero and by erosion of its surface under the action of the external field. During the short plasma existence, the recombination process may be considerable.

Generally, the incident particles (except electrons) yield a density of electron-hole pairs along their tracks higher than the initial one (n_h n_e) and the rate of carrier loss by recombination is

$$-\frac{\mathrm{d}n}{\mathrm{d}t} = \frac{n_x}{\tau_x} + n_h n_e \sigma \nu + \frac{n}{\tau}, \tag{3.149}$$

where three terms in the right hand side represent different mechanisms : recombination via excitons, direct recombination and recombination on recombination centers.

An exciton is a complex resembling the positronium formed by the association of a free electron with a hole. If the incident particle yields n electron-hole pairs, the pair fraction lost by recombination is estimated [M 43] to be $\left(\frac{\delta n}{n}\right) < 0.01‰$ in the limit case when the plasma would last for 10^{-8} s and the exciton lifetime $\tau = 6.5$ μs, so that the recombination via excitons could be neglected.

An upper limit of 1% for the radiative direct recombination is also obtained, hence it is unlikely that this mechanism has a significant contribution to the charge "losses" observed.

The recombination on centers of recombination produces a variation of the carrier density given by the relation

$$\frac{\delta n}{n_T} = \frac{n}{n_T}\frac{\delta t}{\tau} = \frac{n N_R \sigma v}{n_T}\,\delta t, \tag{3.150}$$

where N_R is the density of the recombination centers, σ is the cross section of recombination and v is the thermal velocity of the carriers.

The fractional loss of carriers expected in the general case cannot be estimated, for lack of information about the centers of recombination in Si.

Great losses have been observed in detectors with barrier layer made of N-type homogeneous Si of high purity, particularly for fission fragments.

In diffused junction detectors, the process of recombination takes place in two stages : at first the phosphorus ionized atom, positively charged traps an electron and then the neutral complex resulted traps a hole. The cross-section is expected to be higher in the first stage than in the second because of the initial Coulombian attraction.

Therefore, the second stage limits the recombination rate. The estimated value 5×10^{-14} cm^2 is a reasonable one, in view of the present information on carrier trapping by neutral centers.

It should be noticed that not only the noncompensated part which determines the material conductivity, but the total concentration of phosphorus is of importance. The compensated acceptors too contribute to the process of recombination. This mechanism explains the great variations of the collection efficiency observed in materials with similar ρ and τ.

It may be concluded that the recombination process on centers in plasma can explain both the charge loss value for highly ionized particles and their dependence on the ionization density. This fact also accounts for the bad resolution even in apparently homogeneous devices : any particle trajectory involves a number of recombination centers, their fluctuations controlling the resolution.

3.7. COLLECTION OF ELECTRIC CHARGES BY DIFFUSION FROM OUTSIDE THE DEPLETION LAYER

The pairs created by ionizing particles have finite lifetimes, depending on impurity concentration. The charge collection is complete if the collection time is negligible in comparison with the lifetime. For fast collection, a strong electric field is needed in which the charge-carriers would drift at a high velocity.

The electron-hole pairs created outside the depletion layer of the $P-N$ junction (where no electric field exists) have no significant contribution to the current immediately after their production. They move by diffusion into the region without field and can reach the extremity of the depletion layer where one of the carriers will be collected. The internal field behaves like a barrier for the majority carriers, but collects the minority carriers. The minority carriers cross by diffusion a distance $l = \sqrt{Dt}$ in a time t. D is the diffusion constant of the given carrier and, according to Einstein's relation, has the value

$$D = \frac{kT}{e}\mu. \tag{3.151}$$

The possibility of observing these carriers depends on their lifetime inside the semiconductor material. In Si, at room temperature $D = 10^{-3}\ \mathrm{cm^2/s}$, and thus in 10^{-7} s the carrier diffusion takes place over 10^{-3} cm. The collection by diffusion from a long distance takes place slowly and a long lifetime is required for a high efficiency of the collection, therefore, it is not a useful method for extending the sensitive volume within a junction detector, though in one of the first semiconductor devices for alpha particle detection, only the collection by diffusion was used [G. 18].

Modern devices are the N^+P or P^+N-type junctions with depletion layers, the closest possible to the frontal surface of the detector, so that its "window" should be the smallest possible (Fig. 3.24).

Fig. 3.24 shows three possible cases of heavy charged particle penetration inside the detector volume.

If the particle energy is very small (α_1), the particles stop just in the heavy doped region at the junction surface ($R \leqslant l_w$), but this case is of no interest.

At higher energies the particles (α_2) enter the depletion layer and stop there ($R \leqslant (l_w + d)$). The collection of charges yielded inside the depletion layer has been already studied (it has been assumed $L_e = 0$).

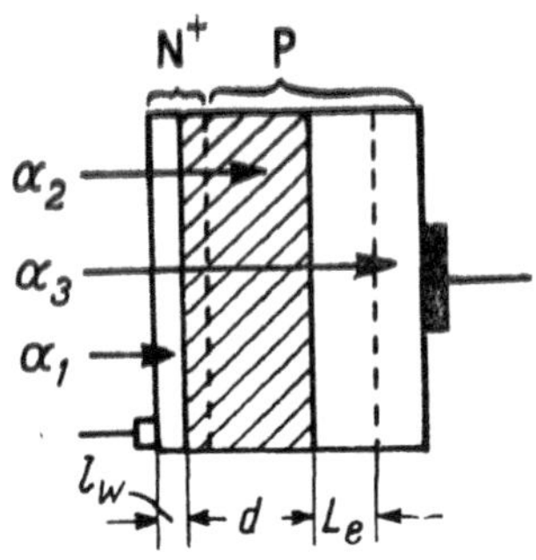

Fig. 3.24 — Penetration depths of alpha particles inside the $N-P$ junction.

At still higher energies (α_3), the particle ranges exceed the depletion layer width ($R > (l_w + d)$) and due to the slow collection by diffusion, the collection time increases.

The rise time of the voltage pulse given by the detector has two components : a fast component determined by the transit time through the depletion layer and a slow one given by the diffusion time from outside the space-charge region, up to it.

Moreover, when the particle range is much longer than the thickness of the depletion layer [M 16], it is not the voltage pulse amplitude which is proportional to the energy, but its integral.

When the particle enters the diffusion zone, the dependence amplitude vs energy is no longer a straight line. Its shape also depends on the time constants of the amplifier which can transmit more or less exactly the slow component of the pulse. The curve has a maximum which is the sharper the smaller the differentiation time constant.

The shape of the pulses observed on a wide band synchroscope and amplified by an amplifier with a 200 MHz band is drawn in Fig. 3.25. The alpha particles and deuterons stop within the barrier, but the protons pass through it. The $t_e = 3$ ns is given by the electronics.

The slow part of the pulse is observed with an amplifier with a differentiation time constant > 10 μs.

It is obvious that for linear (spectrometric) working conditions and for applications where the time characteristics of the device are

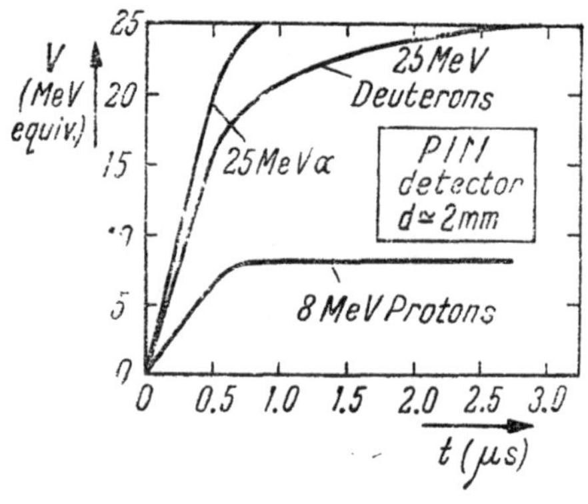

Fig. 3.25 – Pulse shapes for particles stopped inside the depletion region (α, d) and passing through it (p).

concerned, the cases denoted by α_1 and α_2 are incongruous. In applications where the flux of penetrating particles is very strong, it is useful to decrease the lifetime in the diffusion zone, to avoid the background increase due to pulse overlap (the pulses are small as amplitude but numerous and with long durations). In all cases we can escape the influence of the diffusion layer at the frontal surface of the detector, by decreasing its thickness as much as possible.

In the following, the collection of charges from the region l_w [W 20] and then from L_e [K 12], [A 6] will be considered (the charges

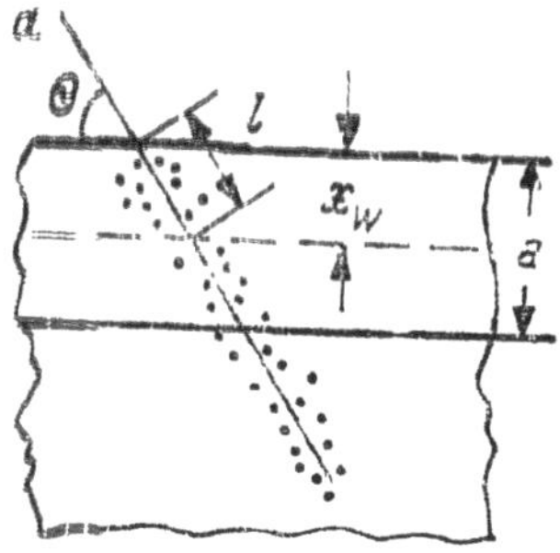

Fig. 3.26 – Surface layer of the N^+P junction.

yielded outside this region, corresponding to condition $R>(l_w + d + L_e)$ do not manage to be collected, due to recombination before reaching the junction).

3.7.1. *Collection of the charge from the surface layer.* If η_2 is assumed to be that fraction of the ionized charge, collected by the junction and coming from the surface layer, then the thickness of the detector window (Fig. 3.26) may be defined as

$$x_w = a(1 - \eta_2),$$

where a is the thickness of the region without field at the frontal surface (denoted by l_w above). Neglecting the lateral diffusion, the problem may be solved as a one-dimensional one.

If τ_h and D_h are constant in the N-layer, the fraction of holes collected in the P-layer [2] will be

$$\eta_2 = \frac{L_h}{a}\left[\frac{\left(\frac{S}{D_h}+\frac{1}{L_h}\right)\exp\left(\frac{a}{L_h}\right)+\left(\frac{S}{D_h}-\frac{1}{L_h}\right)\exp\left(-\frac{a}{L_h}\right)-\frac{2s}{D_h}}{\left(\frac{S}{D_h}+\frac{1}{L_h}\right)\exp\left(\frac{a}{L_a}\right)-\left(\frac{S}{D_h}-\frac{1}{L_h}\right)\exp\left(-\frac{a}{L_h}\right)}\right], \quad (3.152)$$

where

$L_h = \sqrt{D_h \tau_h}$ is the diffusion length and S — the recombination rate at the surface.

If τ_h is long, $L_h \gg a$ and the surface recombination is negligible $(s \to 0)$ and then $\frac{L_h s}{D_h} \ll 1$, $\eta_2 \to 1$; $x_w \to 0$.

Estimation of the window thickness. A lowering of the pulse amplitude has been observed experimentally when the incidence angle of the alpha particles decreased in comparison with the value of 90°.

A particle, normally incident to the surface, loses the energy

$$\left(\frac{\mathrm{d}E_0}{\mathrm{d}x}\right) x_w = \Delta E_0, \tag{3.153}$$

when crossing the layer of thickness x_w, and for an angle θ of incidence the energy loss is

$$\Delta E = \frac{\Delta E_0}{\sin\theta}. \tag{3.154}$$

The energy of the signal at θ angle of incidence is lowered by E' in comparison with the value for normal incidence

$$E' = (E_0 - \Delta E_0) - (E_0 - \Delta E) = \Delta E_0 \left(\frac{1}{\sin\theta} - 1\right). \tag{3.155}$$

The fractional change of the pulse amplitude is

$$\frac{E'}{E_0 - \Delta E_0} = \frac{E_0}{E_0 - \Delta E_0}\left(\frac{1}{\sin\theta} - 1\right). \tag{3.156}$$

It is drawn experimentally as a curve as that given in Fig. 3.27. ΔE_0 can be estimated from the curve slope and from E_0. For $\frac{\mathrm{d}E_0}{\mathrm{d}x} = 140$ keV/μ and $a = 2.4$ μ we obtain a window thickness $x_w = 1.3$ μ. In other cases it has been also observed that $x_w \simeq \frac{a}{2}$.

The hole lifetime in the N-layer at the surface is

$$\tau_h = \frac{1}{N^0 \sigma v}, \tag{3.157}$$

where N^0 is the number of nonionized donors, σ the donor recombination cross-section and v the hole velocity.

N^0 in the diode N-layer doped with phosphorus has been estimated with the relation

$$N^0(x)=N(x)\cdot f[E_f(x)], \tag{3.158}$$

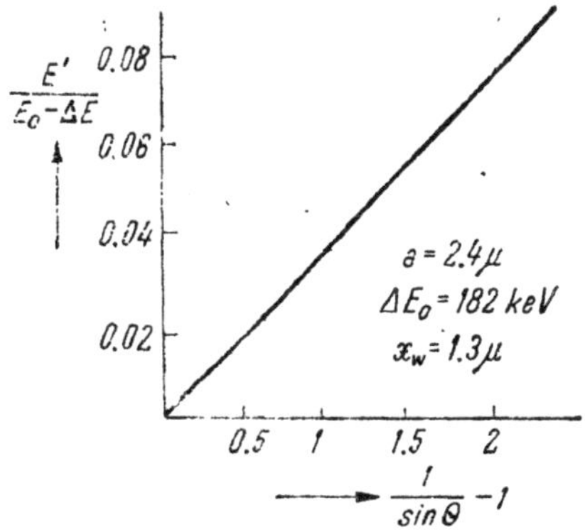

Fig. 3.27 — Pulse amplitude variation with particle incident angle due to window thickness [W 20].

since the probability of a donor level E_d being occupied by an electron is given by the Fermi distribution function

$$f(E_f)=\left[1+\exp\left(\frac{E_f-E_d}{kT}\right)\right]^{-1}. \tag{3.159}$$

Within a diffused layer, the donor concentration varies according to the law

$$N(x)=N^0\cdot \operatorname{erfc}\left(\frac{x}{2\sqrt{Dt}}\right), \tag{3.160}$$

where erfc (y) = complementary error function and

$$y=\frac{x}{2\sqrt{Dt}},$$

with t — diffusion time and D — diffusion coefficient (a function of the temperature used during the diffusion).

Considering $\sigma = 10^{-15}$ cm^2 and $v = 10^7$ cm/s and that inside the volume the lifetime is 10^{-6} s, there results

$$\frac{1}{\tau_h}=\frac{1}{10^{-6}}+N^0\sigma v. \tag{3.161}$$

The results are given in Fig. 3.28.

If the dependence τ vs the distance from the surface is known, the surface layer may be divided into two zones : a region where the charge carriers can move up to the junction ($x > 1.2\ \mu$) and a region

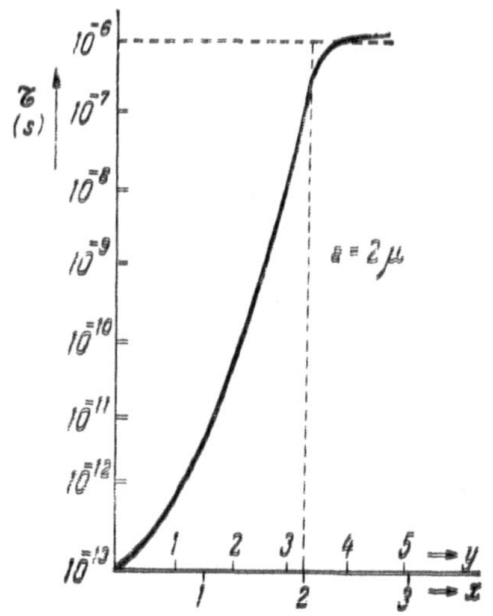

Fig. 3.28 — An example of the minority carrier lifetime variation inside the surface layer of a N^+P junction [W 20].

with a higher recombination ($x < 1.2\ \mu$) determining a thickness x_w. This fact was experimentally checked up.

In the case examined above, $x_w = \frac{a}{2}$ corresponds to $\tau_h \sim 10^{-10}$ s and $N = 5.10^{18}/\text{cm}^2$. If the concentration is kept below this value at the surface, a longer τ_h is obtained and hence a thickness $x_w < \frac{a}{2}$ of the dead layer.

If the thickness of the dead layer is not small enough, it may also influence the detector energy resolution.

3.7.2. *Collection efficiency of diffusing charges yielded by penetrating particles.* If a minority charge carrier is created at a distance l from the junction ($l < L_e$), its diffusion towards the junction in a time t gives rise to a current inside the device [G 18], [G 14], [K 12]

$$i = e \cdot f(t) \cdot \exp\left(-\frac{t}{\tau}\right), \tag{3.162}$$

where

$$f(t) = \frac{lt^{-3/2}}{2\sqrt{\pi D}} \cdot \exp\left(-\frac{l^2}{4Dt}\right), \tag{3.163}$$

with D — diffusion constant and τ — lifetime of the minority carriers.

In the case of N carrier pairs, created by a nuclear particle, supposed to be uniformly distributed along the particle track, between the junction and a plane placed at a distance l from it, the collection current [K 12] is

$$i(t) = \frac{Ne}{l}\left(\frac{D}{\pi t}\right)^{1/2}\left[1 - \exp\left(-\frac{l^2}{4Dt}\right)\right]\exp\left(-\frac{t}{\tau}\right). \tag{3.164}$$

The efficiency of the charge collection depends on the penetration depth l of the particle, compared to the ratio of the depletion layer width to the diffusion length of the minority carriers.

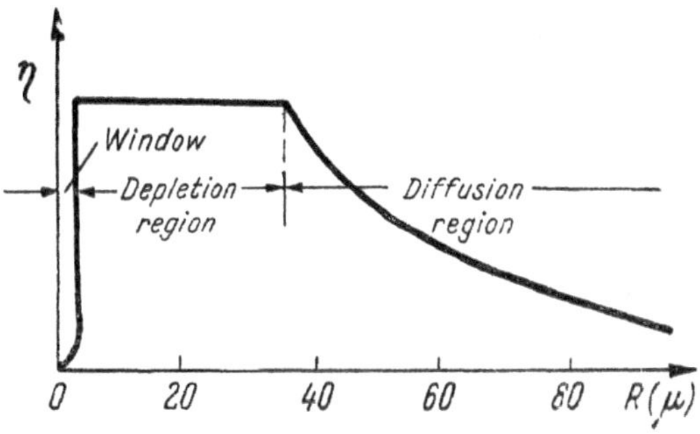

Fig. 3.29 — Typical profile of a $P-N$ junction detector.

The curve representing the variation of the charge collection efficiency vs particle range, is called by Amsel *et al.* [A 6] the "detector profile".

The profile of a proportional counter has a rectangular shape. Fig. 3.29 shows a typical "profile" for a detector with $P-N$ junction [A 6].

The shape of the curve, of the amplitude of voltage pulses vs particle energy is linear for particles stopped within the depletion layer and depends on the time constants of the amplifier, for the diffusion zone, since there the time of charge collection is longer. In the case of small time constants, the curve deviation from a linear one is sharper [A 6].

In practice, the detector profile can be determined from the "linearity" curve amplitude vs energy, obtained with a wide band amplifier and knowing that the total charge (the pulse amplitude expressed in charge units) collected $A(R)$ for particles with range R is

$$A(R) = \int_0^R \eta(x) \cdot q(x, R) \cdot \mathrm{d}x, \tag{3.165}$$

where $q(x, R)$ is the charge yielded by a particle of range R at the depth x and $\eta(x)$ is the collection efficiency.

Usually, $q(x, R)$ is a known function*)

$$q(x, R) = q(R-x) = (R - x)^{-p}, \tag{3.166}$$

and then

$$A(R) \simeq \int_0^R \frac{\eta(x)\,\mathrm{d}x}{(R-x)^p}, \tag{3.167}$$

is an Abellian integral equation for the unknown function $\eta(x)$.

Using the Laplace transform [C 2] we obtain

$$\eta(x) = \frac{\sin p\pi}{\pi} \int_0^\infty \frac{A'(t)}{(x-t)^{1-p}}\,\mathrm{d}t. \tag{3.168}$$

Estimations made with this method [K 12] have shown that the collection efficiency in the diffusion region can be approximately expressed as

$$\eta(r) = \exp\left(-\frac{r}{L}\right), \tag{3.169}$$

where r is the abscissa in the diffusion region, relation which supplies a simplified method for measuring the diffusion length L [K 12].

*) For example, for protons with energies in the range 0.25 and 5 MeV in Si [A 6] the relation

$$-\frac{\mathrm{d}E}{\mathrm{d}x} \cdot 10^{-15}\,(\mathrm{eV/cm}) = 8\,E^{-0.63}\,(\mathrm{MeV}) = 23\,R^{-0.39}(\mu)$$

can be used.

3.8. DETECTOR NOISE

The value of the current amplitude resulting from the collection of the electron-hole pairs created by a particle is not very great. For example [B 45] an alpha particle of 5 MeV creates about 1.5×10^6 pairs that can be collected by a homogeneous detector with 1 mm thickness during a time of about 0.1 μs and with a current pulse amplitude of about 2 μA. In a good detector (with high energy resolution) this current has to be considerably higher than the noise fluctuations of the mean current inside a device. Hence, the semiconductor detectors have to be made from a basic material of higher resistivity. Because of the technical difficulties met with in preparing homogeneous materials with high purity (resistivity of about 10^5 Ω cm), the semiconductor detector with uniform field is very seldom used. The depletion layer of the reverse biased $P-N$ junction is used in order to obtain a device with small leakage currents, because it behaves as a material with high resistivity due to the low density of free charge carriers inside its volume.

The detector with junction may be considered as a solid-state ionization chamber, namely a current generator with large internal resistance shunted by a capacity. It is different from the ionization chamber [G 20] because the resistance of the junction-detector is smaller and its capacity larger. Moreover, through the junction an important leakage current passes, with noise fluctuations worsening the energy resolution of the detection device.

The "leakage current" we have mentioned is in fact the reverse current of the junction and has three important sources :

– the diffusion of the minority carriers, thermally generated outside the depletion region,

– the collection of electron-hole pairs thermally created inside the depletion region,

– surface leakages.

The minority carriers, thermally generated outside the depletion region, reach it by diffusion and are collected. When the bulk material is of P-type (N^+P junction), the diffusion current density is [S 14]

$$J_D = \frac{en_h D_e}{L_e}. \tag{3.170}$$

Usually this current is negligible ($<10^{-2}$ μA/cm²) [M 42] but it increases when the minority carrier lifetime decrease. Langman and Mayer [L 6] calculated the reverse current of the surface barrier semiconductor detector.

Charge generation is the reverse process of the recombination on centers of recombination which have levels in the energy gap. When this process takes place in the depletion layer, the holes and the electrons thus created are collected by the electric field there present and the center is ready to repeat the process. The density of the current produced in this way has the expression [S 14]

$$J_B = G \cdot d_d \cdot e, \tag{3.171}$$

G being the electron-hole pair generation rate on volume unit, d the depletion region width and e the electron charge.

The generation rate G is determined by temperature, nature and location of the generation center levels within the energy gap. Assuming that the recombination center levels determining the carrier lifetime are at the center of the energy gap and have equal electron and hole trapping cross section, it is possible to estimate the maximum current expected to be given by this source. Such an estimation leads to [M 47] a current of about 10 μA/cm² generated in the depletion region in Si for a carrier lifetime of 100 μs.

Deshpande [D 17] calculated the generation-recombination noise and its influence on the *SD* resolution. The shot noise of this current is negligible in some conditions but this noise source may become important in materials with short lifetime of the minority carriers, or devices with large volume of the depletion layer. This noise can be much reduced by cooling the detector.

Generally, the surface leakage current is the principal source of noise in semiconductor detectors. Such currents are usually high and have a larger contribution to the noise than the currents with equal values generated inside the detector volume within the frequency band of interest for us.

The surface currents appear particularly on the lateral surface of the junction in the region where the strong electric field of the depletion layer reaches the surface. The surface leakages at the $P-N$ junction are generally [G 22] assumed to be due to:

— ionic conduction from the external part of an oxide film,

— injection produced by the "channel" inversion layer formed at the surface.

Both the current value and the breakdown voltage of the junction depend significantly on the working and treating of the device surface. A useful control method of the surface conditions of detectors with N gold-silicon barrier-layers is described by Fox and Borkowski [F 13]. They made diodes with N-type silicon of high resistivity, resistent to reverse biases of 4000 V. The leakage current was reduced using a guard-ring construction, similar to that of the junctions obtained by phosphorus diffusion in P-type Si [G 20].

The surface conditions once achieved, the lateral surfaces of the junction have to be protected against later alterations produced either by watery, oil, mercury vapours or other agents of the ambient medium. It is for this reason that these devices are usually tightly closed.

Therefore, the leakage current can be considerably lowered by an adequate technology. The guard-ring construction enables us to obtain negligible leakage current as compared to currents due to carrier generation within the bulk material.

Mention should be made [M 47] of another type of noise due to the contact, specific to surface barrier detectors, and characterized by current pulses resembling the useful signal. They have no influence on the resolution and can be separated by an amplitude analyser in spectrometric applications, and reduced by improving the technology when the detector is used as a monitor. The current intensities J_D and J_B remain then important quantities. For an abrupt junction in Si, assuming that the recombination centers fill in levels lying at the middle of the energy gap and considering that $n_i = 1.5\times 10^{10}\,\mathrm{cm}^3$ one obtains [G 20], [V 2]

$$J_D = \frac{en\,l}{\tau_0} = 2.75\frac{\rho l}{\tau_0}\ [\mathrm{nA/cm^2}] \tag{3.172}$$

and

$$J_B = Gde = \frac{n_i\,de}{2\tau_0} = 38\,\frac{\sqrt{\rho V}}{\tau_0}\ (\mathrm{nA/cm^2}), \tag{3.173}$$

where τ_0 (μs) is the minority carrier lifetime, assuming recombination centers which fill in levels lying at the middle of the energy gap, $\rho(\Omega\ \mathrm{cm})$ is the resistivity of the bulk material, V(V) is the reverse voltage applied on the detector and l (cm) is the distance from which the carriers can diffuse towards the depletion layer (diffusion length in the case of a sufficiently thick device).

By cooling the detector, these currents can be lowered and the *SD* resolution improved [E 10]. In thin junctions with high reverse bias applied, the component J_D is negligible as compared to J_B. The lowest current obtained in a guard-ring device [G 20] varies function of the voltage applied, according to a $\sqrt{V}$ law (Fig. 3.30). This record current corresponds to an effective lifetime of 850 μs.

The mean square fluctuations of the current (noise current) $I_D = (J_D + J_B)\Sigma$ (Σ = junction surface) have the form

$$< I_n^2 > = 2\, e\, I_D \cdot \Delta f. \tag{3.174}$$

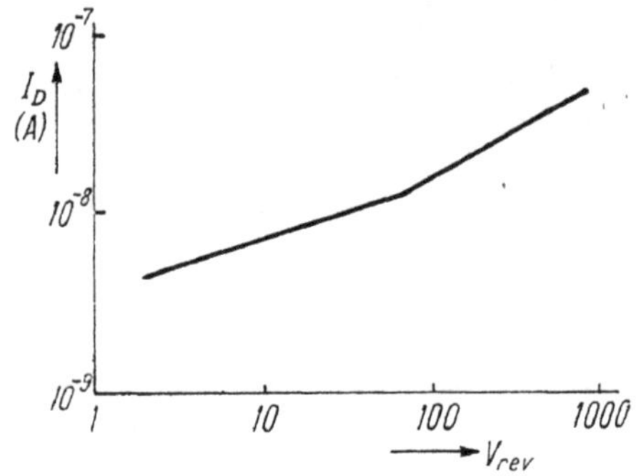

Fig. 3.30 — Reverse current variation with applied voltage.

In the pass band of the amplifier for sufficiently great shunt resistances (parallel connected load resistance of the detector and grid leakage resistance of the input tube), the impedance in which the J_D current fluctuations can give rise to a noise voltage, is almost purely capacitive (due to the total capacity of the detector and of the amplifier) and then

$$< e_n^2 > = \frac{2e\, I_D \cdot \Delta f}{\omega^2 C^2} = \frac{e I_D}{\pi \omega^2 C^2} \cdot \Delta\omega. \tag{3.175}$$

The influence of this noise on the detector energy resolution will be discussed in another chapter, in connection with the noise given by associated electronics.

3.9. DETECTOR ENERGY RESOLUTION

The pulses given by ionizing particles are superposed on the noise, giving rise to a dispersion of the differential $\dfrac{\Delta N}{\Delta V}$ counting rate distribution as a function of pulse amplitudes V.

From the study of ionization chambers [V 2] it is known that this dispersion has a Gaussian distribution (Fig. 3.31)

$$\frac{\Delta N}{\Delta V} = \frac{N_T}{\sqrt{\pi}} \exp\left[-\frac{(V - V_0)^2}{2\sigma_V^2}\right], \tag{3.176}$$

where : N_T is the total counting rate for the radiation with E_0 absorbed energy (the surface under the Gaussian curve) which gives pulses of mean amplitude V_0 and σ_V is the standard deviation.

The dispersion width is usually defined at half maximum value $\frac{\Delta N}{\Delta V}$. In energy units, the *FWHM* (full width at half maximum), i.e., the dispersion produced by the noise is

$$\Delta E_\sigma = 2.36 \sqrt{\sigma_E^2} \tag{3.177}$$

and with another notation

$$\Delta E = 2.36\, E_{eff}, \tag{3.178}$$

E_{eff} being the r.m.s. noise voltage expressed in energy units.

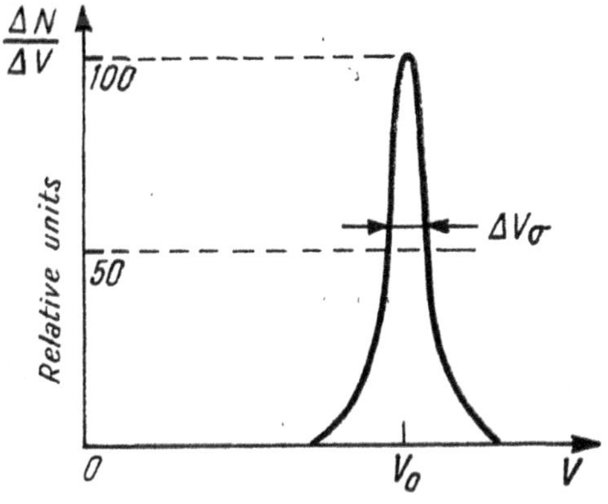

Fig. 3.31 — Amplitude distribution of pulses delivered by the detector.

This is the value of the energy dispersion produced by the noise, but the radiation conversion into electric charges has its intrinsic dispersion related to the mean number N of elementary electric charges created for an energy loss E_0.

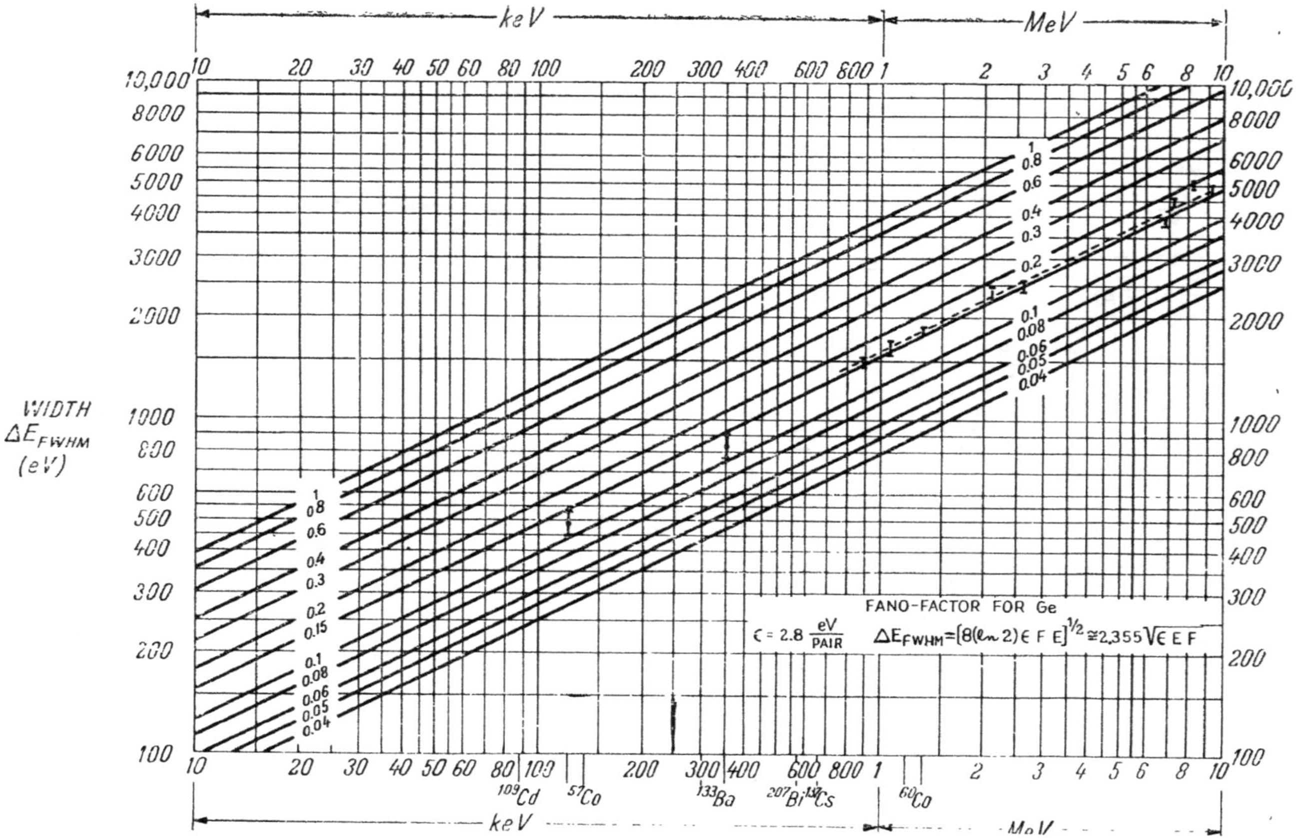

Fig. 3.32 — Energy resolution of Ge semiconductor detectors versus energy [M 10].

The intrinsic dispersion [V 2] of the detector is

$$\Delta E_{int} = 2.36\sqrt{fE_0 \in}, \tag{3.179}$$

f being Fano's reduction factor $\left(\frac{1}{2} \div \frac{1}{3} \text{ for proportional counters}\right)$.

By plotting the resolution vs gamma-rays energy curves for Ge(Li) detectors (Fig. 3.32), Mann *et al.* [M 10] obtained for *SD* a value of 0.15 for the Fano factor. The theory of Fano factor in *SD* was developed by Van Roosbroeck [R 8].

The relative intrinsic dispersion will be then

$$\frac{\Delta E_{int}}{E_0} = 2.36\sqrt{\frac{f\in}{E_0}}. \tag{3.180}$$

In this way, considering $f = 1$ we obtain for an alpha particle of 5.3 MeV

$$\Delta E_{int} = 2.36\sqrt{3.5 \times 5.3 \times 10^6} \text{ eV} = 10 \text{ keV}$$

which means approximately 0.19%. The best resolutions obtained are of 13.5 keV for 5.3 MeV alpha particles [B 26] and of 11 keV for 12 MeV protons [A 9], values close to that estimated above. This proves that for Si the Fano factor is also $f<1$, because it is hard to believe that the detector own noise and the associated electronics noise were so small that practically they had no influence on the resolution.

It is supposed that for heavy particles, the phenomena giving rise to an energy dispersion appear only at the end of their tracks.

It seems that at the end of its track, the particle yields energy to the crystal not by creating electron-hole pairs but by producing a collective displacement of atoms inside the crystalline lattice.

Notice that the same detector has for electrons a better resolution than for alpha particles [S 19].

The energy resolution of the spectrometer is determined by the statistics of electron-hole pair creation, by the Fano factor, the intrinsic noise of the detector and the amplifier noise.

The charge collection efficiency depends on the voltage applied and becomes 100% at sufficiently high voltage. The voltage increase on the detector leads to the increase of I_D and thus of the noise, but at the same time the capacity C decreases and the signal increases.

There is then an optimum working voltage at which the signal to noise ratio is maximum. The *FWHM* spectrometer energy resolution is given by the relation

$$\Delta E^2 = \Delta E^2_{int} + \Delta E^2_{det} + \Delta E^2_{amp}, \tag{3.181}$$

with

E_{int} — the intrinsic energy dispersion,
E_{det} — the energy dispersion due to the detector own noise,
E_{amp} — the energy dispersion due to the amplifier noise.

The spectrometer resolution will be analysed in connection with the signal to noise ratio at the amplifier output.

In the case of a considerable amount of recombinations, the resolution becomes worse due to "multiple peaks" (resolved or not).

The detector nonhomogeneity can also contribute to the resolution worsening, especially for wide surface devices.

Another factor leading to a worse resolution especially for short range particles, is the finite thickness of the detector window giving rise to a supplementary energy dispersion.

CHAPTER IV

CHARACTERISTICS OF SEMICONDUCTOR DETECTORS

The working of different types of semiconductor detectors (SD) can be characterized by electrical and "physical" (detection) parameters, which determine the characteristics of these detectors.

The correct working conditions of the detector can be chosen only by taking into account these characteristics. The purpose of this chapter is to review such characteristics.

4.1. ELECTRICAL CHARACTERISTICS

Since the most frequently met with SD are the reverse biased $N-P$ junctions, their characteristics will be here analysed i.e., the volt-ampere characteristic ($i_D - V_a$), the barrier capacity and the depleted region thickness vs applied voltage, ($C_D - V_a$) and ($d - V_a$), respectively.

4.1.1. *Volt-ampere characteristic.* The shape of the voltage-current characteristic of the semiconducting diode used in radiation detection is represented in Fig. 4.1. The sensitive volume of the detector corresponds to the depleted region and in order that it has the largest possible thickness, we have worked with the "reverse biased diode"; the positive voltage is applied on the N region of the junction and the negative one on the P region.

It follows that the reverse region of the diode characteristic which, as shown in § 2.11, can be described up to breakdown by an equation of the form

$$J_D = J_S \left[\exp\left(\frac{eV_a}{kT}\right) - 1\right]. \tag{4.1}$$

is much more interesting for us.

We notice that for $V_a < 0$ and $|V_a| \gg \frac{kT}{e}$, i.e. at high reverse voltages, $J_D = -J_s$. The value of this saturation current is given by

$$J_s = e\left(\frac{D_h n_{hN}}{L_h} + \frac{D_e n_{eP}}{L_e}\right). \tag{4.2}$$

However, no clear saturation is found experimentally; the reverse current increases proportionally to the applied voltage (Fig. 4.2). This fact is due to surface leakages and to a current given by thermally created carriers inside the depleted region which is proportional to the thickness of this region

$$J_B \sim \sqrt{\rho V_a}. \tag{4.3}$$

This current takes considerable values, especially in silicon at room temperature and very thick SD. A low noise is obtained when

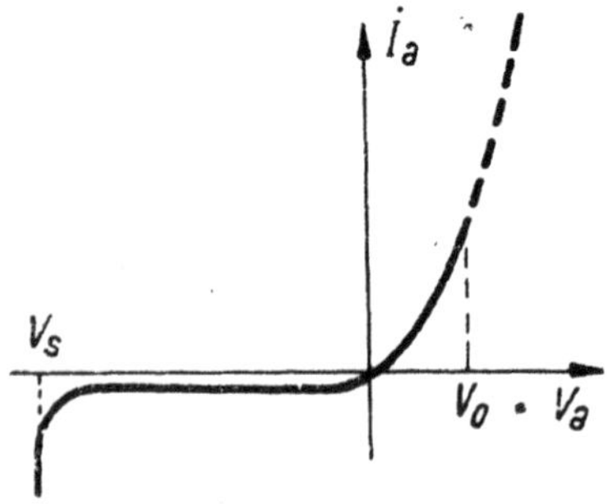

Fig. 4.1 — Volt-ampere characteristic of the $P-N$ junction.

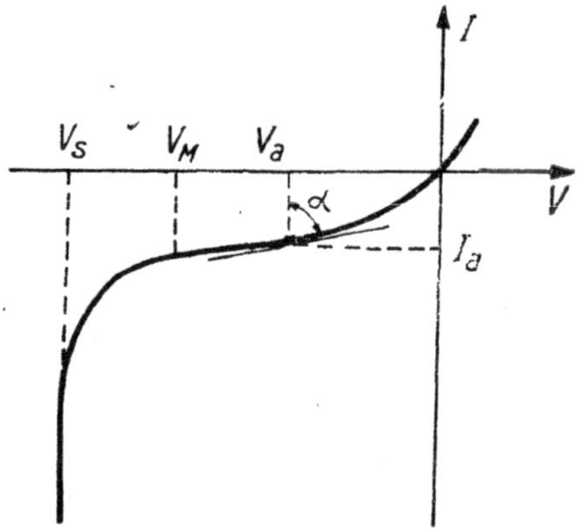

Fig. 4.2 — Static characteristic of a $P-N$ junction and its principal parameters.

the reverse current at the working point has the smallest possible value. A good detector has $I_D < 1\ \mu A$ (Fig. 4.2.).

Thus, before reaching the region of the barrier breakdown voltage (V_{bd}), the maximum voltage (V_M) that may be applied on the detector is limited by the appearance of an intense noise.

Current pulses appear in the neighbourhood of the breakdown voltage, the number and amplitude of which increase at the same time with the bias voltage increase. These pulses may also occur far from the breakdown voltage (e.g. [W 20] at 45 V though $V_{bd} = 75$ V). Usually V_{bd} does not vary during the working of the detector. But the reverse current can increase if due precautions are not taken (the junction should not be illuminated, the electrons emitted by the target should be prevented to reach the detector surface [W 20] a.s.o.).

At the chosen static working point (M) the differential resistance of the barrier layer $R_b = \tan\alpha$ (Fig. 4.2) is usually of the order of tens of megohms.

The shape of the characteristic for direct bias supplies information about the series resistance of the detector (the series resistance is high if the current does not increase sharply at small values of the direct voltage). A great series resistance gives great thermal noise and increases the rise time of the detector pulses.

All electrical tests are carried out in darkness. The photocurrent is also measured. If it is smaller than some tenths of μA at 2—3 V reverse voltage at the room light, it means that the detector window is thick.

4.1.2. *Barrier capacity*. Besides the reverse current, the barrier capacity, which determines also the amplitude of the voltage signal, influences the noise given by the detector at the working point.

The variation of the barrier capacity vs reverse voltage applied, shows whether the diode is an abrupt ($C_D \sim V_a^{-1/2}$) or a linear ($C_D \sim V_a^{-1/3}$) junction. If the device capacity does not vary with the applied voltage for sufficiently high reverse voltages, the diode is a *PIN* one.

The P^+N diodes used as *SD* are generally of abrupt junction types and in this case, as shown in section 2.11, they have a surface unit capacity (C_D) related to the applied voltage V_a and the resistivity (ρ) of the bulk material by the relation

$$\frac{1}{C_D^2} = \frac{2\mu\rho}{\varepsilon}(V_0 + |V_a|), \qquad (4.4)$$

where μ is the carrier mobility, ε is the material permitivity and V_0 the internal diffusion voltage of the junction

$$V_0 = \frac{kT}{e}\ln\frac{n_{eN}\cdot n_{hP}}{n_i^2} \qquad (4.5)$$

The high reverse voltage region of the characteristic $C - V_a$, experimentally obtained, enables us to determine by extrapolation the internal voltage of the barrier (V_a). In this way we can also determine the silicon resistivity after the detector manufacturing and make a comparison with the initial resistivity of the material.

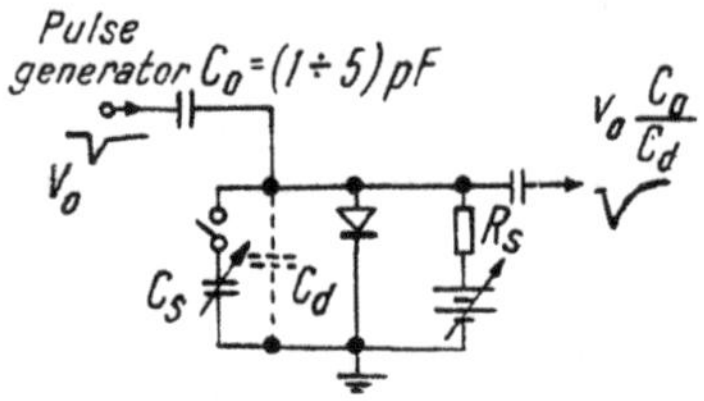

Fig. 4.3 — Circuit used for detector capacity measurements.

A convenient method for estimating the detector dynamic capacity consists in establishing the influence a variable standard capacity C_s, connected in parallel with the barrier capacity C_D (Fig. 4.3), would have on the amplitude of the voltage pulse applied on the detector.

The capacity C_S which connected in parallel with C_D produces the lowering of the pulse amplitudes to half their height, has a value equal to the detector capacity.

More accurate estimations require the taking into account of the more complex equivalent circuit of the detector (Fig. 4.4).

In addition to the elements C_b and R_b of the depleted region, the resistance (R_s) and the capacity (C_s) of the neutral region as well as the contact resistance R_k also intervene. The resistances R_k and

Fig. 4.4 — The equivalent circuit of a $P - N$ junction semiconductor detector.

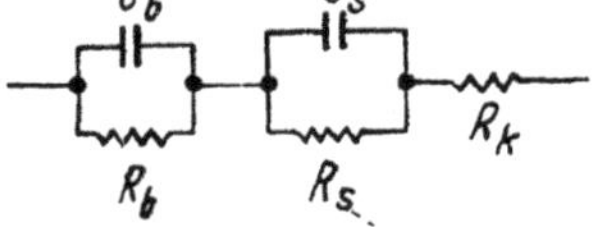

R_s of a good detector have small values, so that the circuit reduces to the capacity C_b in parallel with the resistance R_b.

Fig. 4.5 is a plot of some theoretical characteristics $C-V$ [M 31]. The higher the resistivity of the bulk material, the lower is the value of the capacity, at a given applied voltage, and the larger the amplitude of the voltage signal pulse

$$\Delta V = \frac{E}{\epsilon} \cdot \frac{e}{C} \tag{4.6}$$

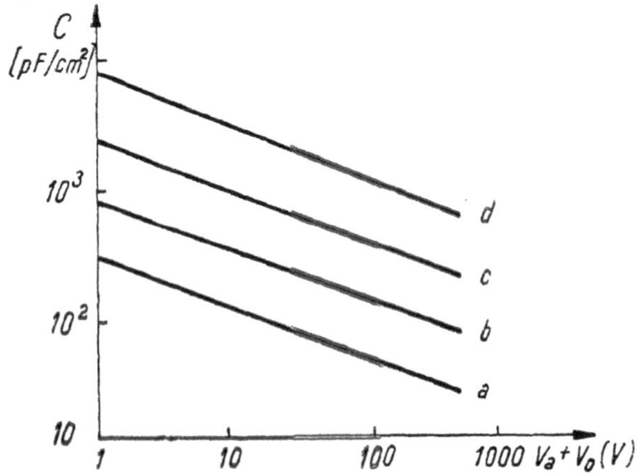

Fig. 4.5 — Barrier capacity [M 31] for a Si $P-N$ junction.
a) $N = 10^{12}$ carriers/cm^3; $\rho_N = 3470\ \Omega$ cm; $\rho_P = 15600\ \Omega$ cm.
b) $N = 10^{13}$ carriers/cm^3; $\rho_N = 347\ \Omega$ cm; $\rho_P = 1560\ \Omega$ cm.
c) $N = 10^{14}$ carriers/cm^3; $\rho_N = 34.7\ \Omega$ cm; $\rho_P = 156\ \Omega$ cm.
d) $N = 10^{15}$ carriers/cm^3; $\rho_N = 3.47\ \Omega$ cm; $\rho_P = 15.6\ \Omega$ cm.

In the first approximation, simple relations [D 12] can be used in calculations

$$\frac{C}{\Sigma} = c(\text{pF/cm}^2) \simeq \frac{18\,000}{\sqrt{\rho(\Omega\text{cm})\, V_a(V)}}, \tag{4.7}$$

for a silicon P^+N diode, and

$$c \simeq \frac{30\,000}{\sqrt{\rho V_a}} \tag{4.8}$$

for silicon N^+P junction.

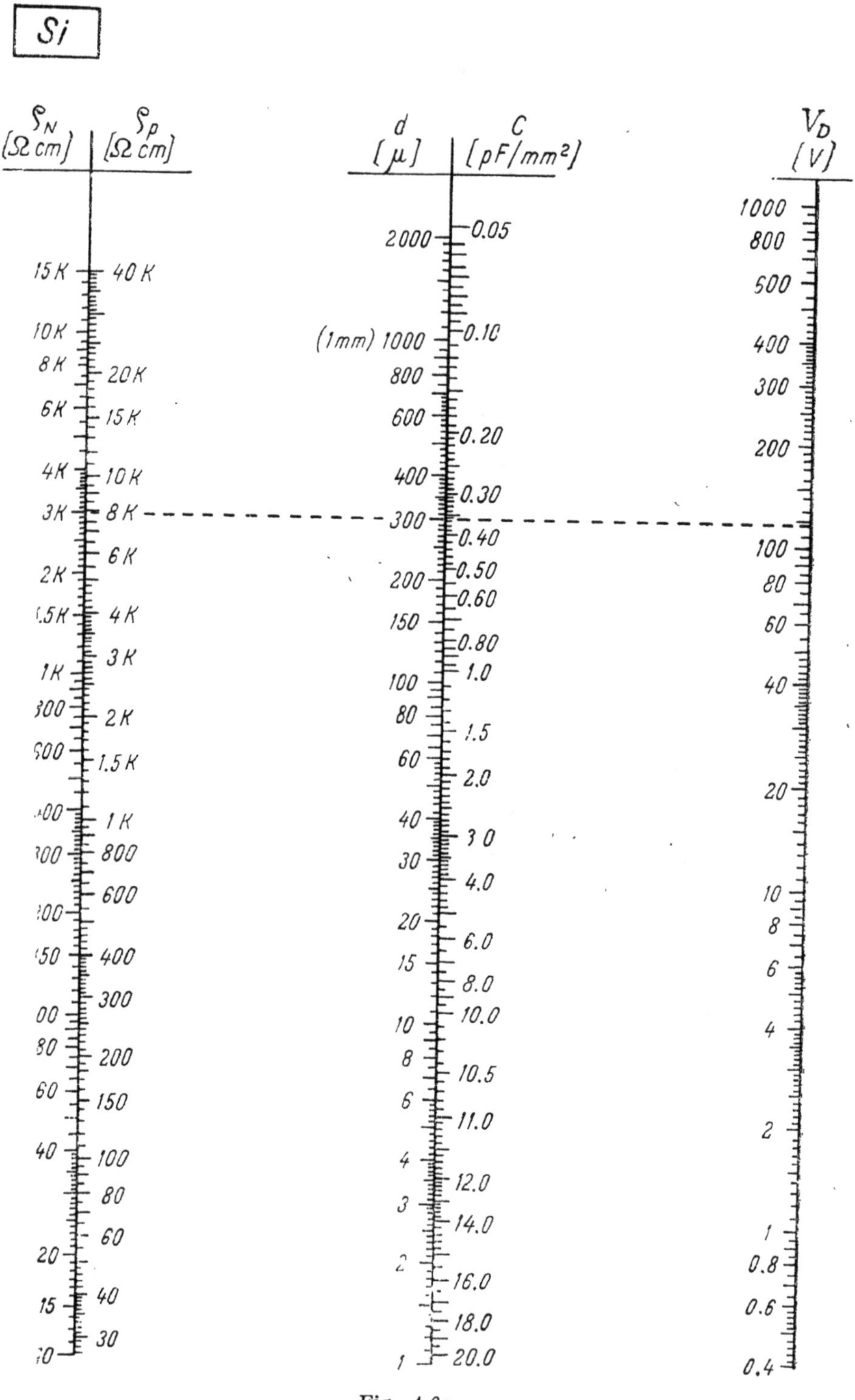

Fig. 4.6a

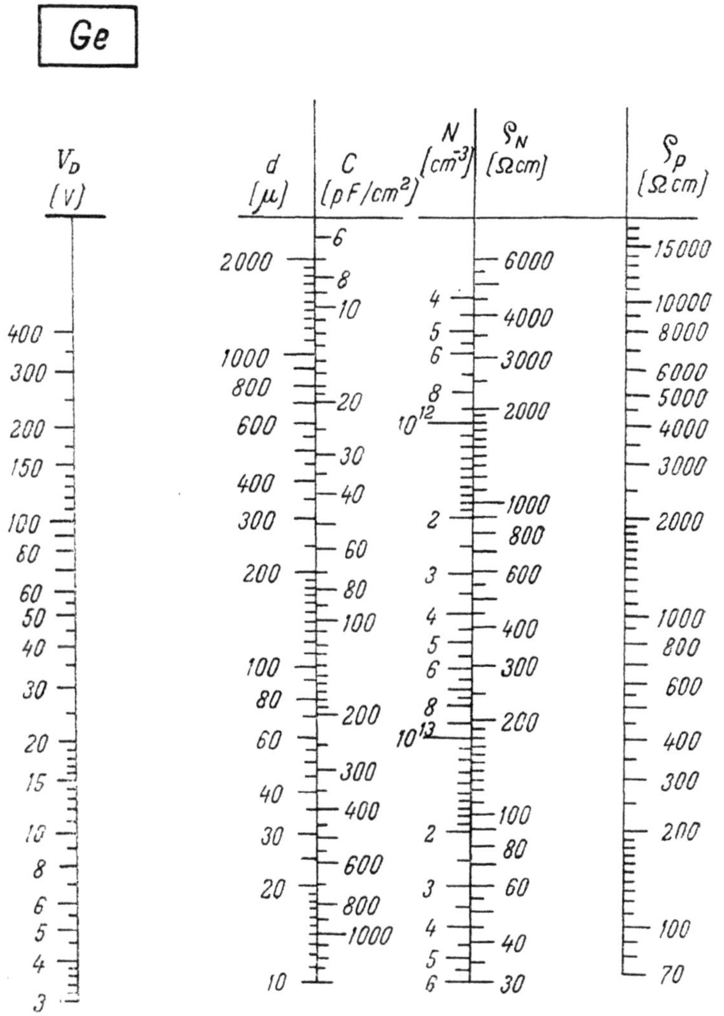

b

Fig. 4.6a, 4.6b — Nomographs used to calculate d and C in function of V_a for different resistivities of the bulk material. One assumes $\mu_e = 1200$ cm²/V·s and $\mu_h = 450$ cm²/V·s, $\varepsilon = 12$ [B 26] in Si and $\mu_e = 3500$ cm²/V·s; $\mu_h = 1500$ cm²/V·s, $\varepsilon = 16$ in Ge [T 8].

4.1.3. *Thickness of the depleted region.* From the $C-V$ characteristic, the $d-V$ characteristic may be deduced by knowing that

$$d = \frac{\varepsilon}{C} = \sqrt{2\varepsilon\mu\rho(V_0 + |V_a|)}. \tag{4.9}$$

Simple relationships [D 12] can be used for approximate calculations in the case of silicon detectors

$$d[\mu] \simeq 0{,}5\sqrt{\rho[\Omega\text{cm}]\cdot V_a[\text{V}]}, \tag{4.10}$$

for a P^+N junction (or barrier layer on N-type Si), and

$$d \simeq 0{,}3\sqrt{\rho V_a} \tag{4.11}$$

for a P^+N junction.

This is a characteristic of particular importance because it gives the "detector profile", hence the maximum particle energy at which linearity in spectrometry is still obtained.

The thickness of the depleted layer and the barrier capacity in Si and Ge can be easily estimated from the nomographs in Fig. 4.6 [T 8] for a given voltage, when the resistivity of the detector bulk material is known. We notice again the advantage of materials with higher resistivity if thicker depleted layers are desired.

4.2. DETECTION CHARACTERISTICS

The main detection characteristics of the SD are : the maximum energy, linearly detected as a function of the applied voltage $E_M(V_M)$, the linearity characteristic $V_s(E)$ at V_a = const., the efficiency characteristic $\eta(V_a)$ at E = const., the detector profile $\eta(R)$ at V_a = = const, the energy resolution $\Delta E\ (V_a)$ at E = const. and the time response.

4.2.1. *Maximum energy linearly detected.* The characteristic $E_M(V_a)$ can be easily established if the $d(V_a)$ characteristic and the range-energy relation $R\ (E)$ are known and we consider that the maximum energy that can be linearly detected E_M corresponds to a range $R = d$.

In this way, the curves in Fig. 4.7 were calculated for alpha particles, protons and electrons and for the semiconductor doping conditions shown in Fig. 4.5.

From the curves shown in Fig. 4.7 we can estimate the minimum bias voltage V_{min} that has to be applied on a detector with given resistivity, in order that the energies of the detected particles be within the linearity range of the characteristics $V_s(E)$.

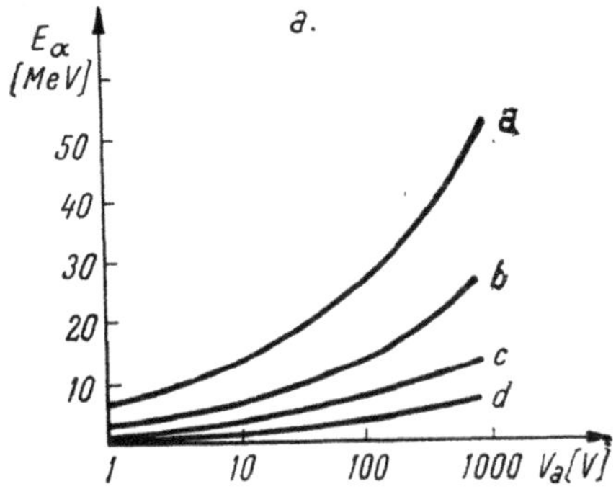

Fig. 4.7 — Maximum energy linearly detected in $P-N$ junctions; a — for alpha-particles, b — for protons, c — for electrons [M 31].

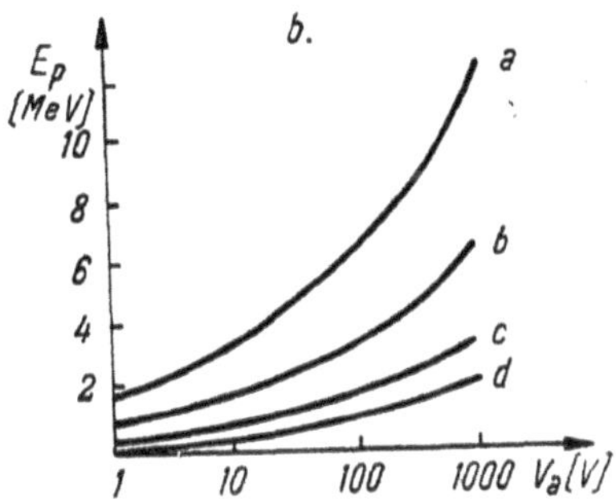

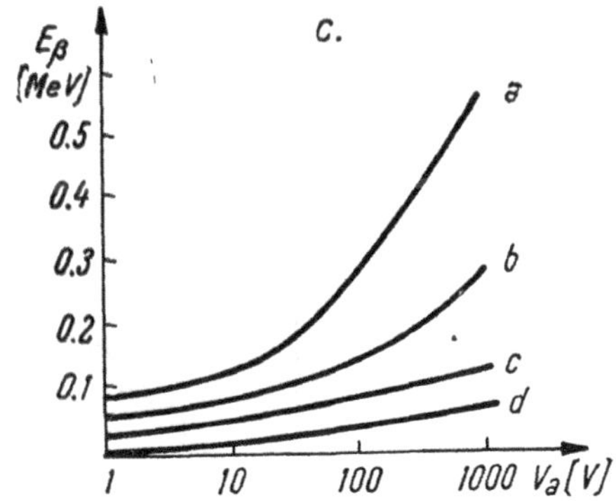

4.2.2. *Linearity characteristics*. The maximum energy linearly detected can be estimated more accurately from the linearity characteristic $V_s(E)$ at V_a = const., experimentally obtained with natural *) alpha sources or heavy charged particle beam with known energy (given by an accelerator) and using the absorber method.

*) E.g. ^{238}Pu (5.495 MeV); ^{239}Pu (5.147 MeV); ^{210}Po (5.305 MeV); ^{212}Pb (6.04; 6.08; 8.78 MeV); ^{241}Am (5.375; 5.435; 5.477; 5.530 MeV); ^{244}Cm (5.759; 5.801 MeV). A comprehensive list of alpha sources is given in Appendix A.II.

Such a set of characteristics for different voltages applied on a given detector is plotted in Fig. 4.8 [D 6].

The very good linear dependence of the pulse amplitude on energy

$$V_s = \frac{E}{\epsilon} \cdot \frac{e}{C} \tag{4.12}$$

is due to the fact that the mean energy required for the creation of an electron-hole pair (ϵ) is independent of the particle energy.

The range of linearity is limited in the upper part by the thickness of the depleted region, the maximum energy linearly detected corresponds to a particle range equal to the barrier width.

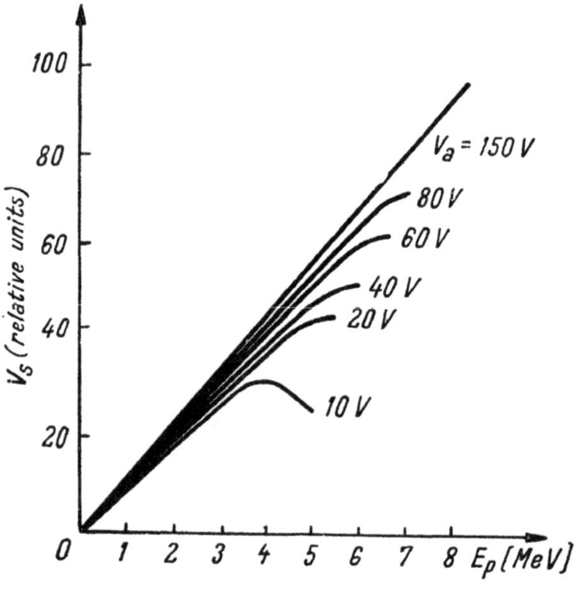

Fig. 4.8 — The linearity characteristic [D 6].

4.2.3. *Collection efficiency characteristics.* The linear regions of the curves $V_s(E)$ plotted for different voltages applied on the detector, have not the same slope because of the variation of the collection efficiency (η) and barrier capacity with the bias voltage applied on the detector (V_a).

The collection efficiency at low voltages is smaller than unity (Fig. 4.9), particularly because :

— the thickness of the depleted layer at low voltages applied on the junction can be smaller than the particle range and collection losses arise because the charge carrier diffuse slowly towards the barrier ;

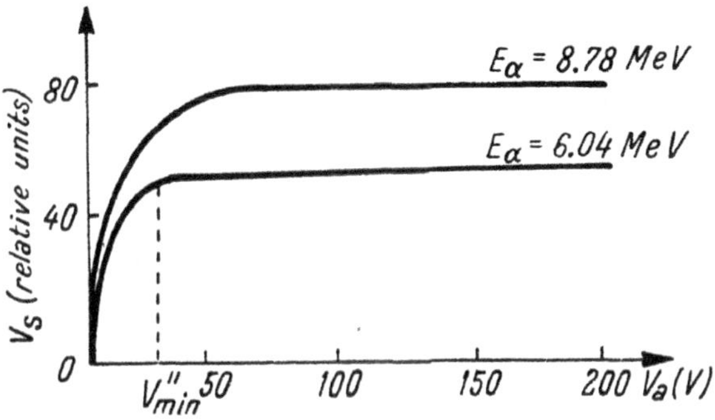

Fig. 4.9 — The collection efficiency characteristic.

— the carriers have a low drift velocity due to the weak electric field and many recombinations happen. In the case of many traps, the shape of the characteristic depends significantly on the time constant of the amplifier.

The variation with applied voltage of the pulse amplitude may be also due to the variation of the junction capacity (if charge sensitive amplifier with sufficiently great input dynamic capacity is not used).

For particles with small range ($R \ll d$) it was shown that

$$\eta = \frac{2}{2+\gamma} \text{ where } \gamma \sim \frac{\rho}{\mu\tau}, \tag{4.13}$$

whence it follows that the bulk material with smaller resistivity provides a better collection efficiency. Such junctions can be used [V 7] without applying any external bias ; for charge collection, the internal electric field of the junction due to the fixed space charge (ionized donors and acceptors) is strong enough. With a charge resistance $R_s \ll R_{bo}$, (e.g. $R_s = 250$ kΩ), the required "short circuit" regime is ensured. In these conditions the reverse current of the junction is zero and the intrinsic noise of the detector is negligible.

When a V_a bias voltage is applied and a good efficiency, unaffected by variations of the detector feeding voltage is desirable, the working point should be chosen on the plateau of the characteristic

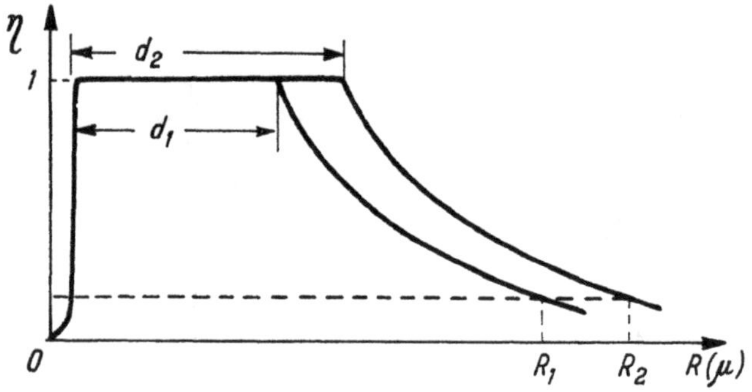

Fig. 4.10 — The profile of the detector.

v_s (V_a). In this way the efficiency characteristic gives the lower limit of the voltage applied on the detector (V_{min} in Fig. 4.9).

4.2.4. *The detector profile.* A useful characteristic for penetrating (range longer than the depleted layer thickness) particle detection is the variation of collection efficiency vs particle range [W 20] (Fig. 4.10).

The efficiency lower than unity for charges collected from outside the barrier ($R>d$) is due to losses by recombination occurring during the minority carrier diffusion towards the depleted layer. In this region.

$$\eta(R) \sim \exp\left(-\frac{R}{L}\right), \tag{4.14}$$

where $L = D$.

The longer the minority carrier lifetime, the wider the diffusion region of the detector.

Inside the region of the "detector window" the carrier lifetime is very short (due to surface recombinations) and thus the collection efficiency is zero. The window thickness for diffused junctions is usually about half the thickness of the junction heavily doped region.

It is desirable that the detector window be as small as possible such that not to worsen the energy resolution.

The thickness of the detector window can be measured with methods the principles of which were presented in section 3.7.1. (variation of the pulse amplitude with particle incidence angle), or measuring the spectral dependence of the junction photocurrent within the range $\lambda = 4000-6000$ A, drawn taking into account the absorption constant for each wave length.

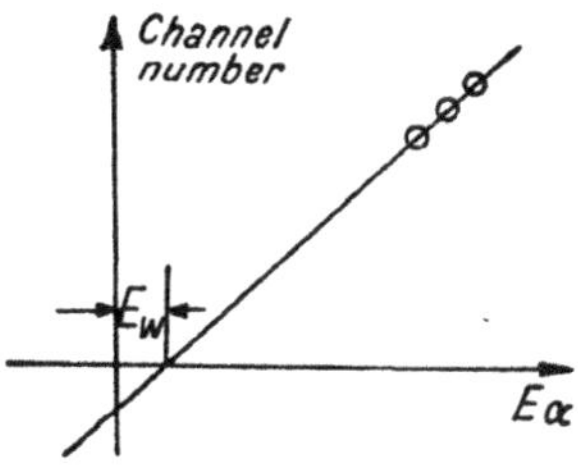

Fig. 4.11 — Determination of energy lost inside the detector window.

Another method, less accurate, consists in the estimation of the energy lost inside the window by extrapolating the straight line-channel number vs E (Fig. 4.11) drawn through some points obtained with alpha sources of close energies.

Windows thinner than 0.03 μ can be obtained in surface barrier detectors.

In nonspectrometric working conditions the useful working range of the detector can be much widened on the energy scale (particularly for a long lifetime of the minority carriers) due to the diffusion region. Then, all pulses with amplitudes greater than that corresponding to the discrimination threshold would be counted.

We notice that for this region, the pulses given by particles with different energies have different time characteristics and two pulses with the same amplitude but coming from particles with different energies can be distinguished. In the case $E_2 > E_1$ (Fig. 4.12), the pulse has a slow component due to diffusion and by derivation it gives a smaller amplitude than the fast pulse corresponding to the energy E_1 when the range is shorter than the thickness of the depleted layer.

4.2.5. *Energy resolution.* The volt-ampere characteristic ($i_D - V_a$) of the diode imposes a limit on the maximum voltage (V_M) that may be applied on the detector. The characteristic ($E_M - V_a$) limits the minimum voltage that can be applied (V'_{min}) in order that the particle range should not exceed the thickness of the junction depleted region. The minimum energy (V''_{min}) for particles with short ranges can be limited by the efficiency characteristic.

It follows that using the characteristics described above we can determine only the voltage range ($V_{min}-V_M$) that may be applied on the detector (V_{min} is the highest voltage between V'_{min} and V''_{min}).

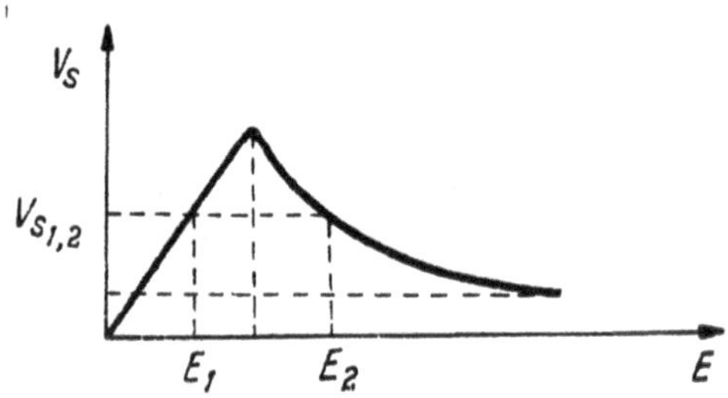

Fig. 4.12 – The linearity characteristic extrapolated to the diffusion region [F 18].

In spectrometric applications the energy resolution of the system is of interest. The resolution characteristic ΔE (V_a) (Fig. 4.13) permits the determination of the optimum voltage applied on the detector, a value belonging to the range V_{min} and V_M and situated on the curve plateau. In this manner working conditions with minimum energy resolution are obtained.

The shape of the curve shown in Fig. 4.13 can be explained in the following way :

– On the linear part the resolution improves when the applied voltage increases, due to increase of the charge collection efficiency [Z 3].

Under such working conditions, the charge carriers are collected by diffusion, the fluctuations of the number of recombining carriers are large and the resolution is worse than in the case of total carrier collection.

– At high voltages the resolution worsens due to the appearance of noise.

The resolution limit is determined by the statistical fluctuations of the number of electron-hole pairs "injected" by the ionizing particle. It has been found quite recently [M 23] that the Fano factor in *SD* is smaller than unity; values of about 0.3 or still lower were measured. From the point of view of pair creation statistics, the percentage energy resolution improves when the particle energy increases. On comparing the resolutions for electrons and alpha particles of a given *SD*, we notice that the energy dispersion (*FWHM*) has a larger value in the case of a particle with larger mass. This fact is thought to be due to interactions occurring towards the end of the particle track where collisions with the solid lattice can lead to significant fluctuations of the energy transferred by nonionizing processes.

Fig. 4.13 — The energy resolution characteristic.

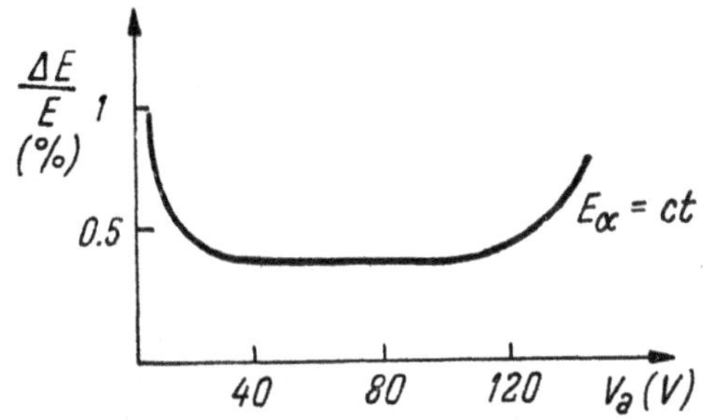

Another phenomenon, still unexplained [M 23] is the asymmetry of the amplitude spectrum given by monoenergetic particles; the number of pulses in the lower energy region of the spectrum is $\sim 1\%$ larger than on the right side of the peak. This fact makes difficult the identification of a weak spectrum superposed over the intense one.

For detectors with considerable thick windows, the widening of the amplitude distribution may be assigned to Landau scattering in the dead layer. This fact was checked up for several *SD* with window thickness within 0.5 and 3μ [M 23].

In detectors with large volumes, the possibility of partial charge collection has to be considered. A homogeneous distribution of trapping and recombination centers does not spoil significantly the resolution if the particle tracks are parallel to the lines of force of the electric field inside the *SD*.

Some of the detectors may have another fault: a double peak — due to nonuniformities produced for instance by a bad ohmic contact which gives birth to a nonuniform field whithin the depletion layer or to the nonuniform lifetime of the minority carriers, a.s.o. Radiation damages can also produce such effects, which appear only at high doses.

The total resolution of the spectrometer with *SD* depends on the noises given by the detector and the input tube of the preamplifier (see chapter VI). We may say that the optimum total resolution can be obtained only with a preamplifier designed to take into account the capacity and the reverse current of the *SD*.

The resolution characteristic must be obtained in conditions as close as possible to that of the experiment. A source with a weak alpha activity and small thickness has to be used. Both the detector and the source are in vacuum and darkness and the collimation is made such that all particles fall practically perpendicular to the *SD* surface. Amsel and co-workers [W 20] used a clock ruby which had two important advantages : the slit edges were perfectly polished and its diameter was known with an accuracy of microns. A good collimation worsens very much the efficiency of detectors with large surfaces. In such a case we can use a honeycomb collimation net covering the whole surface of the detector and providing a sufficiently good detection efficiency.

A suitable alpha source for checking both the resolution and the linearity of the *SD* is the ^{241}Am source which has three close intense lines : 5.477 MeV (85%), 5.435 MeV (12,5%) and 5.375 MeV (1,7%).

An energy spectrum of the alpha particles given by the ^{241}Am source in *SD* [G 24] is plotted in Fig. 4.14.

If the energy E_α is accurately known we can also estimate the charge collection efficiency injecting at the input known charge quantities VC_0 (by means of a pulse generator) such as to obtain at the output pulses with the same amplitude as that given by alpha particles with known energies E_α. The collection efficiency is

$$\eta = \frac{VC_0}{\frac{E}{\epsilon}e} \tag{4.15}$$

with the same notations as above.

Usually this estimation confirms the resolution to be the expected one from electric measurements and the collection efficiency equal to 100%.

4.2.6. *Time characteristics. Fast responses.* Semiconductor detectors with fast response (a short rise time of the pulse) are required in several applications, such as the measurement of charged particle fly time, experiments of multiple coincidence used in the case of scarce events, detection of charged particles in the presence of a high gamma ray background, fast storing of data in experiments with pulsed accelerator, a.s.o.

The fast response of the junction semiconductor detectors is limited by the electric field that can be applied on the junction and by the limiting velocity of the charge carriers. In the $P-N$ junction detectors made of a bulk material with low resistivity and in thin

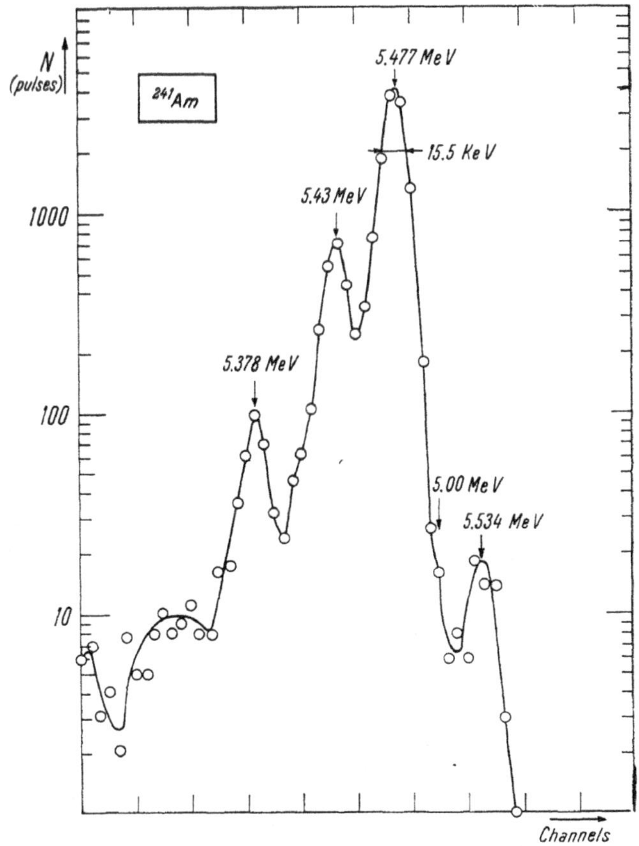

Fig. 4.14— Energy spectrum of alpha particles emitted by the ^{241}Am source, and measured with a semiconductor detector [G 24].

PIN detectors with depletion region extended in the whole volume, rise times of about 1 ns [K 5] and tenths of ns were obtained for Si and cooled Ge (the carrier mobility of which is higher), respectively.

In the case of thick depletion regions ($P-N$ junctions detectors made of high resistivity basic material and thick PIN detectors) the pulse rise time can take values up to hundreds of ns.

The parameters of the detector that has to be used in a time analysis are chosen such that to reduce the collection time of charges and at the same time to preserve a sufficiently great signal to noise ratio. The charge collection time is given by the relationship

$$t_c \simeq \frac{d^2}{\mu V_a}. \tag{4.16}$$

A nomograph for the estimation of electron transit time in silicon is drawn in Fig. 4.15.

Three types of SD are of interest in the time analysis :

— The thin (dE/dx) detector with depletion region extended over the whole volume : thickness $d = 100\ \mu$, $t_{CM} = 2$ ns at 100 V [K 5]. But the signal is affected by the relatively high capacity of the SD and the low quantity of energy lost by the particle inside such a detector (for relativist particles it is only ~ 40 keV).

— The thick PIN detector. The large thickness d slows down the charge collection. (Fig. 4.16).

— The detector with surface barrier layer or junction. In the case of abrupt junctions, $d \sim \sqrt{\rho V_a}$, such that the voltage vanishes in equation (4.16). The use of such detectors, permits the variation of the thickness d between the limits required by the stopping power we expect from the particle under consideration when a relatively constant rise time is kept. For Si we obtain [M 23]

$$t_{c0.1-0.9}[s] = 10^{-12} \cdot \rho\ [\Omega\text{cm}]. \tag{4.17}$$

More rigorous calculations made under the assumption that the carrier mobilities do not depend on the electric field value permitted to construct the nomographs of Fig. 3.8, from which the material resistivity required in order to obtain a certain rise time may be easily estimated. The same calculations show that the best results may be obtained with P^+N junctions when the particle range is $\leqslant \frac{d}{3}$.

In this case the electron-hole pairs are created near the negative electrode and the electrons have to travel the whole thickness d of the depletion region. The electron mobility being higher than that of the holes, a shorter rise time results.

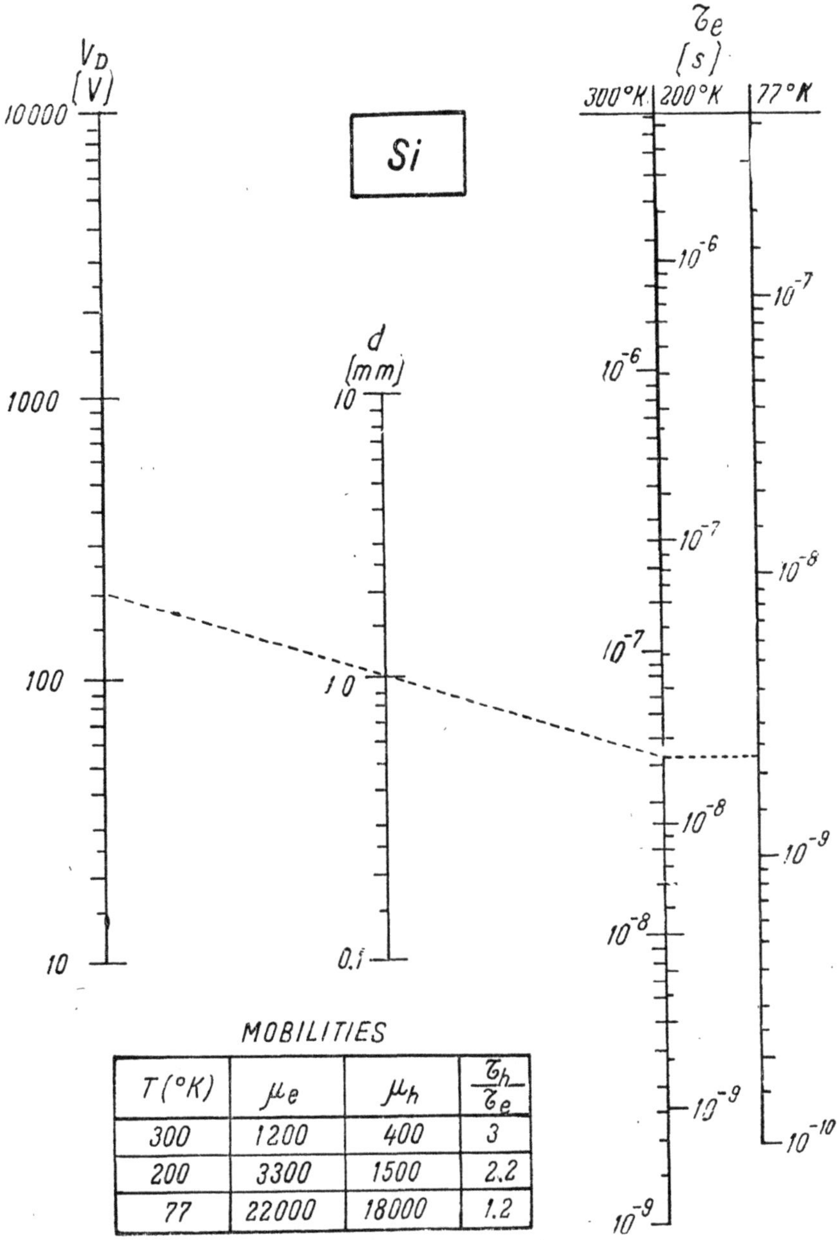

T(°K)	μ_e	μ_h	$\frac{\tau_h}{\tau_e}$
300	1200	400	3
200	3300	1500	2.2
77	22000	18000	1.2

Fig.4.15 – Nomograph used to determine the charge-carrier transit time in Si *NIP*-type semiconductor detectors.

In the case of long range particles, pairs are also created in the vicinity of the positive electrode. A worsening of the voltage pulse rise time follows due to the lower mobility of the holes.

The experimental curves plotted in Fig. 4.17 [M 23] show the increase of the rise time as a function of the particle range.

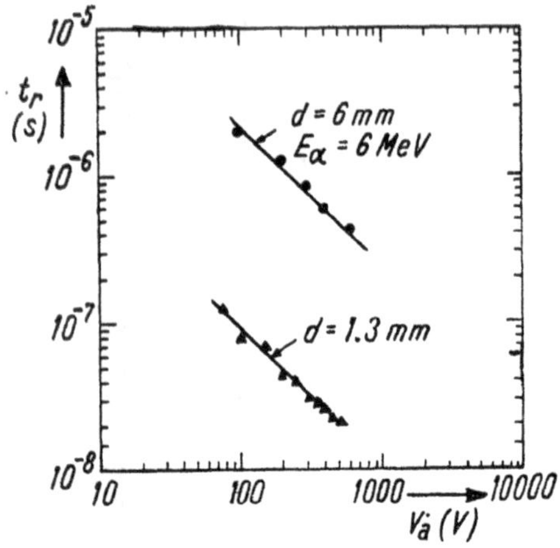

Fig. 4.16 — The effect of V_a and d on t_r for a *PIN* detector [M 23].

If the range exceeds the thickness of the depletion region, the rise time becomes still longer because of a slow component given by the diffusion of the minority carriers created outside the detector depletion region.

The rise time of the pulses given by fission fragments is longer though their range is short. This fact appears as an effect of the charge collection delay due to the screening of the thick plasma created around the particle track ("plasma time").

The considerable thicknesses of the neutral silicon (outside the depletion region) constitute the shortcoming of the *SD* with surface barrier layer or diffused junction. Thus, the resistance (R_s) and the capacity (C_s) of the bulk region (Fig. 4.18) should be also taken into consideration as bound in series with the barrier capacity (C_B) and resistance (R_B) in the equivalent circuit of the *SD*.

Usually, in a junction *SD* $R_B \gg R_s$ and $C_s \gg C_B$, so that in the equivalent circuit of the detector only the capacity (C_s) anp the resistance (R_s) are considered and the time constant determining the pulse rise time is [M 23]

$$\tau = R_s \frac{C_A \cdot C_B}{C_A + C_B} \tag{4.18}$$

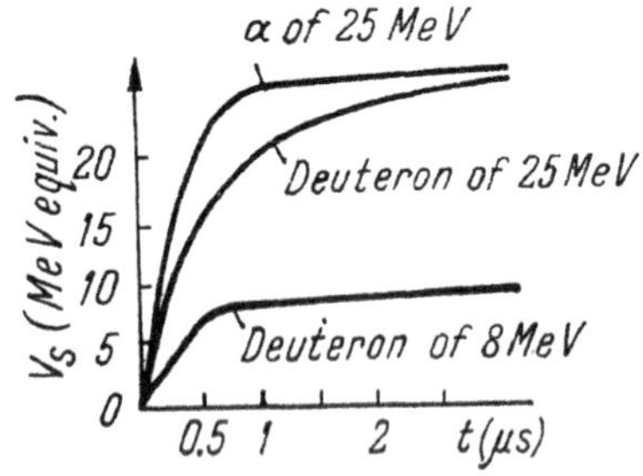

Fig. 4.17 — Variation of t_r with the particle range [M 23].

which becomes

$$\tau \simeq R_s C_B \tag{4.19}$$

for charge sensitive preamplifiers which have a large input dynamical capacity $C_A \gg C_B$.

The influence of the *SD* equivalent circuit elements on the shape of the voltage pulse given by the detector was calculated [P 14] and the results are presented in section 3.5.

Reference [S 17] gives the results of the checking of 60 detectors; their resistance R_s and pulse rise time t_r were measured. The series

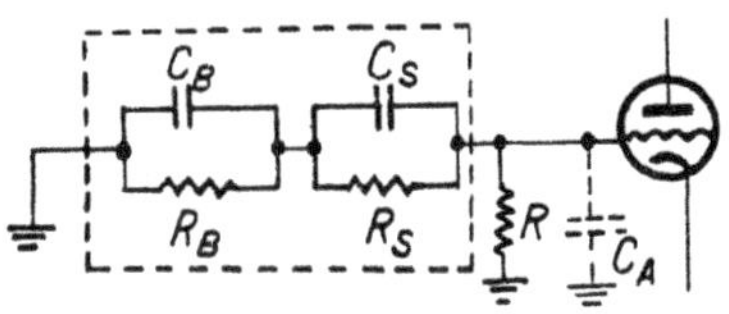

Fig. 4.18 — Equivalent circuit of *SD* and preamplifier input.

resistance (Fig. 4.19) was measured with a *RF* bridge working with a small amplitude signal of 1 MHz superposed over the detector bias. The total capacity C' and the conductance G' of the diode were also measured with the bridge and R_s was estimated with the assumptions presented in that paper, according to the relation

$$R_s = \frac{G'}{\omega^2 C'^2}. \tag{4.20}$$

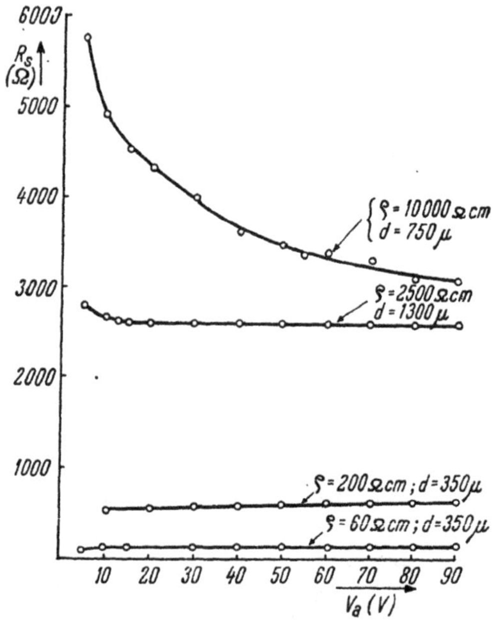

Fig. 4.19 — Series resistance versus reverse voltage for semiconductor detectors of different thicknesses made from materials with different resistivities [S 17].

It may be noticed from Fig. 4.20 that the pulse rise time is influenced by the material resistivity (which determines the transit time of the carriers) and the detector series resistance (which determines the time constant of the equivalent circuit).

The effect of the bulk material resistance, depletion region thickness and electric current intensity on the pulse rise time was experimentally studied [K 10] at a pulsed electron accelerator with 0.5—1.5 ns pulse duration and 60 Hz repetition frequency. The pulse total energy was varied (by varying the number of electrons) within

75 and 150 MeV. Every electron had 75 keV energy and a 45 μ residual range in silicon.

The pulse shape from Fig. 4.21 was obtained with a diffused junction *SD* with a small surface (1.2 mm^2) made of silicon with $\rho = 300\,\Omega$cm and 1 mm thick having a capacity of 10 pF at 225 V;

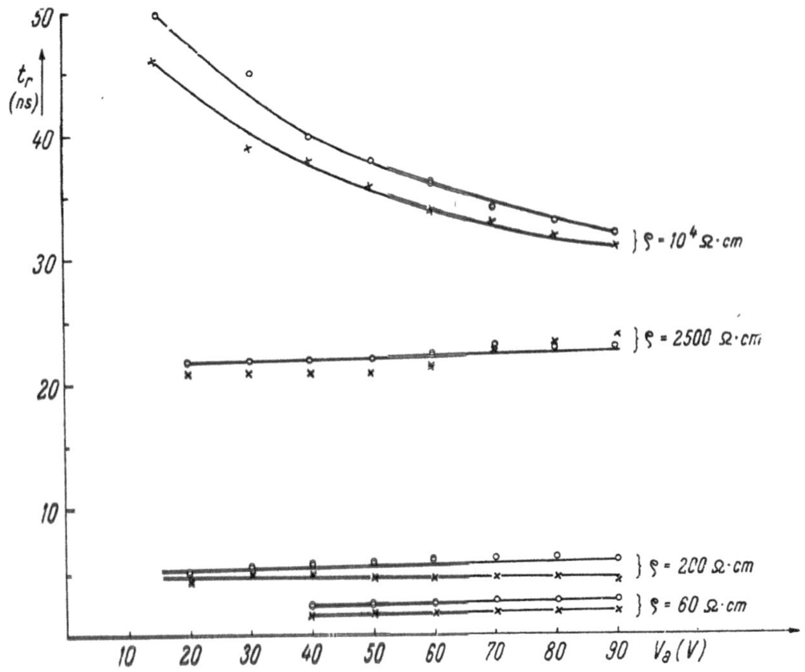

Fig. 4.20—Pulse rise-time for detectors with resistances given in Fig. 4.19 [S 17].

the result presented in Fig. 4.22 was obtained with a silicon *PIN* detector having $d = 2.7$ mm. At the same time with the field intensity decrease, the charge collection efficiency also decreases, pointing out the presence of a relatively high density of recombination centres.

As one may see from Fig. 4.22 the pulse rise time in thick *PIN* detectors is rather long. It may be shortened by lowering the *SD* temperature. In this way the carrier mobility increases. An example of the rise time variation with the temperature is given in Fig. 4.23, for a *PIN* detector with a 1 cm^2 surface and $d = 3$ mm thickness.

An increased voltage applied on the detector at room temperature does not put an end to the problem of the rise time shortening, since, as seen from Fig. 4.24, at a given voltage, t_r no longer decreases when the voltage increases because the carrier mobility is no longer constant for electric fields above 10^4 V/cm ; the carrier velocity reaches a limiting value.

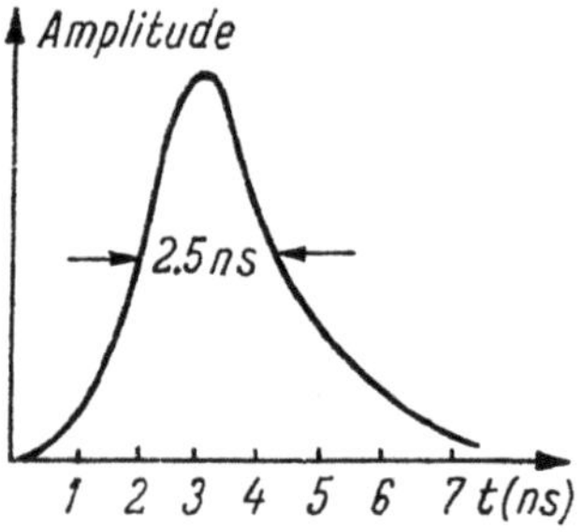

Fig. 4.21 – Pulse shape for a diffused junction *SD* [M 18].

By carefully chosing the *SD* parameters and taking a series of precautions, pulses with rise times of 1 ns can be obtained in Si (and rise times shorter than 1 ns in Ge where the carrier mobility is higher), but unfortunately, with low amplitudes, of the order of millivolts, from low energy particles.

The amplification of these fast signals raises many difficulties as concerns the signal to noise ratio. It is suggested to use slow-fast coincidence circuits, in which the detector signals are sent on two

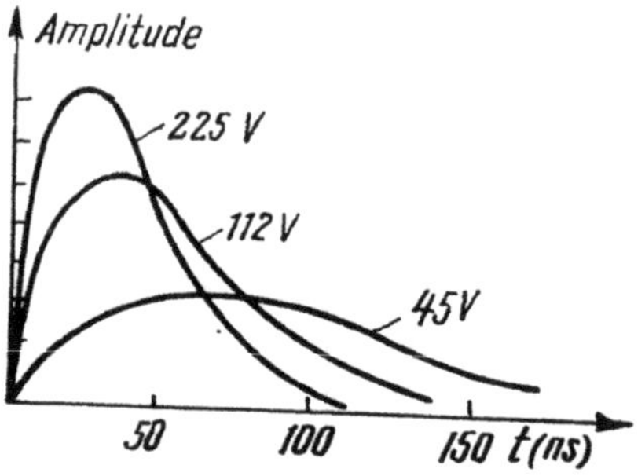

Fig. 4.22 – Thick *PIN* detector pulse shape [M 18].

lines, i.e. a slow line of narrow band and small noise and a fast line on which the signal to noise ratio is sacrificed but good fast responses are obtained [V 8].

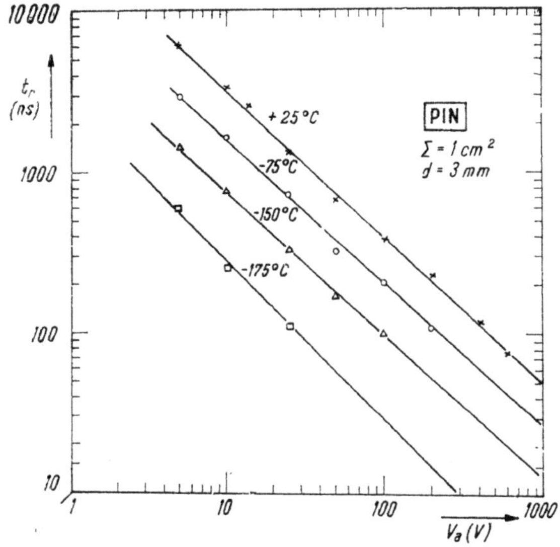

Fig. 4.23 — Variation of t_r with temperature for a *PIN* detector [K 5].

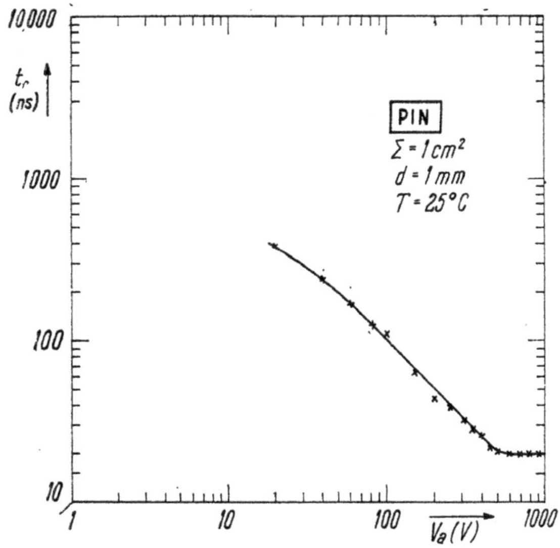

Fig. 4.24 — Variation of t_r with the electric field intensity for a *PIN* detector.

In fact, the problem may be also solved in two other ways, that is either to work out detectors with internal gain, able to give pulses with large amplitudes and good fast responses, or to achieve circuits with semiconductor devices (tunnel diodes, transistors) steadily working at low signal values.

Both solutions seem attractive, but their carrying out implies at the present time great difficulties.

4.3. EFFECTS OF TEMPERATURE, MAGNETIC FIELD AND LIGHT ON THE SEMICONDUCTOR DETECTOR CHARACTERISTICS

The intrinsic concentration of the charge carriers decreases and the material resistivity increases very much at a considerable decrease of the temperature. Under such conditions, we may no longer neglect the voltage fall on the detector volume outside the depletion region; the working of a junction detector resembles that of a detector with homogeneous semiconducting crystal.

An increase of sensitivity to gamma-rays has been also observed in external magnetic fields [W 6]. In magnetic fields of about 1 kilogauss parallel to the frontal surface of the detector, the pulses given by alpha particles in *SD* with barrier layer, were thoroughly smothered at the liquid helium temperature [W 4].

In all cases [G 31], [W 6] an improvement of the detector energy resolution was observed at lower temperatures, related to the reverse current lowering, particularly for Ge, which has a narrower energy gap than Si and thus cannot be practically used in ionizing particle detection unless it is not cooled.

Some anomalies were also observed in the semiconductor detector behaviour, such as a fast decrease of the pulse amplitude and energy resolution of *PIN* silicon semiconductor detectors between 40 and 35°K [D 19].

A decrease of about two orders of magnitude of the reverse current [W 6] was also observed in case of Si cooled at −20°C.

The results obtained in the study of the variation of energy resolution with the temperature within −72 and +125°C, for two semiconductor detectors with diffused N^+P junction on a bulk material of great resistivity [G 31] are presented in Fig. 4.25.

When the temperature increases we observe a worsening of the energy resolution and a narrowing of the range of voltages which applied on the detector preserve the best resolution.

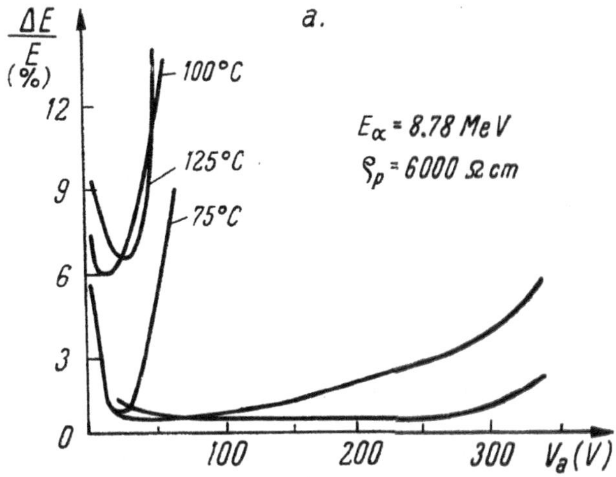

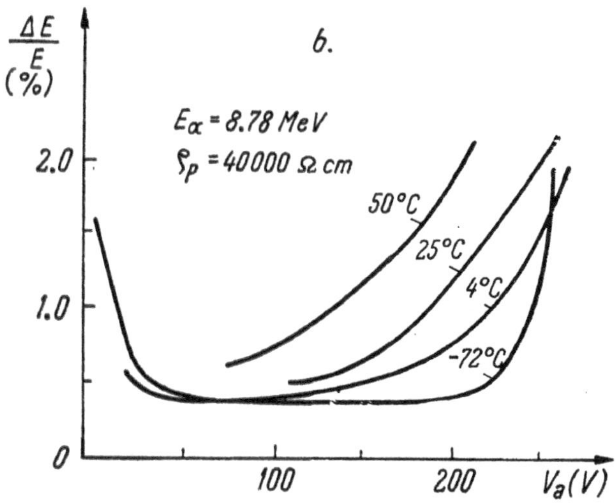

Fig. 4.25 — Variation of the energy resolution characteristic with themperature for a diffused $N + P$ junction [W 6] *a*) $\rho = 6000\ \Omega$ cm ; *b*) $\rho = 40\,000\ \Omega$ cm.

It should be noticed [G 31] that the low temperature should not be applied suddenly because the detector can be put out of order (the contacts come off, a.s.o.). The cooling is easily performed when using special detector designs with cooling wings [W 6].

The improvement of charge collection efficiency at lower temperatures can be explained [B 8] by admitting a lowering of the mean energy ($\in$) required by an electron-hole pair creation. It is also necessary to take into account the charge carriers of the surface inversion layer in the calculation of the electric charge distribution at the junction. Then, a dependence on temperature as shown in Fig. 4.26 is obtained for the internal voltage (V_0). But a detailed understanding of the phenomena occurring there requires further experimentation.

Some applications, such as the neutron flux measurement inside a reactor, require devices [H 11] working at higher temperatures. Detectors with $P-N$ junctions obtained by aluminium diffusion on N-type SiC monocrystals brought up by sublimation technique and doped with nitrogen have been designed [C 1].

These detectors could work at temperatures up to 700°C and allowed integral fluxes of charged particles two orders of magnitude higher than the Si detectors without important damages.

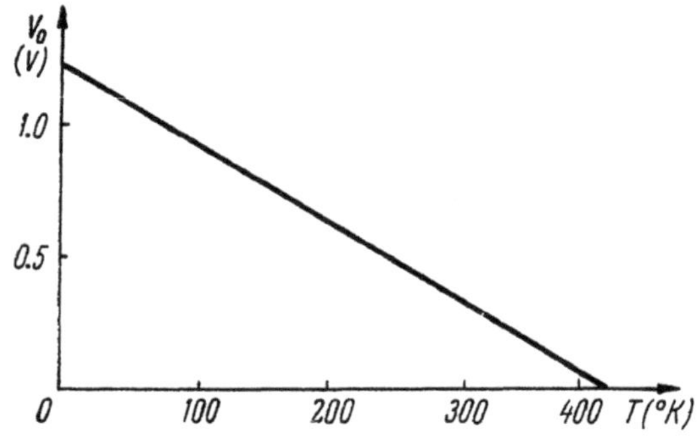

Fig. 4.26 — Variation of internal voltage with temperature for a diffused $N-P$ junction [B8] $\rho = 1000\ \Omega$ cm.

The magnetic field changes the electron and hole trajectories. When the magnetic field is perpendicular to the electric field a deflection upon an arc of circle takes place and when they are parallel the electron trajectory becomes a spiral.

In both cases, the recombination probability is increased (in the former case, the carriers are deflected towards the surface, where the recombination rate is higher and in the latter case the time taken by the carriers to travel a given distance increases).

However, Shockley showed that no important effects are to be expected in magnetic fields

$$H \leqslant \frac{10^8}{\mu}, \tag{4.21}$$

hence in silicon at room temperature for $H \leqslant 5.10^4$ Oersted. In germanium at 78°K, $\mu \simeq 3.10^4$ cm²/V·s and at 4°K $\mu \simeq 10^6$ cm²/V·s, so that magnetic fields of about 3.10^3 Oe, respectively produce effects, a fact experimentally proved [W 20].

The light is strongly absorbed by semiconductors, with an absorption constant of about $10^4 - 10^6\,\mathrm{cm}^{-1}$ [D 11] and creates electron-hole pairs.

Any light penetrating inside the depleted layer of the device contributes to the increase of the "darkness" current, and consequently to the detector noise, such that the detector should be kept and used in darkness. If the window is thick, the light can be thoroughly absorbed by it. The surface barriers with thin gold film have a high photosensitivity and the measurements achieved resulted in [D 11] a 0.13 μA/μW sensitivity at 5460 Å wave length in the case of a detector covered with a 50 Å gold film.

Amsel reported that by covering the surface barrier semiconductor detector with tin oxide instead of gold, the photosensitivity decreases.

4.4. DETECTOR SENSITIVITY TO NEUTRONS AND GAMMA-RAYS

The detectors used as spectrometric devices in the study of nuclear reactions are usually working in conditions of an intense gamma-ray or sometimes neutron background.

The gamma-rays give rise to photoelectrons, Compton electrons and electron-positron pairs inside silicon and the surrounding material. Because of the small atomic number of the silicon, the photoelectric and pair creation processes are not favoured. The electrons lose a small energy inside the narrow barrier of the $P-N$ junction so that the gamma rays give small pulses with low efficiency [D 4]. In spite of a higher atomic number, in Ge, things are not much worse.

The superposition of unexpected pulses that could lead to unpleasant effects can be prevented by using a small detector load resistance and a reduced time constant for the amplifier. The amplitude of electron pulses decreases when working with a minimum thickness of the depleted layer (exactly that required in stopping the particles under study).

In detectors with Au-Si surface barrier layer, the neutrons can produce unexpected pulses by three mechanisms:

— (n, p) and (n, α) reactions in Si. The following reactions are produced

^{28}Si (n, p); $Q = 3.86$ MeV. The isotopic abundance is 92%. The cross-section increases from 0.02 barns at 5 MeV to 0.4 barns at 8 MeV, decreasing at 0.22 barns for 14 MeV neutron energy [D 4].

^{28}Si (n, α); $Q = -2.66$ MeV.

^{29}Si (n, p); $Q = -3.20$ MeV, isotopic abundance 4.7%, 0.05 barns cross-section at 14 MeV.

^{30}Si (n, α); $Q = -4.19$ MeV, isotopic abundance 3.1%, 0.05 barns cross-section at 14 MeV.

From all these reactions, the most important are those induced in ^{28}Si because of the great isotopic abundance and cross-section.

— Reactions inside the gold layer. The charged particle emission is inhibited by the great atomic number. The only observable effect is the capture followed by emission of the 412 keV radiation, but the productivity is very low for thickness about 50 μg/cm^2 or smaller.

— Recoiled protons yielded in the elastic scattering on hydrogen and water vapours absorbed by the detector surface or by the neighbouring ones. The gold with its very low absorption is appropriate from this point of view.

The use of Si detectors of any type is limited (if due precautions are not taken) in beams of fast neutrons with ~ 8 MeV energy at which the cross-section of the ^{28}Si (n, p) reaction has a maximum [D 4]. At this energy the effect of ^{28}Si (n, α) reaction with identically great cross-section, is also felt.

From this point of view, Ge with an atomic number (32) higher than that of Si (14), has the advantage to prevent charged particle emission.

Such detectors, with Ge, have a smaller response to fast neutrons than the Si detectors with identical thickness of the depletion region, though the reactions induced by fast neutrons in Ge isotopes have lower thresholds than in Si:

^{70}Ge (n, p); $Q = -0.82$ MeV, isotopic abundance $= 20.5\%$

$$\sigma = \begin{cases} 0.014 \text{ barns at } 1.8 \text{ MeV} \\ 0.09 \text{ barns at } 14 \text{ MeV} \end{cases}$$

$^{72}Ge\ (n, p)$; $Q = -3.2$ MeV, isotopic abundance $= 27.4\%$
$\sigma = 0.06$ barns at 14 MeV

$^{73}Ge\ (n, p)$; $Q = -0.71$ MeV, isotopic abundance $= 7.6\%$
$\sigma = 0.13$ barns at 14 MeV.

Other isotopes, present in the natural element are ^{74}Ge (36.7%) and ^{76}Ge (7.7%), but the thresholds of the (n, p) reaction for these neutron rich isotopes are much above 5 MeV. The fact that the Ge is affected to a less extent by fast neutron beams, represents for Ge a considerable advantage as compared to Si.

4.5. EFFECTS OF RADIATION DAMAGE ON DETECTOR CHARACTERISTICS

After a long exposure to intense beams of charged particles and especially to that of fast neutrons, a modification of the semiconducting device properties is observed, since they depend on small concentration variations of impurity atoms and defects in crystal. The radiation damages can limit the useful life of the detector and it is important to take every precaution to minimize them. But on the other hand the nuclear reactions on some Si isotopes can be used to create donor and acceptor elements inside the material bulk ; high resistivities are thus obtained by nuclear means.

The problem of radiation damages on semiconductors can be approached in two stages [C 16]. In a first stage the spatial distribution and density of defect centres are studied and in the second stage the electric effects of these centres on electric and detection characteristics of the devices are considered.

Before examining the irradiation effects on *SD* characteristics, we shall briefly present the processes leading to defect formation inside the silicon lattice under radiation influence.

4.5.1. *Formation of defects inside the lattice.* The main mechanism of defect formation under heavy charged particle action consists in collisions displacing the Si atoms from their places inside the lattice. The result is the appearance of a "vacancy" and of an interstitial atom (Fig. 4.27) breaking down the crystal periodicity. This defect is called "Frenkel pair".

The minimum energy required for displacing the Si atom from its place inside the lattice is $E_d = 15-30$ eV, depending on the radiation energy and type [B 34].

The number and the energy distribution of the recoiled atoms depend on the value and the energy dependence of the interaction cross-section, so that different types of radiation yield different defect

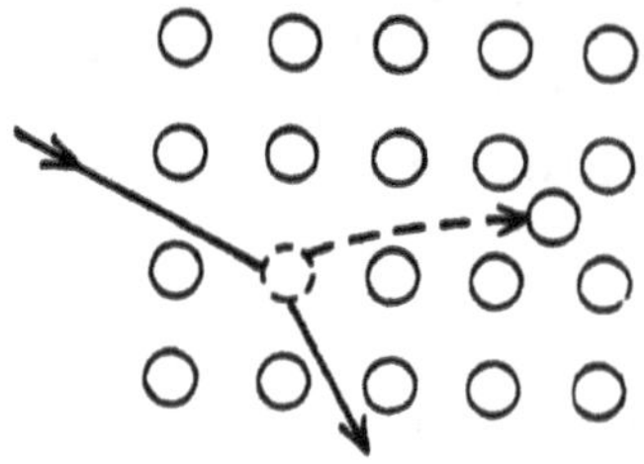

Fig. 4.27 — Frenkel-pair production.

distributions. Therefore, we shall here consider separately three radiation types i.e. neutrons, heavy charged particles and electrons (including gamma-rays which produce defects especially by the Compton electrons or photoelectrons they usually yield).

4.5.1.1. *Neutron induced defects.* The maximum energy $\hat{E}_{Si}$ given to the Si nucleus in an elastic collision between a neutron with energy E_n and a Si nucleus [D 11] is

$$\hat{E}_{Si} = 0{,}133\ E_n. \tag{4.22}$$

The recoiled atoms have energies E_{Si} up to this maximum value according to the function $W_d\ (E_n,\ E_{Si})$ which gives the probability that in a given collision the energy of the primary recoil is E_{Si}.

For fast neutrons $\hat{E}_{Si} \gg E_d$, so that many collisions lead to displacements and the cross-section of the process (σ_d) is equal to the diffusion cross-section. In the case of neutrons from a graphite reactor it is shown that the mean energy $\overline{E}_{Si}$ of the recoiled atoms is

$$\overline{E}_{Si} \simeq \frac{1}{2} \cdot \frac{\hat{E}_{Si}}{\ln\ (\hat{E}_{Si}/E_d)}. \tag{4.23}$$

The primary recoiled atoms yielded in this way can produce, in their turn, secondary displacements; every primary event yields a number $\nu\,(E_{Si})$ of defects

$$\nu\,(E_{Si}) \simeq 0.56\,\frac{E_{Si}+E_d}{E_d}. \tag{4.24}$$

The mean number of displacements $\bar{\nu}$ is given by

$$\bar{\nu} \simeq \frac{E_{Si}}{2E_d} \tag{4.25}$$

For neutrons of 1 MeV we get $\bar{\nu} \sim 1000$. However, some of the recoiled atoms with energies higher than 7 keV can lose energy by ionization and not by secondary displacements. If displacements produced by atoms with energies above the threshold 7 keV are neglected, a lower limit $\bar{\nu} \simeq 150$ is obtained. The Si ions can have energies high enough to produce secondary displacements; their range is however short, so that the displacements occur within a small volume.

Neutrons with energies lower than ~ 200 eV cannot yield displacements in Si because $\hat{E}_{Si} < E_d$. But they can yield transmutations. In ^{30}Si (4% isotopic abundance) the neutron capture is followed by β emission by which the ^{31}Si becomes the stable ^{31}P with a donor-like behaviour. The slow neutron capture cross-section in ^{30}Si is only 0.12 barns.

4.5.1.2. *Defects induced by heavy charged particles.* The interactions among heavy charged particles and Si nuclei occur almost entirely through Rutherford diffusion.

For protons with E_p energy

$$\hat{E}_{Si} = 0{,}133 E_p\,. \tag{4.26}$$

Supposing $E_d = 30$ eV and expressing E_p in MeV, the cross-section of the displacement process is

$$\sigma_d\,(E_p) = \frac{1{,}5.10^{-20}}{E_p}\,\mathrm{cm}^2. \tag{4.27}$$

From this relation (4.27) we observe that defects are important mainly at the end of the particle range. This fact is shown in Fig. 4.28.

But defects are not concentrated inside small volumes as in the neutron case.

The mean energy of a primary recoiled atom is

$$\overline{E}_{Si} \simeq E_d \cdot \ln\left(\frac{\hat{E}_{Si}}{E_d}\right), \tag{4.28}$$

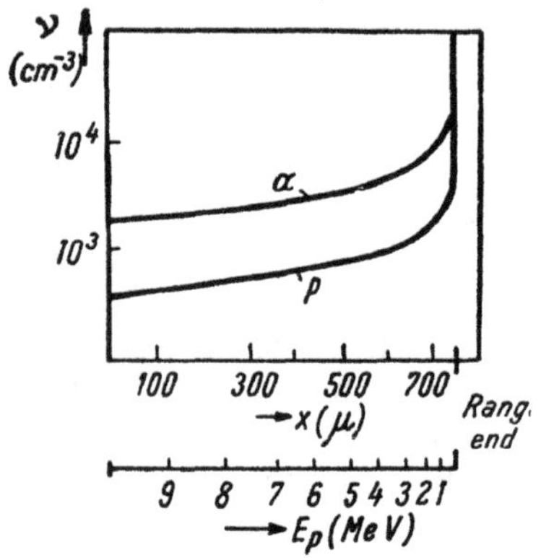

Fig. 4.28—Density distribution of Frenkel defects in Si [D.13].

and the mean number of displacements yielded by a primary recoiled atom is

$$\overline{\nu} \simeq 0.12 + 0.56 \ln\left(\frac{E_p}{2.25 \cdot 10^{-4}}\right). \tag{4.29}$$

The corresponding formulae for alpha particles of energy E_α (MeV) in silicon are

$$\hat{E}_{Si} = 0.437\, E_\alpha, \tag{4.30}$$

$$\sigma_d \simeq \frac{2.36 \cdot 10^{-19}}{E_\alpha}\ \text{cm}^2, \tag{4.31}$$

$$\overline{\nu} \simeq 0.12 + 0.56 \ln\left(\frac{E_\alpha}{6.9 \cdot 10^{-5}}\right). \tag{4.32}$$

The total number of Frenkel defects, dN_F, yielded by a particle travelling a distance dx inside a material with N_0 nuclei per cm^3, is

$$dN_F = N_0 \cdot \sigma_d \cdot \overline{\nu}\, dx, \tag{4.33}$$

so that at an energy loss dE, the incident particle yields a defect number

$$\mathrm{d}N_F = \frac{N_0 \sigma_d \overline{\nu}\, \mathrm{d}E}{\frac{\mathrm{d}E}{\mathrm{d}x}}, \tag{4.34}$$

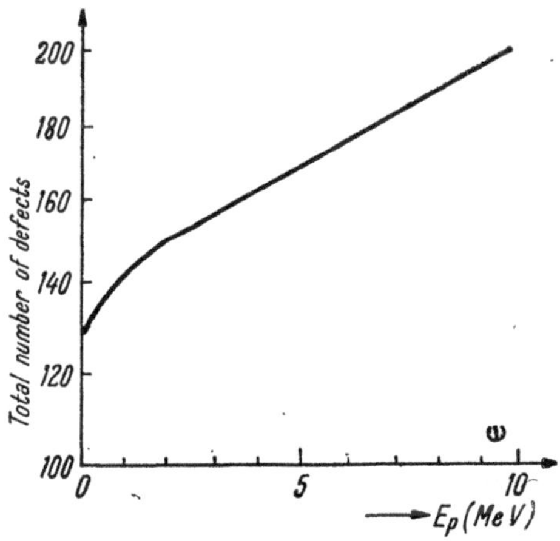

Fig. 4.29 — Total number of Frenkel defects yielded during the proton slowing down [D 11].

A lower limit estimated for the total number of Frenkel defects yielded at the slowing down of a proton with energy E_p is plotted in Fig. 4.29.

The number of defects yielded within every unit interval of the particle track is

$$\frac{\mathrm{d}N_F}{\mathrm{d}x} = N_0\, \sigma_d \overline{\nu} \tag{4.35}$$

In the case of fission fragment bombardment, almost all atoms along the track can be displaced. It is almost impossible to make a practical density estimation of the defects yielded in such conditions.

4.5.1.3. *Defects induced by electrons and gamma-rays.* Considering relativistic effects, the electrons can transfer to recoiled atoms a maximum energy

$$\hat{E}_{\mathrm{Si}} = 2\frac{(E_e + 2m_0 c^2)}{MC^2} E_e \tag{4.36}$$

where E_e, m_0 are the kinetic energy and the rest mass of the electron, respectively and M is the mass of the Si nucleus.

In silicon, an electron cannot transfer enough energy to produce a displacement below $\sim$250 keV, and in germanium below $\sim$600 keV. The energy transferred to recoiled ions is low even for electron energies up to 10 MeV, such that very few displacements are induced. As distinguished from the defects induced by fast neutrons, the electrons yield relative uniform distribution of the defects.

In the case of gamma-ray irradiation, the Compton diffusion is the prevailing phenomenon in silicon within $E_\gamma = 60$ keV$-$15MeV. The electrons yielded in this way are responsible for the defects produced by gamma-rays.

The cross-section of the displacement process in the case of gamma-ray irradiation is very small. For instance, the probability for a gamma-ray of 1 MeV to yield a single defect when crossing a silicon layer of 1 mm thick is only 0.001.

The defect yielding rate for reactor gamma-rays was estimated for silicon at 9×10^{-3} Frenkel pairs per cm^3 for every photon per cm^2. This rate is so low in comparison with the defects induced by neutrons that in most cases it may be neglected.

4.5.2. *Irradiation damage effects on semiconductor detector characteristics.* Frenkel defects arise within small volumes around primary recoiled atoms. The interstices thus created have a donor impurity behaviour and the vacancies an acceptor-like one.

Accordingly, donor and acceptor levels arise in the energy gap. The levels induced next to the middle of the energy interval between the bottom of the conduction band gap and the top of the valency band are very effective in promoting electron and hole recombinations by their alternate capture.

Consequently, defects lower the carrier concentration by majority carrier capture and shorten the minority carrier lifetime by alternate capture of electrons and holes. In this way the resistivity of the material subject to irradiation increases and the carrier lifetime decreases.

With a material with high resistivity undergoing irradiation, the resistivity can significantly increase without introducing important changes in the minority carrier lifetime. The silicon is the only semiconductor tending towards an intrinsic conductivity when being irradiated. Most materials tend towards a limit concentration of the carriers about 10^{16}/cm^3. Samples of P and N-type germanium tend towards a P conductivity with low resistivity under irradiation.

In the case of silicon detectors, the reverse current is expected to decrease, the energy resolution to worsen and the charge collection efficiency to become lower (corresponding to the decrease of carrier lifetime).

The effect of carrier concentration change given by irradiation induced defects is more important when the initial carrier concentration is low i.e., in materials with high resistivity. In such cases, the resistivity can increase very much under irradiation before a significant change of the lifetime occurs. In detectors made of a low resistivity material, the modification of the lifetime is expected to be the prevailing effect under irradiation.

Exposing for several days detectors with barrier layer on Si with $\rho = 1000\,\Omega$ cm to a 10^{11} alpha particle/cm^2 beam from an alpha source of ^{241}Am ($E = 5.48$ MeV), an increase of the leakage current and a worsening of the energy resolution were observed after an irradiation with $10^8 - 10^9$ alpha particles/cm^2 [D 8].

A characteristic example is given in Fig. 4.30. After $10^9 - 10^{10}$ alpha particles/cm^2, multiple peaks can be observed in the spectrum of a monoenergetic source at $V_a = 2$ V. The satellite peaks broadened with the dose increase and sometimes merged in a wider one.

For higher voltages applied to the detector (20 V), the effect on the energy resolution is smaller because the electric field increase reduces the carrier transit time.

In *NIP* detectors the effect is stronger, due to the weaker electric field and the longer path the charge carriers have to travel. A spectrum before and after irradiation with $\sim 10^{10}$ alpha particles/cm^2 is drawn in Fig. 4.31. The particle energy is 43 MeV.

A heating at 120°C, for 24 hours, does not produce a considerable improvement of the resolution. The multiple peaks have a tendency to fuse and the leakage current decreases slightly.

In other experiments, after ten days of irradiation with protons, deuterons and alpha particles of 0.20 MeV (yielded by a thin boron target bombarded with Li ions) of a Si detector with $\rho = 3000\ \Omega$ cm, $\rho = 12\,000\ \Omega$ cm was obtained with a still good energy resolution.

In the case of fast neutron irradiation [K 10], [B 2] the effect of the defects induced is not significant [D 8], [M 19] up to about 5.10^{11} fast neutrons/cm^2. Beyond this dose the resolution becomes worse, multiple peaks [K 10] appear in the spectrum of monoenergetic alpha particles and after about 10^{13} neutrons/cm^2 (Fig. 4.32) the pulse amplitudes are much reduced.

In such measurements of irradiation effects the increase of the pulse rise time during irradiation should be taken into account when choosing the amplifier time constant.

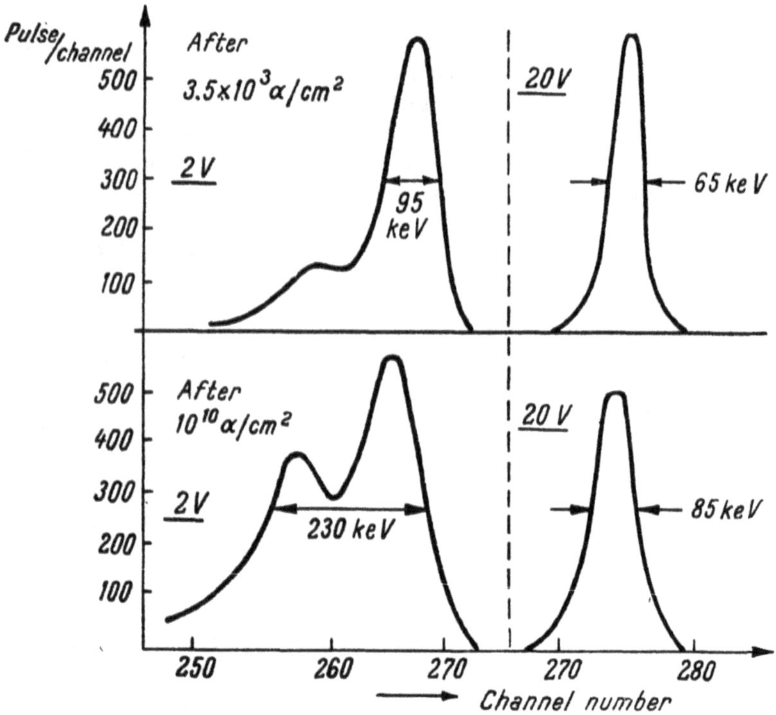

Fig. 4.30 — Effect of alpha-particle irradiation on the energy resolution of a Si detector with $\rho = 1000\ \Omega$ cm [D 8]

Few measurements were made of defects produced by irradiation with gamma-rays or electrons of silicon detectors. Kramer exposed two detectors to a total dose of 10^8R of gamma-rays yielded by a ^{60}Co source. No modification of pulse amplitudes was observed, though the carrier lifetime had decreased.

Lately, besides the effects expected from previous studies, that is the reverse current increase, the detector noise increase and energy resolution worsening, and the decrease of charge collection efficiency, particularly at low voltages, a decrease under proton irradiation and an increase under electron irradiation of the N^+P junction capacity were also observed [L 15]. The capacity change can be explained by a variation of ionized acceptor density ($C \sim \sqrt{N_a}$). N_a decreases under proton irradiation and increases under electron irradiation.

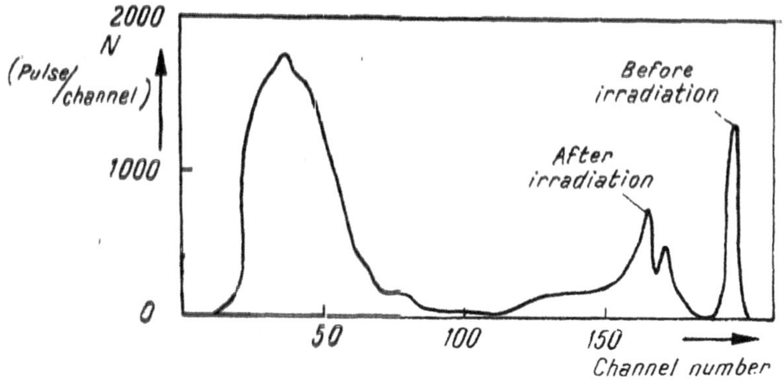

Fig. 4.31. — Response of a *NIP* detector to 43 MeV alpha-particles before and after exposure to $\sim 10^{10}$ particles/cm² [D 13].

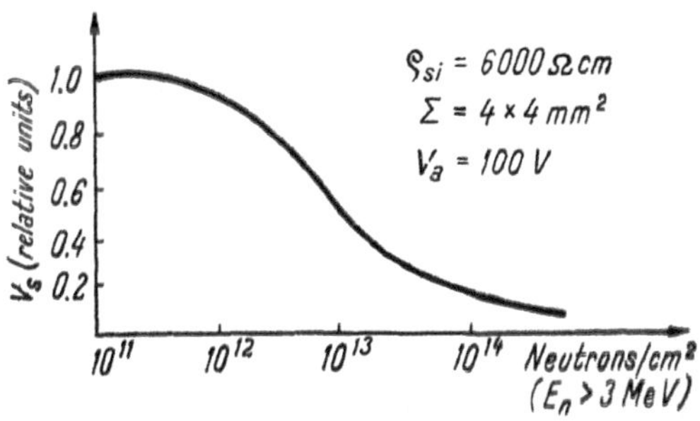

Fig. 4.32 — Effect of fast neutron irradiation on the amplitude of pulses given by the junction [M 19].

In the case of *NIP* detectors with large sensitive volumes, the irradiation effects are considerable, since the collection field is weaker. In paper [B 36] it is shown how electron trapping length in *NIP* junction on Si changes under irradiation with gamma-rays from a ^{60}Co source. Significant changes of collection efficiency and collection time were also observed at doses of about 10^5 R considered to be negligible for P—N junctions. For this low dose the reverse current and noise variation were small.

Other experiments confirming the higher sensitivity of NIP detectors to radiations are presented in [G 21]. The detectors used had $\varnothing = 8$ mm—2 cm and $d = 2-4$ mm. After 10^8 protons (of 4 MeV/cm^2) the detector resolutions become worse due to the worsening of charge collection efficiency. A similar result was obtained after irradiation with alpha particles of 43 MeV. After about 2×10^{11} alpha particles/cm^2, the multiple peaks appeared without affecting the rectifying properties. The rectifying properties were quickly lost during experiments with fast neutrons ($E_n = 1-10$ MeV) and a flux of 10^9 neutrons/cm^2.s. At this intensity the detector was out of order in some minutes. During exposure at $V_a = 308$ V the reverse current was increasing linearly with the dose at a rate

$$\frac{\Delta I_D}{\Phi\, \Delta t} = 3.2 \cdot 10^{-11}\, \mu\text{A/n.cm}^2.$$

It was checked up that the induced defects had an acceptor-like character since they disappeared after a new Li drift at a temperature of 130—150°C and $V_a = 100-500$ V.

Table 4.1

Orientative values of allowed doses in some typical cases [D 13]

Radiation type	Surface barriers or diffused junctions	NIP detectors
Alpha particle of 5—50 MeV	10^{10} cm^{-2}	10^8 cm^{-2}
Protons of 5—10 MeV	10^{11} cm^{-2}	10^8-10^9 cm^{-2}
Electrons of 2—5 MeV	$10^{13}-10^{14}$ cm^{-2}	—
Fast neutrons	$10^{12}-10^{13}$ cm^{-2}	10^{11} cm^{-2}
Gamma-rays	$>10^8$ R	10^5 R
Fission fragments	10^8 cm^{-2}	—

Approximate values of allowed exposures to different nuclear radiations are given in Table 4.1. [D 13]. They are only a general guide since the maximum dose allowed varies from a detector to another, and for the same detector depends on its working conditions (especially on the electric field intensity).

The experimental results obtained, emphasize the importance of a strong electric field in preventing the energy resolution worsening during irradiation.

For a given spectrum of particles, we can estimate the density of Frenkel defects introduced during irradiation and provide the required protection or apply a correction to the data obtained [D 8].

In any case, when working with semiconductor detectors it is proper to get from time to time a spectrum of amplitudes with a monoenergetic particle source to check up the detector resolution.

CHAPTER V

SEMICONDUCTOR DETECTOR TYPES

At the present time there exists a large number of semiconductor detector types, resulting from the manifold requirements of practical applications.

After a brief examination of the considerations which led to the choice of certain types of semiconducting devices to work as radiation detectors, the main types classified according to their design particularities will be described. Typical examples of applications to nuclear physics experiments will be then given. Other examples can be found, for instance, in [D 21].

5.1. METHODS FOR OBTAINING HIGH ELECTRIC FIELDS IN SEMICONDUCTORS

The plane-parallel geometry with uniform electric field can be easily achieved in semiconductor detectors but the requirement of a high electric field (for fast collection) implies problems of noise and dissipation.

The detector noise is low if a low direct current leaks through the detector, i.e. when the material resistivity is great enough. A silicon with $\rho = 1000\ \Omega$ cm is considered to have a great resistivity. A device made of such a material with 5 mm diameter and 1 mm thickness has a resistance of about 500 Ω. If a uniform field of 10^3 V/cm, required for charge collection in good conditions, is achieved the leakage current through the device will be 0.2 A and a 20 W power will be dissipated. It follows that a 1000 Ω cm resistivity is still small.

In Si of N-type with a $10^{13}/\text{cm}^3$ donor concentration, a number of 10^{13} conduction electrons/cm^3 will exist at room temperature and the material will have $\rho = 500\ \Omega$ cm. By purification and compensation the electron net concentration can be lowered within 10^{10} and 10^{11} electrons/cm^3, hence $\rho \sim 10^5\ \Omega$ cm can be obtained, corresponding to the intrinsic material at room temperature.

The product $n_e n_h$ depends on the semiconductor type and varies considerably with the temperature but is independent of impurity concentrations (n_e or n_h can be changed by adding impurities but their product remains constant). For Si at room temperature $n_e n_h \simeq \simeq 10^{20}/\text{cm}^6$ and the total concentration $(n_e + n_h)$ of the carriers is minimum at $n_e = n_h =$ intrinsic concentration at this temperature. The prevailing term in the dependence on temperature of the product $n_e n_h$ is

$$\exp\left(-\frac{E_g}{kT}\right).$$

At low enough temperatures, e.g. at 20°K, the electron concentration is about $10^5/\text{cm}^3$ and $\rho = 10^{10}\ \Omega$ cm. This method of obtaining a high resistivity is rendered difficult by an avalanche ionization phenomenon occurring in high-electric fields ($>10 \div 100$ V/cm). A satisfactory method of lowering the electron concentration consists in using impurities which give rise to energy levels at the middle of the energy gap. For instance, the gold is an acceptor ; it receives electrons from the conduction band and from donors. Concentrations of about 10^{10} electrons/cm^3 and 10^5 electrons/cm^3 can be obtained at room temperature and at 78°K, respectively. Electrons captured by gold cannot be ionized by impact because they have a high ionization potential. Impurities from lower levels have an undesirable effect : they behave as trapping and recombination centers, limiting the electron and hole lifetime, particularly at low temperatures. Thus a $\tau_e \simeq 10^{-2}$ μs was measured for gold doped silicon at 78°K. In spite of this fact, there are some results obtained with gold doped silicon semiconductor detectors.

With a reverse biased $N-P$ junction, strong electric fields can be achieved in semiconductors, without requiring a bulk material with extremely high resistivity. The silicon is generally used since it works at room temperature, while germanium has to be cooled in order to give a low noise.

Since the window thickness must be minimum, junctions made by diffusion at the surface and leading to a N^+P structure, are used.

The P-type silicon of high resistivity is more easily obtained than the N-type silicon due to boron impurities (acceptors) implying great difficulties in purification. However, the barrier layer devices analogous to P^+N junctions are also often used. The inversion layer formed at the surface is caused by the chemical treatment of the silicon surface. The contact at the inversion layer is made through a thin gold film laid down by evaporation in vacuum. In fact many experiments point out the active rôle played by the metal in obtaining rectifying characteristics; it is not only a mere contact.

Devices with large sensible volumes are made of a bulk material with comparatively low resistivity ($\sim 100\ \Omega$ cm), when a *PIN* construction is used.

A very thick depletion region is important not only because it provides a large sensible volume but lowers the device capacity, hence increases the signal and the signal to noise ratio.

The main characteristics of the semiconductor detectors are: the depletion region thickness which determines the sensible volume and the transit time of the carriers; the barrier capacity and the series resistance of the bulk material which control the fast responses and the energy resolution; the reverse current which affects the noise level.

Assuming the depletion region thickness as a classification criterion, the semiconductor detectors can be divided into two categories, namely

— thin detectors: $P-N$ junctions achieved by superficial diffusion or surface barriers,

— thick detectors: *PIN* junctions and homogeneous semiconductor detectors.

In the following we shall examine the most frequently used semiconductor detectors, classified from the constructive point of view.

5.2. HOMOGENEOUS SEMICONDUCTOR DETECTORS

Potentially, such detectors supply means for obtaining a large sensible volume (which cannot be reached in junction detectors). For this reason they are suitable to penetrating particle detection.

Moreover, they allow a straightforward experimental checking up of the simplified theoretical model suggested for the description of the detector working [S 21] principle.

These detectors are made of materials with high resistivities ($\rho \geqslant 10^8\ \Omega$ cm) cut as parallelepipeds or cylinders. Ohmic contacts are achieved on two opposite faces of the parallelepiped; they should permit the application of an electric field, without altering the charge carrier concentration inside the detector.

The good working of the device depends on the material high resistivity and long carrier lifetime. In other words, materials for which high E_g and low densities of traps and recombination centres are ensured during the technology are to be preferred.

From this point of view, the quantity $\mu_h \tau_h \mathcal{E}$, also called hole trapping length, represents the figure of merit of the material. It follows that for practical values of $\mathcal{E}$ only Si and Ge are adequate from all semiconductors [D 12]. Other materials, such as for instance semiconducting compounds, will become useful when their preparation methods will be improved.

The main difficulties in the technology of such detectors of Si and GaAs are related to the obtaining of a material with high ρ without lowering τ below allowable limits (10^{-7}—10^{-8} s).

Gold doped N-type silicon has a unique property: if gold impurities have a concentration of 5×10^{14} atoms/cm^3, three times greater than the concentration of the material impurities (e.g. $2\times\times10^{14}$ atoms/cm^3 in phosphorus), the crystal behaves as an intrinsic semiconductor. With this type of silicon, cooled at 170° K, $\rho > 10^9\ \Omega$ cm was obtained. The doping reduced τ to 10^{-7} s, but in this case $\mu_h \tau_h \mathcal{E} \simeq \simeq 1$ cm was also obtained, such that the collection efficiency was good for a crystal of 2.5 mm thickness. The semiconductor detector obtained in this way has a good collection efficiency for electrons and a low one for holes [D 11].

The high resistivity silicon ($4\times10^6\ \Omega$ cm) required for the construction of homogeneous semiconductor detectors was also obtained [V 6] by irradiating the N-type silicon with gamma-rays from a 7 kCu Co^{60} source. The detector with 1.8 cm diameter and 2.8 mm thickness was exposed to an integral dose of 1.5×10^7 R.

In order that the detector has a good working the electric field applied should be strong enough, so that the transit time of the carriers injected by ionization should be shorter than the carrier lifetime inside the given material

$$\frac{d}{\mu\mathcal{E}} = \tau_c < \tau. \tag{5.1}$$

In a 2000 V/cm electric field, the mean free path of the charge carriers is about 1 cm at a temperature close to 120°K; hence a detector can be designed, thick enough such that a minimum ionizing particle loses at least 1—2 MeV energy while crossing the crystal.

The fast responses of these detectors are not too good. For instance in a detector of 1 cm thickness, the carrier transit time is limited to 10^{-7} s by the maximum velocity of the carriers inside Si (10^7 cm/s).

Pulses with 1 mV amplitude from minimum ionizing particles (π^- mesons) [V 3] were obtained in a crystal of 0.5 cm thickness and 2 cm diameter. The pulse rise-times were shorter than 0.1 μs. An energy resolution of 1.5% for 30 MeV protons was reached with a cubic detector with 6 cm side [D 11]. One of the largest volume so far constructed is $4\times2\times2$ cm³ [M 19].

The energy resolution of these detectors is limited by noises. The main noise components [S 21] are : the thermal noise, the noise given by the detector current, the effects of the strong electric field and the contact noises. Because of them, lonely particles cannot be observed in Si, but only if SD is cooled at least at the temperature of the liquid nitrogen.

With GaAs which has $E_g = 1.43$ eV [D 11] one can work at room temperature. Resolutions of 2% were obtained for 5 MeV alpha particles [M 19], but the collection field cannot be much increased without the appearance of cumbersome oscillations. Since the GaAs has a higher Z than the Si (31—33), it is more fitted for gamma-ray detection. But from this point of view the most convenient seems to be the CdS with its $Z = 48$. Exposed to a gamma-ray flux of 1 Roentgen/minute, a conventional, commercial detector [D 11] gives 0.2 μA with 5—10 s response time and 0.02 μA "dark current".

Unfortunately, in these crystals there appear several effects of slow collection of carriers formed by ionization. Such an effect called "polarisation", some time after the voltage application on the detector, makes the pulse amplitudes to decrease continuously. This is due to trapped charges giving rise to a field opposite to the collection field.

An internal multiplication effect can be also obtained and explained in the following manner. As a result of the unequal collection due to a preferential trapping of a certain carrier type, a net charge arises which can induce the injection carriers of opposite polarity from bordering electrodes. During the time required for reaching the equilibrium, an additional "secondary" current flows and its integration in the amplifier circuit gives a total charge much greater than the ionization charge. The multiplication factor can reach 10^3 in CdS, but at the expense of losing the linearity and the time resolution.

The time, required by trapped charges to establish the field value inside the whole crystal is given by the "dielectric relaxation time" τ_d equal to the product of capacity by resistance of a dielectric block of resistivity ρ (Ω cm) and dielectric constant ε [D 12]

$$\tau_d = 10^{-12} \frac{\rho \varepsilon}{4\pi}. \tag{5.2}$$

The charge multiplication occurs when the carrier trapping time exceeds the relaxation time.

Polarisation phenomena were eliminated by supplying the detector at *HF* [B 37]. The detection is performed in this case by observing the influence of the losses inside the crystal on the self-oscillations of an oscillating circuit mounted in parallel with the *SD*.

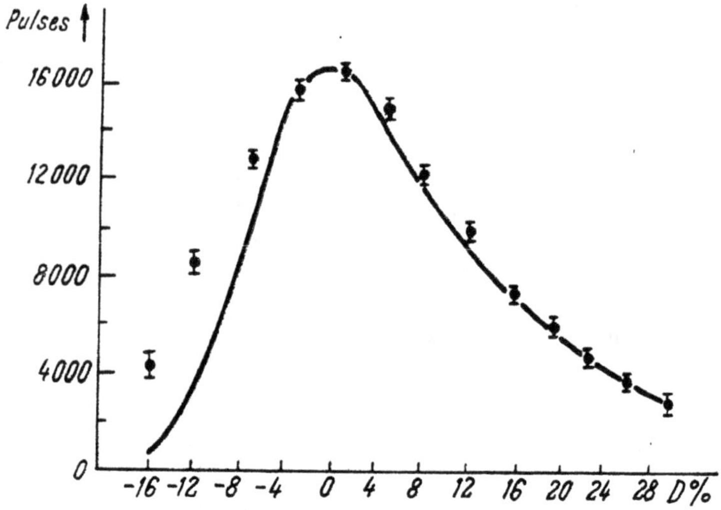

Fig. 5.1 — Percentage deviation about the most probable energy loss value of mesons with momentum 1.5 BeV/c [V 3].

The crystal detector working in conditions of a non-uniform electric field (due to the space charge created) was studied in [T 9].

Because of the great difficulties in obtaining the necessary material, homogeneous semiconductor detectors are very seldom used.

An example of application in the high energy physics is given in Fig. 5.1. A Si crystal compensated with Au with 20 mm diameter and 2.5 mm thickness was used [V 3] in measuring the energy loss fluctuations for minimum ionized particles. The distribution of energy losses for π-mesons with 1.5 BeV/c momentum given in percents referring to the most likely value of these losses is given in Fig. 5.1. The experimental points are rather well fitted to the Landau curve.

5.3. DIFFUSED N—P JUNCTION DETECTORS

An N^+P or P^+N junction used as radiation detector is different from junctions used in rectifying circuits because it has a thicker depletion region closer to the surface.

The former characteristic is imposed by the necessity of a large sensible volume and the latter by the condition that the thickness of the detector "window" be negligible.

A cross-section of such a junction is shown in Fig. 5.2. Inside a P-type bulk material (Si) of high resistivity ($\rho = 100 - 10000\ \Omega$ cm) and $l_N + l_p$ thickness, a donor (usually P) is diffused so that a thin

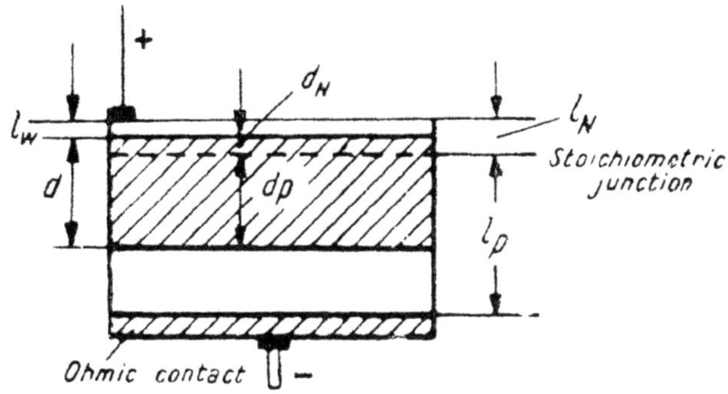

Fig. 5.2 — Cross-section through a diffused junction SD.

film of $l_N < 1\mu$ thickness, heavily doped (N^+) should result at the surface. The condition $\rho_N \ll \rho_P$ extends the depletion region particularly in the P part of the junction, since

$$\frac{d_N}{d_P} = \frac{\rho_N}{\rho_P}. \tag{5.3}$$

The diffusion depth determines a detector window thickness of about $l_F/2$. In this heavily doped region, the minority carrier life-time is very short (the recombination rate is great) and practically there is no collection of ionization charges.

In the depletion region (of d thickness) the electric field intensity may be $\sim 10^4$ V/cm ; the electric field is maximum near the surface and decreases linearly with the depth in the depletion region. When applying a reverse voltage this thickness increases advancing mainly inside the weakly doped region (the *P*-type in the above example).

The junction manufacturing by diffusion implies the following stages, most of them representing a pure transistor technology [B 40]. The Si monocrystal is cut under the desired form (disc or parallelepiped), grinded with an abrasive and then treated with an etching substance. The desired impurity (e.g. *P*) is diffused inside the silicon at a temperature of 800 — 1250°C in an atmosphere of inert gas to prevent oxidations. The achievement of a thin window requires a diffusion temperature of 900°C and 10 minutes duration. A diffusion depth of 0.3 μ and a surface dead layer of 0.15 μ are obtained in this way. After diffusion contacts are applied on both sides of the junction.

The detector technology with phosphorus diffusion in *P*-type silicon is described in [M 31] or [M 42], [M 43], [M 39], [M 30], [I 2]. Though this method is generally used in the technology of diffused junction semiconductor detectors, results obtained with Ga [W 20] and B [D 19], [K 20] diffusion in *N*-type silicon were also reported.

The heating required for diffusion reduces the minority carrier lifetime [D 4]. A junction made by double diffusion at high temperature is a bad detector of fission fragments which give rise to important recombinations [M 43]. Devices with better performances were obtained [M 43] for gaseous diffusion at lower temperatures (600°C).

A detailed study of the window thickness as a function of the diffusion cycle used [F 19] permitted an estimation of the diffusion length shorter than 0.05 μ (within which alpha particles of 6.02 MeV lose an energy below 6 keV). Window thicknesses below 0.5 μ can be usually obtained. Detectors with windows of 0.013 μ were also made [J 1] and a method for measuring small diffusion depths (below 0.1 μ) was elaborated.

A thin window was also made [M 12] by injecting boron at the surface of *N*-type Si with $\rho = 5000\ \Omega$ cm, by bombarding it with B ions of 100 keV and 500 keV energies (with 0.14 and 0.7 μ ranges in Si, respectively) accelerated at Van de Graaff.

Diodes with $i_D \simeq 10$ μA at 100 V and a breakdown voltage below 400 V were obtained. The best sample had 50 keV resolution (*FWHM*).

Better results obtained with this "ion-implantation technique" are presented in [M 37], [M 13].

Electrical contacts at the frontal surface of the semiconductor detector are easily achieved due to the high conductivity of this surface. A gold spiral or band fixed with a conducting cement usually makes up a satisfactory contact.

The ohmic contact in the region of high resistivity must not inject minority carriers. For this purpose, the surface recombination rate in the region of this contact is increased through a rough mechanical processing, or yielding a large excess of majority carriers by a corresponding doping.

The value of the surface leakage current and the breakdown voltage of the junction depend significantly on the chemical treating of the surface and particularly of the edge where the junction reaches the crystal surface. Effects of different treatments on the reverse current and the breakdown voltage were also studied [B 47]. The surface conditions once achieved, the junction edges must be protected against subsequent alterings caused by water or oil vapours, mercury vapours, a.s.o. A method of sealing the junction edges is described in [J 1]. Silicon oxide films brought up on bulk silicon before or after the phosphorus diffusion were proposed as protections [M 42]. Glass with As (it has a low melting temperature) [D 12] or teflon [I 2] are also used. A transistor like sealing meant to ensure safety during bad operation conditions was achieved for diffused junction detectors used on artificial earth satellites [B 49].

With such detectors, depletion layer thicknesses up to 600 μ [D 12] ($\rho = 10{,}000\ \Omega$ cm with 500 V applied voltage) can be usually achieved. A thin window and a good energy resolution at room temperature make them useful in the detection of fission fragments, alpha particles (up to 40 MeV with 15 keV resolution for a useful surface of about 1 cm^2), beta particles (8 keV resolution) and heavy ions. When the detector is cooled the resolution improves. With a 50 mm^2 surface detector the best resolution obtained was [D 12] 13 keV for 5 MeV alpha particles.

As compared to other types of semiconductor detectors, the diffused junctions are now considered [G 21] the most reliable and useful semiconductor detectors. The diffusion technique is suited for mass production. Special structures can be also manufactured, e.g., a multichannel mosaic (100 channels) on silicon support was produced [G 21] and used for two years in spectrometry.

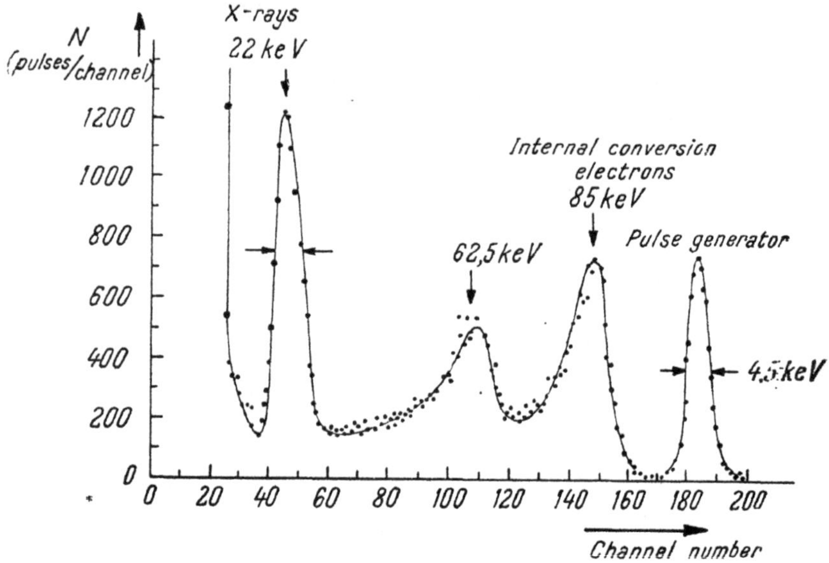

Fig. 5.3 — X-ray and internal conversion electron spectrum of the ^{109}Cd source obtained with a diffused junction on Si with $\rho_p = 16$ k Ω cm [M 9].

Figure 5.3 gives an example of using a diffused junction detector [G 4] of *P*-type silicon with $\rho = 16{,}000$ Ω cm. One may see the peaks of 22 keV (X rays), 62.5 and 84.5 keV (internal conversion electrons) given by a ^{109}Cd source.

5.4. SURFACE-BARRIER DETECTORS

In competition with the diffused junctions, the barrier detectors are of a simpler construction and can be easily manufactured in laboratory. For this reason this detector type has a wider use, though so far the comparison of these two junction types does not categorically plead for one or the other. Opinions do not agree. It is said [G 21], for instance, that the barrier parameters can be hardly checked because the barrier layer theory is not sufficiently developed and, upon the authors' experience, there is a very low efficiency in getting good detectors, though the technology is simple.

Another disadvantage of barrier *SD* consists in the fact that their bulk material (*N*-type silicon) can be more hardly obtained with the purity degree required (than the *P*-type silicon) because of boron impurities (acceptors), which cause most of the troubles in purification. However, from this point of view, the *N*-type silicon with $\rho = 3500\ \Omega$ cm is comparable with the *P*-type silicon with $\rho = 10{,}000\ \Omega$ cm which, due to an electron to hole mobility ratio equal with three, has the same density of non-compensated carriers, i.e., at a given voltage it presents the same barrier thickness. Methods of working out surface barriers on *P*-type silicon have recently been described [I 2].

These semiconductor detector types were also made on other materials, e.g. GaAs [K 11] and GaP [G 11].

Because the method for the preparation of surface-barrier semiconductor detectors implies no heating of the material (therefore no shortening of minority carrier lifetime) it is stated [M 19] [D 4] that these detectors have potentially better performances than diffused junctions and are preferable. Because of their smaller reverse currents, they are preferred when larger surfaces are required [D 11]. Generally these semiconductor detectors do not need sealing as diffused junction *SD*. The main argument remains the one stated above, that is their simple construction. Methods of manufacturing barrier layer semiconductor detectors are described in detail in literature [H 3], [A 16], [H 4], [D 6], [B 26], [A 8].

On an *N*-type silicon monocrystal sample whose surface underwent fine mechanical processing and etching, a *P*-type inversion layer appears under oxygen action (when being kept in ambient room conditions for 2 or 3 days, or more quickly by chemical treatment with oxygen rich substances as H_2O_2, humid oxygen, ozone and watery vapours, a.s.o). The contact with this inversion layer is made through a thin gold film deposited through vacuum evaporation. The ohmic contact on materials with high resistivity is achieved through the method described for diffused junctions.

In fact, the exact mechanism of rectifying junction formation through this method is not well known; there is no theoretical model as for diffused junctions [S 11]. The junction rectifying properties are influenced not only by the energy levels (given by the band theory of conductivity) resulting from the breaking down of the crystal periodicity and by the existence of surface impurities [H 12], but also by the nature of the metal achieving the contact with the inversion layer. For instance, instead of gold, tin and nickel were successfully used, but with materials as Al, Be and Mg rectifying contacts on N-type Si were not obtained [M 42], pointing out that the metal plays a more important part than a simple contact on the inversion layer obtained through chemical methods. Recent experiments [S 18] pointed out that the metal presence is required in "slow states" occupation (energy levels induced by surface absorbed gases). It was also shown [G 3] how rectification could be explained as a function of the Fermi level relative position in metal as to the conduction and valency band in semiconductor.

The gold layer deposited on the frontal surface cannot be made too thick since it determines the thickness of the detector window. A transparent film corresponding to 75% light transmission (20—40 $\mu g/cm^2$) is satisfactory [D 4].

The ohmic contact surface on the opposite side must be small in order to inject as few carriers as possible. If on both sides of the Si, the contacts are rectifying, one junction is reversely biased and the other directly biased representing an important noise source [D 27].

Barrier layer detectors were also built on Ge surface [D 4], [Z 3]. Since Ge has a stopping power twice that of the silicon, germanium detectors with the same thickness of the depletion region provide linearity at higher energies. Ge has the disadvantage of requiring a liquid nitrogen working temperature. At this temperature the carrier mobility is very great i.e., $\mu \simeq 1.5 \times 10^4$ cm^2 /Vs so that pulse rise times of about $\sim 5 \times 10^{-10}$ s are expected [D 4].

As in the case of diffused junctions, the electric field has a maximum value close to the surface (at the x_0 distance). In a first approximation [F 13] it can be made equal to a field generated in the neighbourhood of a uniform sheet of electric charges with surface density (eN_s) and thickness $x_a - x_o$ (the thickness of the depletion region)

$$\mathcal{E}_M = \frac{eN_s}{2\pi\,\varepsilon_r\,\varepsilon_0}\,\ln\frac{x_a}{x_0}, \tag{5.4}$$

where ε_0 is the vacuum dielectric constant, ε_r the relative dielectric constant (for Si, $\varepsilon_r \approx 12$).

If the surface charge density (eN_s) is high enough in comparison with the volume space charge, breakdowns will occur at edges. At the edge of the gold contact bright spots ("microplasma") appear when the applied voltage exceeds a certain value.

In order to obtain surface barrier semiconductor detectors able to work at high applied voltages, methods of sealing the detector sides have been studied [F 13], [B 29]. The surface conductivity was

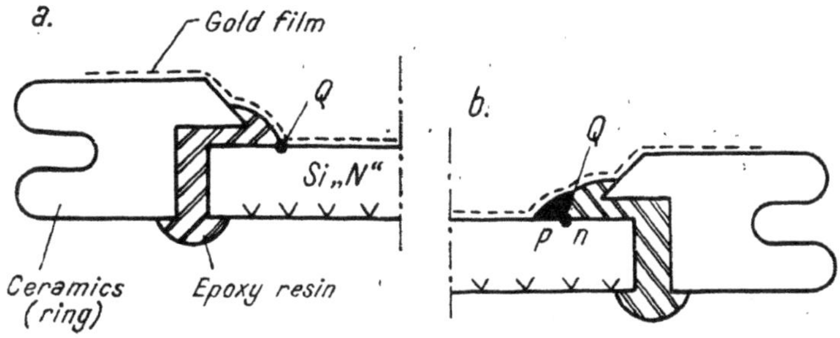

Fig. 5.4 — Cross-section profile of a barrier-layer detector: *a* — before mounting the iodine epoxy resin ring, *b* — after mounting it.

measured for silicon in contact with several materials used in the detector manufacture. In contact with nitrogen rich epoxy resins the Si surface had a pronounced *N*-type conductivity.

In order to reduce the great potential gradient at edges (the point *Q* of the cross-section in Fig. 5.4 *a*), an 0.8 mm thick ring of a lightly iodine doped epoxy resin without nitrogen (Fig. 5.4 *b*) was put around the gold boundaries, before it being deposited.

This resin gives birth to a *P*-type inversion layer on Si. Therefore, the junction edge remains in *Q* too, and the limit of the gold film is removed away from the maximum electric field of the junction (there is a gradual transition).

In the case of *N*-type Si with $\rho = 6000\ \Omega$ cm, voltages of 4000 V could be applied without the occurrence of breakdowns. However, for voltages higher than 1000 V the noise was excessively great unless special construction with guard-ring was used. In a guard-ring construction [B 19], thicknesses up to 1.5 mm were obtained for the depletion layer, with 9.2 keV (FWHM) energy resolution for 1 MeV electrons. The detector surface had 20 mm^2. Usually, the thickness *d* up to 0.7 mm can be obtained when applying reverse voltages up to 500 V, without taking special measures. The thickness record [M 23] is that of Klema who got 3.5 mm on Si of 20,000 Ω cm at 2000 V. A good detector must have a reverse current of $< 0.3\ \mu A/cm^2$ at 50 V voltage [D 4].

Resolutions of 14 keV were measured for 5 MeV alpha particles [B 26]. In a detector with $d \simeq 120\ \mu$ (72 V applied on surface-barrier detector of Si with $\rho = 1000\ \Omega$ cm) risetimes $t_r \simeq 3 \times 10^{-9}$ s [D 4] could be measured for pulses given by 5 MeV alpha particles.

Systematic measurements [W 15] made with a series of diffused junction and surface barrier detectors emphasized the better resolution of barrier layer semiconductor detectors. Besides the energy resolution and pulse amplitude, the reverse current, the noise level as a bias function for detectors in air and in vacuum, the window thickness and the long duration stability were also measured.

A detector mosaic can be easily made up on a common Si support, on which the Au is evaporated through a mask. Multiple devices were thus built, only 0.1 mm spaced [D 4] without neighbouring detectors interacting among them. Another example of special construction is a thin $\frac{dE}{dx}$ semiconductor detector combined with an integral one and making up a detector with barrier layer on an N-type layer deeply diffused inside a P-type silicon of high resistivity.

In the last years the barrier layer semiconductor detectors have been frequently used in nuclear physics experiments at low energies.

The fission fragments and heavy ions can be easily detected because they have short ranges; they do not require very thick depletion layers.

A characteristic energy spectrum for fission fragments yielded in the ^{252}Cf spontaneous fission obtained with a spectrometer using two barrier layer semiconductor detectors in coincidence, is drawn in Fig. 5.5 [J 6].

The fission fragment detection does not require semiconductor detectors with depletion layers thicker than 50 μ. With greater thicknesses, charged particles yielded in nuclear reactions can be linearly detected. For instance, in Fig. 5.6 a complex energy spectrum is shown [D 4] obtained in the bombardment of a thin target containing C, B and O with 5 MeV deuterons. For details about fission fragment detection with *SD*, see [S 6], [K 9], [M 45] [F 12] a.o.

Similar results are obtained with diffused junctions.

The minimum energy detected is limited by the noise.

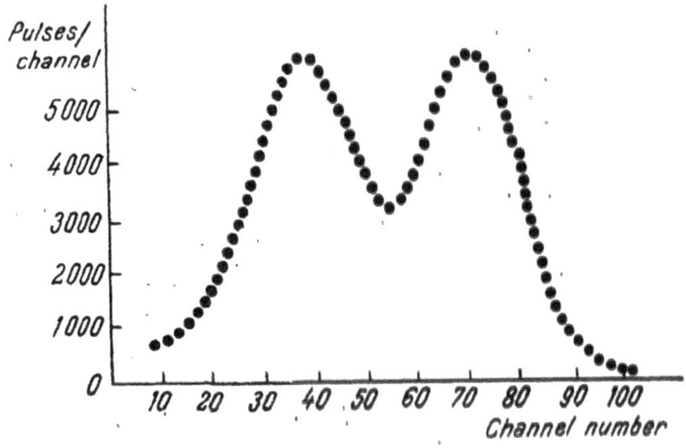

Fig. 5.5 — Energy spectrum of fission fragments yielded in the spontaneous fission of ^{252}Cf [J 6].

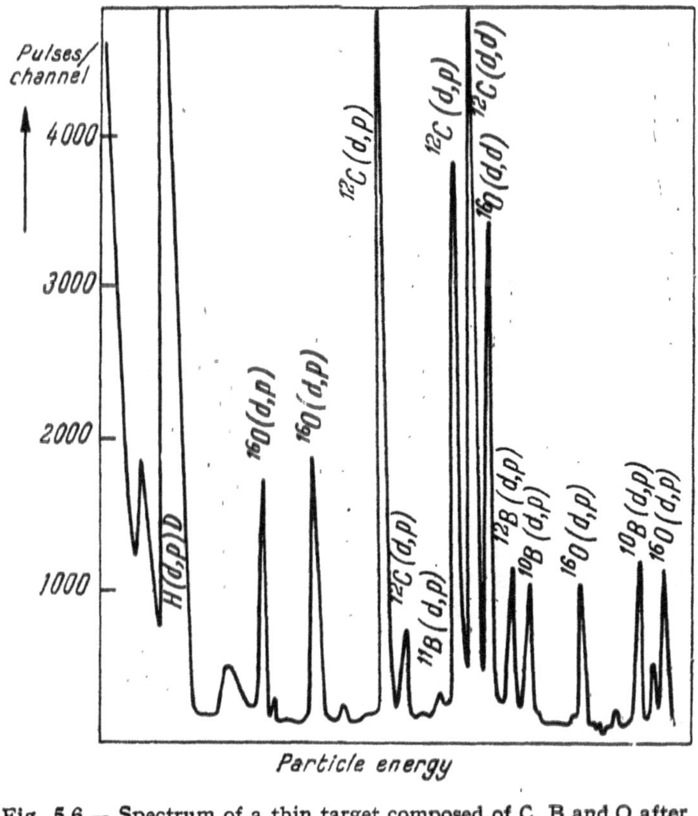

Fig. 5.6 — Spectrum of a thin target composed of C, B and O after it being bombarded with 5 MeV deuterons [D 4].

Ewing [E 15] managed to detect 18 keV protons with a surface barrier *SD* made of *N*-type silicon with $\rho = 300\ \Omega$ cm. The spectra he obtained for 18 and 25 keV protons are given in Figs. 5.7 *a* and *b*.

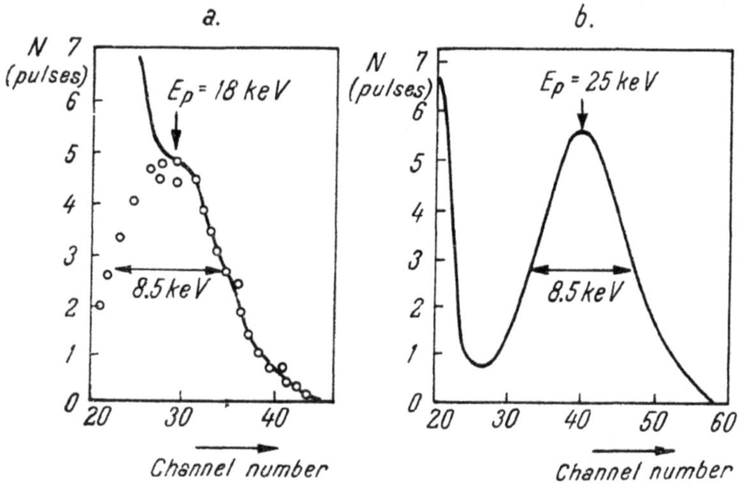

Fig. 5.7 — Spectra of 18 and 25 keV protons. The spectrum after subtracting the noise is represented by dots.

Surface barrier semiconductor detectors were also obtained at I.A.P. [B 4]. Figure 5.8 is a picture of such a detector.

5.5. GUARD-RING DETECTORS

The best performances of the diffused junction or surface-barrier semiconductor detectors can be achieved in the guard-ring variant. The guard-ring reduces very much a component of the detector current which is always prevailing and depends on the ambient medium conditions i.e. the surface leakage current. Hence, only the two volume components remain : the diffusion current and particularly the current given by thermally generated carriers inside the depletion layer, the sum of which is small enough in the case of Si at room temperature.

The decrease of the current by an order of magnitude enables the application of a higher voltage and so the thickness of the depletion layer can be very much increased.

The guard-ring semiconductor detectors have the advantage of being very suited to experimental checkings of some theoretical junction models, since the components of the volume reverse current can be easily estimated. In this way a careful designing is made for associated electronic apparatus which ought to provide the best working conditions of the spectrometer employing a guard-ring semiconductor detector [G 20].

The construction of guard-ring detectors is not too complicated. Methods used in the case of both diffused junctions and surface barriers are presented in [G 20], [F 19] and [G 11].

After diffusing the phosphorus inside N-type Si of high resistivity (1500 Ω cm), the N-type circular surface of the device is divided into two regions by etching a narrow circular portion (tenth of milimeters) (Fig. 5.9).

Fig. 5.8 — Surface barrier-layer *SD* manufactured at IAP [B 4].

The central region represents the detector itself and the bordering region the guard-ring which collects the surface leakage currents, particularly considerable there. In the case of surface barrier semiconductor detectors [L 15] the gold film is evaporated in vacuum, through a mask. Between the central region and the guard-ring there is only a low voltage (the volume currents are small and do not give a significant voltage drop on the load resistance R_l; Fig. 5.9 *a*) so that an average surface current cannot flow. Besides, both electrode voltages

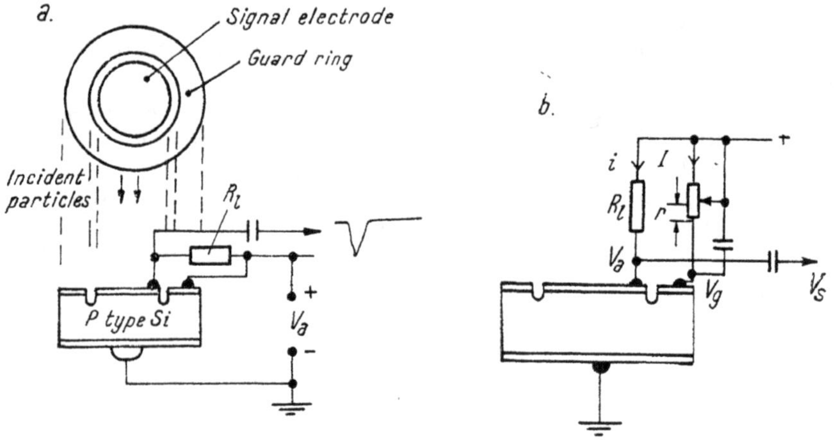

Fig. 5.9 — Guard-ring detector : *a* — circuit configuration for low bulk currents, *b* — circuit configuration in which $V_a = V_S$.

can be rigorously equal in the circuit shown in Fig. 5.9 *b*, where the resistance r is chosen to satisfy the relation

$$\frac{r}{R_l} = \frac{i}{I}. \tag{5.5}$$

This is not the only condition for eliminating the noise given by the surface currents. If the spacing between these two surfaces is too large, the surface noise given by the borders of the junction central region may be increased in the presence of the guard-ring, though the mean surface leakages outside the central region could be zero. This anomaly could be accounted for by the presence of currents with opposite signs which give a zero average and summing up fluctuations. The spacing between the guard-ring and the central region of the detector must be thinner than the depletion layer at the working voltage of the detector.

In this case the depletion layer of the detector extends over the whole space between the ring and the central region when the voltage applied exceeds a certain value, producing an increase of the resistance between electrodes. It was observed [G 20] that a surface treatment giving rise to a stable surface with reduced *N*-type conductivity in the separation region is required for obtaining the best performances.

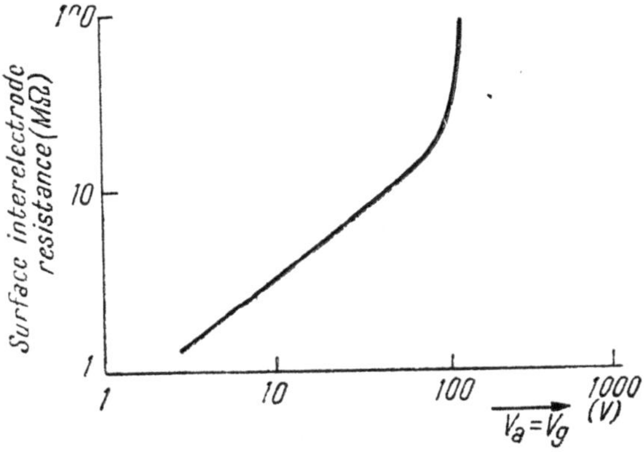

Fig. 5.10 — Variation of the guard-ring detector insulation resistance [G 20].

A variation of the inter-electrode impedance with the voltage applied of the form drawn in Fig. 5.10 was obtained [G 20] for a separation ring thickness of 62 μ, achieved through a photo-engraving method. A guard-ring surface barrier detector [C 11] with 1 mm depletion layer, at 1000 V applied voltage, has 0.3 mm spacing thickness.

The equivalent circuit of the guard-ring detector is given in Fig. 5.11 *b* [F 19] and, for comparison, the equivalent circuit of the detector without guard-ring is given in Fig. 5.11 *a*. We notice that the guard-ring section takes over the resistance R_Σ of surface leakages. In this figure the following notations have been used : R_d — diode resistance (given by volume currents), R_Σ — resistance of surface leakages, R_s — detector series resistance, C_d — junction capacity, C_{gr} — guard-ring capacity, Z_c — insulation impedance and CG — a charge generator.

An efficiency checking method of the insulating spacing properties consists in observing the system noise dependence on the guard-ring voltage V_g, the voltage V_a on the detector being kept constant. Such a curve is drawn in Fig. 5.12. For small time constants (0.1 μ s) in the detector-amplifier system, the shot-noise of the preamplifier input tube prevails, and hence the *FWHM* is not influenced by the detector current. For large time constants ($\tau = 10\mu s$) when the noise is determined by the detector current, it was observed that the noise was minimum at $V_g = V_a$ and increased for $V_g \neq V_a$, due to the

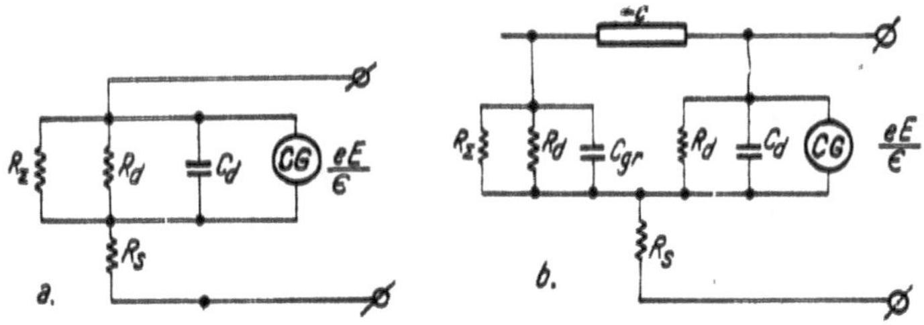

Fig. 5.11 — Equivalent circuit of the *SD* without guard-ring (*a*) and with guard-ring (*b*) [F 19].

surface leakages generated by voltage inequality between the two diodes. Small differences ($\pm 10\%$) between V_g and V_a do not affect significantly the resolution.

The leakage current can be measured by means of a potentiometer circuit [G 20] as shown in Fig. 5.13. Thus the measurement is made at the same potential of the detector surface and the guard-ring. The current decrease given by the guard-ring is shown by the curves in Fig. 5.14 [C 11]. This detector has a 20 mm² useful surface and a 540 V working voltage.

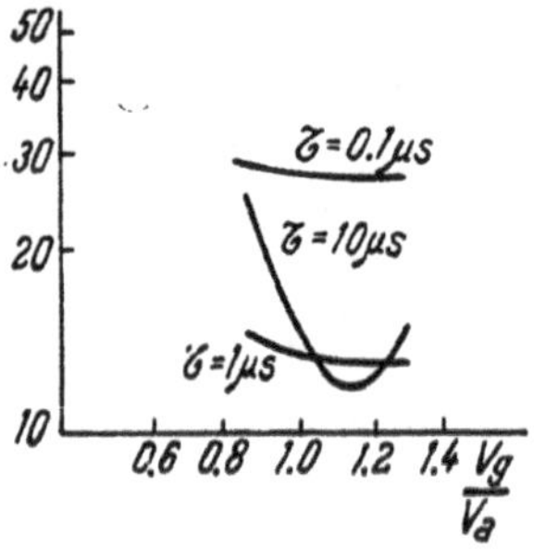

Fig. 5.12 — *SD* noise versus guard-ring voltage. $V_a = 100$ V; $\rho = 2000\ \Omega$ cm; $\Sigma = 0.4$ cm² [F 19].

Guard-ring semiconductor detectors are suited for electron detection, since they generally have depletion layer thicknesses of the order of 1 mm².

Fig. 5.15 shows the spectrum of ^{207} Bi conversion electrons obtained with a guard-ring detector [F 13]. The detector surface is 20 mm^2 and at a voltage of 1800 V, one obtains $d = 1.5$ mm.

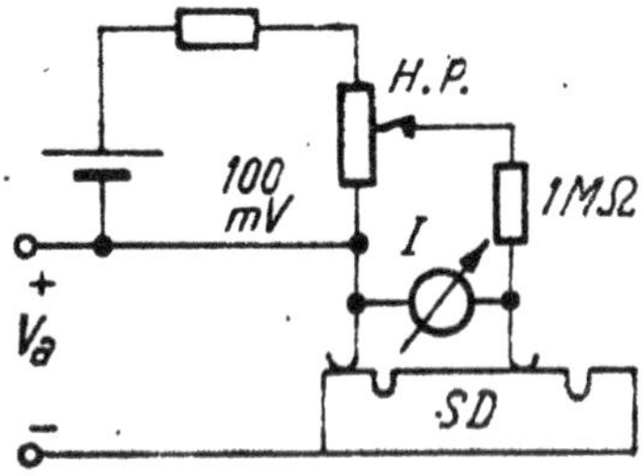

Fig. 5.13 — **Potentiometric circuit configuration used to measure the *SD* leakage current. *HP* = helicoidal potentiometer; *I* = zero instrument with high internal impedance.**

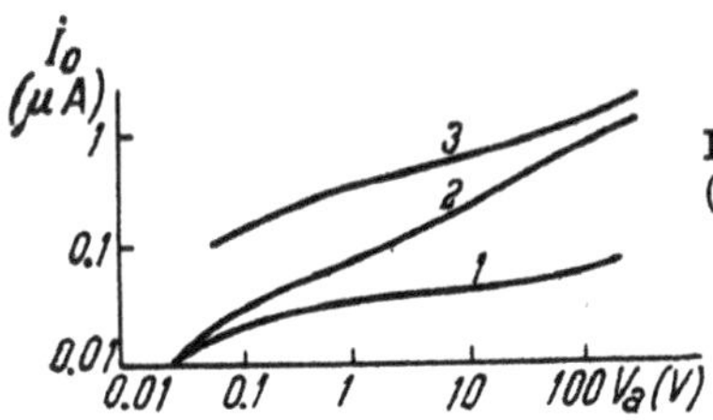

Fig. 5.14 — **The characteristic of a guard-ring *SD*, (*1*) connected guard-ring; (*2*) disconnected guard-ring; (*3*) guard-ring characteristic [C 11].**

Because of their small masses and succesive collisions with electrons of Si atoms, the track of detected electrons is not, generally, a straight line. For this reason, with a 1 mm thick semiconductor detector (equivalent to the range of 540 keV electrons), a linearity up to 1.8 MeV was obtained [S 7].

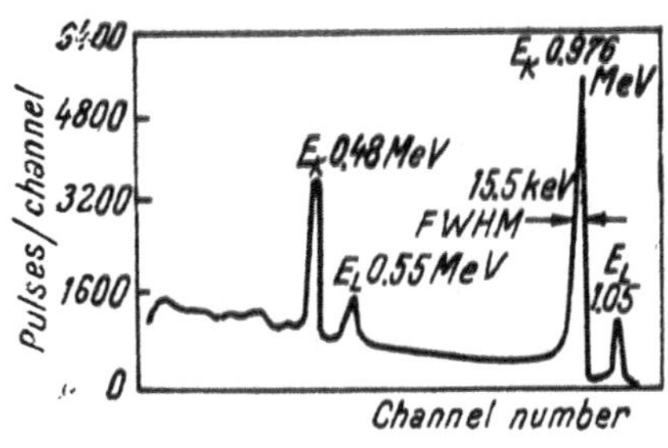

Fig. 5.15 — **Spectrum of the ^{207}Bi conversion electrons obtained with a guard-ring *SD* at V_a = 1800 V [F 13].**

Different spectrum shapes can be obtained function of the detector thickness to electron range ratio. Figure 5.16 illustrates this fact [D 27].

The *SD* had a thickness equal to the 200 keV electron range. Electrons with energies below 200 keV give a spectrum of the form *a* (Fig. 5.16). If the electron energy increases beyond 200 keV, part of electrons yield their whole energy in multiple diffusions and produce

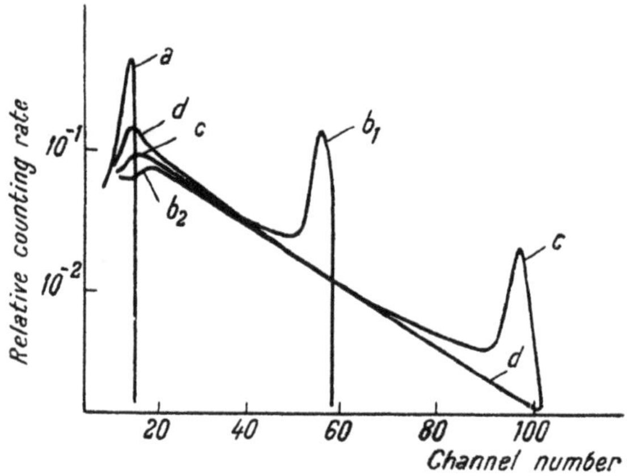

Fig. 5.16 — Electron spectrum at energies of 80 keV (*a*); 396 keV (*b*); 790 keV (*c*) and 1.22 MeV (*d*), given by a *SD* [D 27].

a peak of total absorption (b_1) and another part undergoing a minimum number of diffusions gives a "minimum ionization peak" (b_2). In addition, there are electrons diffused backwards which give rise to the middle region of the spectrum.

5.6. TOTALLY DEPLETED DETECTORS

When the depletion layer of a diffused junction or surface-barrier semiconductor detector extends all over its volume (for applied voltages of the order of tens of volts), we call it a $\mathrm{d}E/\mathrm{d}x$ sensitive detector.

Thin detectors can be used in systems for charged particle identification at low and medium energies, according to the relationship

$$E\frac{\mathrm{d}E}{\mathrm{d}x} \sim Mz^2e^2, \tag{5.6}$$

where M is the the particle mass and ze the electric charge of the particle.

The minimum thickness of the detector is restricted by the electric noise and the dispersion increase of the ionization losses when they become very small in comparison with the particle energy.

In Figures 5.17—5.22 are given plots of the variation of ionization loss with the particle energy in Si detectors of different thicknesses [G 21], in the case of electrons (Fig. 5.17), protons (Fig. 5.18) deuterons (Fig. 5.19), tritons (Fig. 5.20), ^{3}He (Fig. 5.21) and alpha particles (Fig. 5.22).

Considerable construction difficulties are related to the obtaining of silicon discs with thicknesses of tens or hundreds of microns.

By diffusion on *P*-type silicon with $\rho = 6000\ \Omega$ cm [W 11] thin detectors were obtained with thicknesses of about 50 μ and 5 mm diameter. A thin phosphorus layer was diffused and then covered with a gold film on which gold wire contacts were fixed with silver paste.

All attempts to achieve thin detectors from *N*-type silicon by gold evaporation on both sides failed because of a hole injection which appeared on the direct biased face and gave rise to high reverse currents and a prohibitive noise.

However, thin surface-barrier detectors were made [A 10] from *N*-type silicon with a gold surface and the ohmic contact obtained by Al, Be or Mg evaporation. A disc of 1 cm diameter and 125 μ thickness was cut in a bulk material with $\rho = 1000-2000\ \Omega$ cm. After etching, the detector thickness became 50—100 μ. A non-injecting ohmic contact [I 2] was also made from gold deposited by vacuum evaporation after a previous Si oxidation in air (one hour at 300°C). The detector was enclosed in teflon.

Very good results [M 2], recently reported, were obtained with a combination of known techniques i.e,. phosphorus diffusion for achieving the junction, the ohmic contact technique used for surface barriers, and the passivating technique with oxide, leading to a high stability in time of the characteristics. The use of a plane etching method provided a good uniformity of the detector thickness : the thickness variations inside a circle with 1.5 cm diameter were less than 0.75 μ in comparison with the 25 μ thickness of the detector.

The thickness uniformity is especially required in dE/dx measurements. A good uniformity is obtained if the samples are cut with

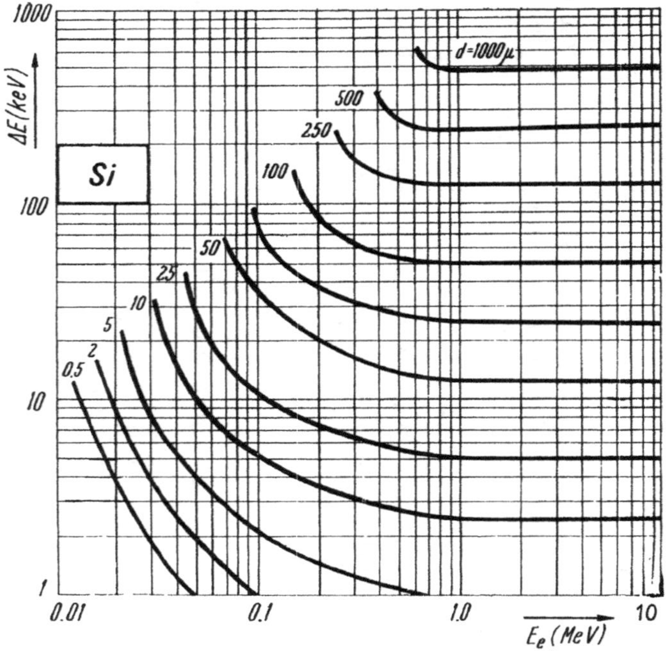

Fig. 5.17 — Electron energy losses in Si with different thicknesses [G 21].

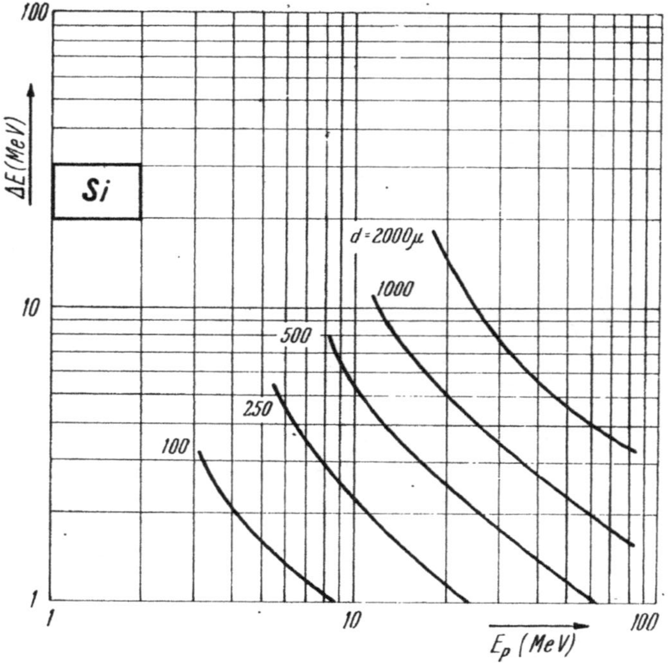

Fig. 5.18 — Proton energy losses in Si with different thicknesses [G 21].

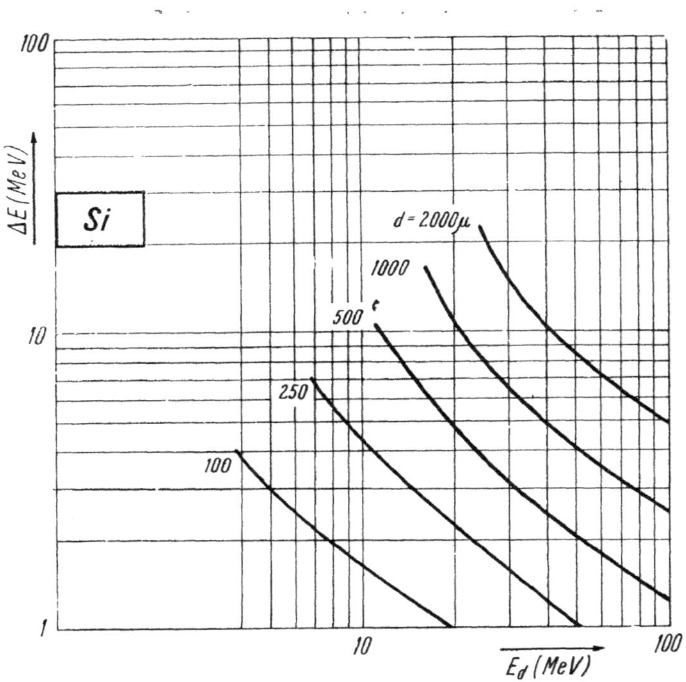

Fig. 5.19 — Deuteron energy losses in Si with different thicknesses [G 21].

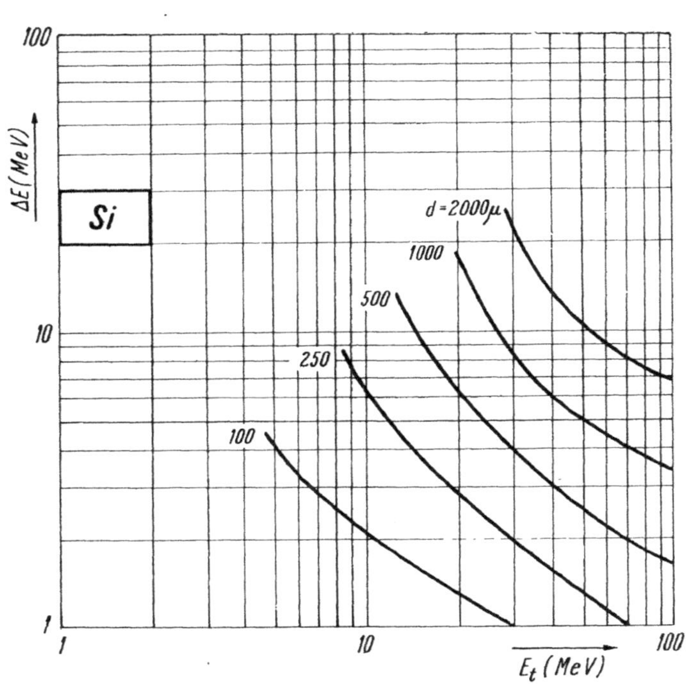

Fig. 5.20 — Triton energy losses in Si with different thicknesses [G 21].

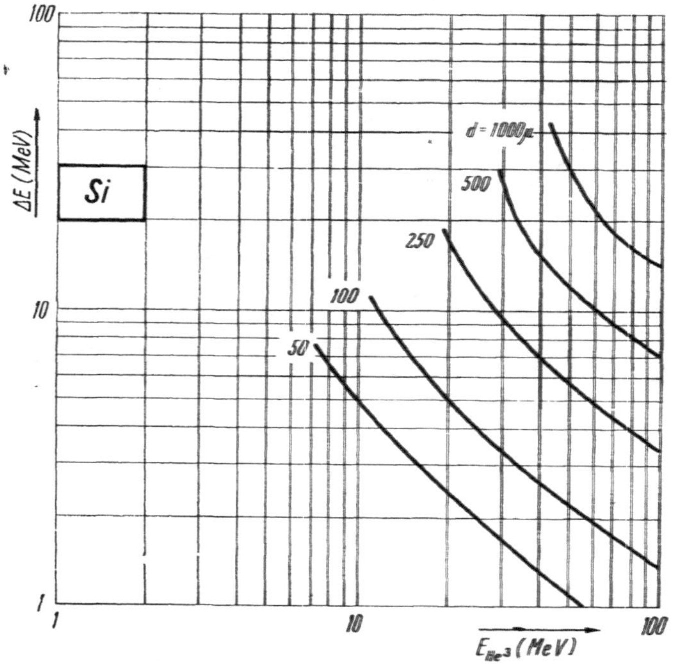

Fig. 5.21 — ^{3}He-ion energy losses in Si with different thicknesses [G 21].

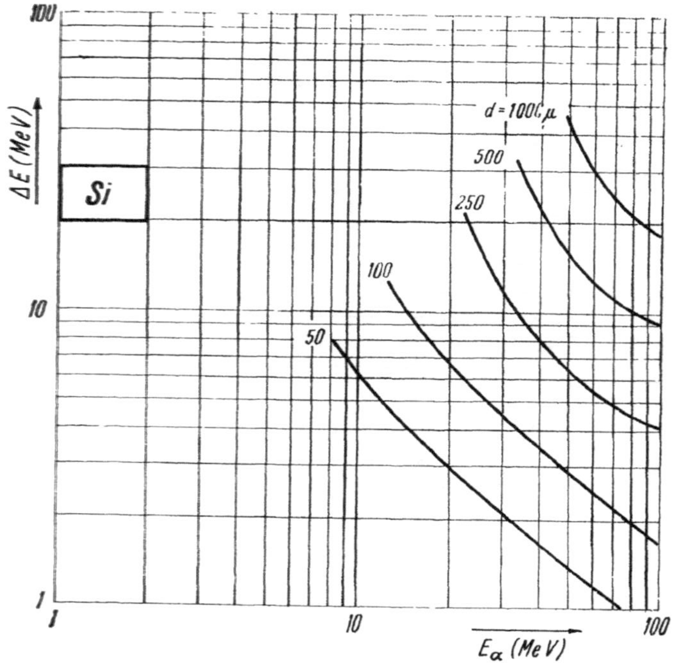

Fig. 5.22 — Alpha-particle energy losses in Si with different thicknesses [G 21].

a diamond and no more than 20μ [A 10] are removed during etching because non-uniformities increase due to this treatment.

Figure **5.23** shows the pieces used in mounting the detector [M 2] and Fig. **5.24**, the characteristics (capacity, current and energy resolution as a function of voltage) of a typical detector with 25 μ thickness and 0.4 cm² surface made of *P*-type silicon with $\rho = 400\,\Omega$cm.

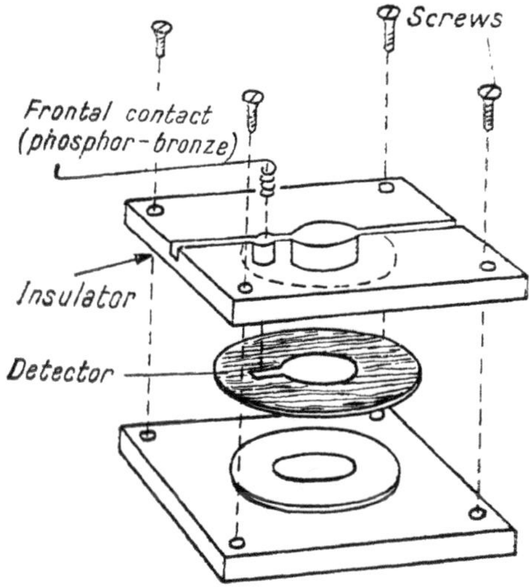

Fig. 5.23 — Mounting set-up of a dE/dx detector [M 2].

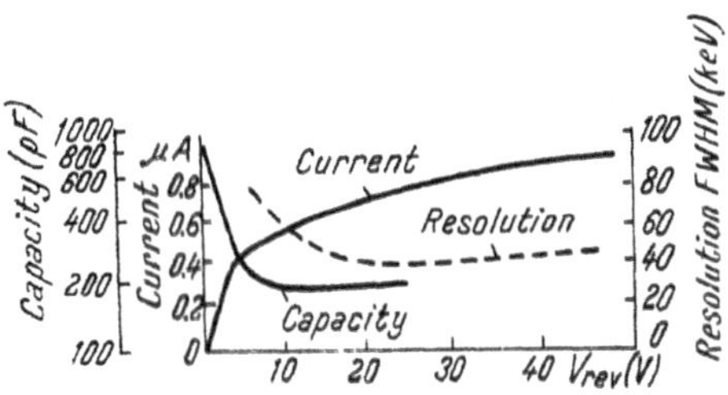

Fig. 5.24 — The characteristics of a dE/dx detector [M 2].

A ceramic ring with a dilatation coefficient close to that of the silicon can be also used as a detector support [C 16].

From Fig. 5.24 we notice that the voltage at which the depletion region extends all over the volume up to the ohmic contact is 7 V.

The ohmic contact has a good behaviour : the current does not increase suddenly when the depletion region reaches the ohmic contact, which means that it does not inject minority carriers. This current has an exponential behaviour with the temperature [A 10].

If the device resistance is great when the voltage of direct polarity is applied, the back contact is a rectifying one.

The detector capacity varies according to the law $V_a^{-1/2}$, function of the applied voltage until the whole volume becomes sensitive and then remains constant for determined thickness, surface area and dielectric constant of the device material.

With this detector, a "line" of 17 keV *FWHM* was obtained in the ^{241}Am alpha spectrum at —60° C. The response was the same for incident particles on both faces of the detector.

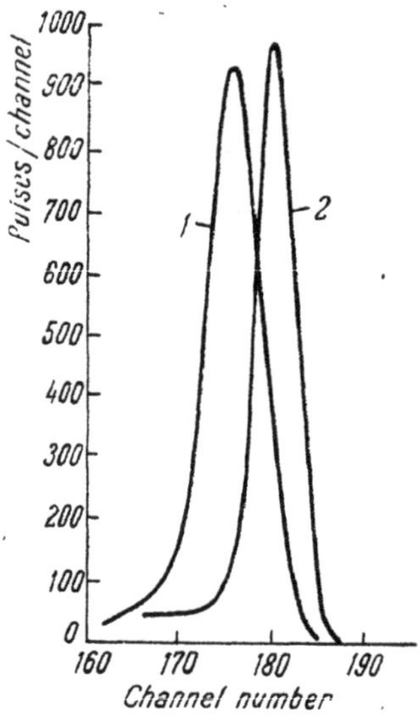

Fig. 5.25 — ^{210}Po-source spectrum given by a dE/dx detector [W 11], 1 — The particle is incident on the front face; 2 — The particle is incident on the back-surface.

The ^{210}Po alpha particle spectrum obtained by Wegner [W 11] is plotted in Fig. 5.25. The small difference between the spectra obtained from the irradiation of both faces of the detector points out the small thickness (1.25 μ) of the gold layer used as ohmic contact.

The 50 μ dE/dx detector had 7.9% resolution (*FWHM*) for ^{3}He ions of 21 MeV. The amplitude dependence on the applied voltage of the pulses given by this detector is drawn in Fig. 5.26. A charge sensitive preamplifier was used.

Very thin detectors were obtained. For instance, [S 27] for 11 μ thickness, alpha particles from a Th (C + C′) source with energies of 8.78 and 6.07 MeV lose inside the detector an energy of 1.06 and 1.5 MeV, respectively. They were detected as two separate peaks with 8.5 and 6.5% resolution, respectively. Other achievements are described in [W 13], [B 39], [I 1], [I 2], [D 12], [B 41], [E 6]. Awcock and Young [A 17] elaborated a theory of electron-hole pair collection in these types of semiconductor detectors.

An interesting problem [D 9], [G 6], [W 14], [C 13] is the influence of particle track orientation against the crystal anisotropy axes on the range of these particles inside the crystal. It was observed (mainly at low energies) that heavy ions (e. g., ^{85}Kr and ^{125}Xe) with energies of about 20—80 keV, have a longer range when penetrating the crystal along a principal axis than in the case of incidence along other direction. The phenomenon was explained as an ion "channelling" along this direction due to a series of collisions with the lattice atoms. The crystal structure being more "open" along a principal axis, the heavy ions of low energy "focus" the track in a more linear way than the random collisions.

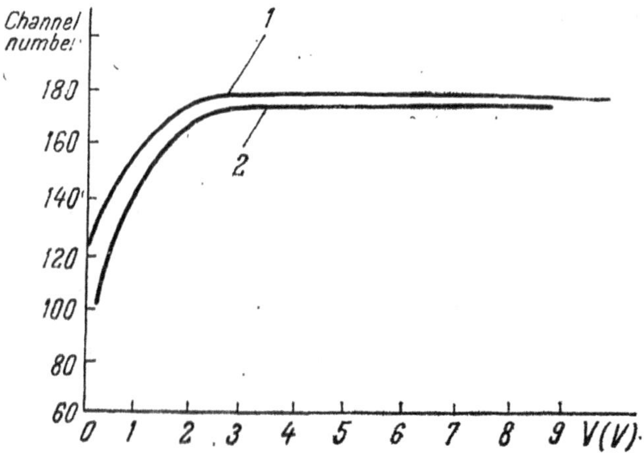

Fig. 5.26 — Pulse amplitude variation with the detector bias [W 11].

A first practical conclusion for dE/dx detectors is the necessity of cutting the Si wafer at an angle of 10—20° from the plane [1, 1, 1].

Lately, the bombarding energy range between 10 and 30 MeV has been increasingly used in the low energy physics. Because this energy

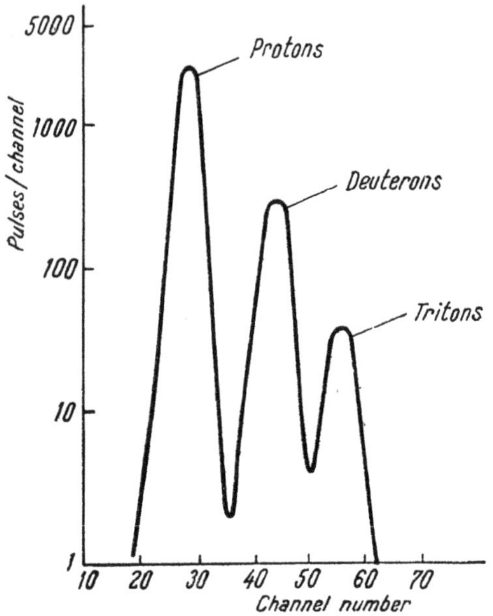

Fig. 5.27—Mass spectrum obtained with an $E.\frac{dE}{dx}$ system [G16]

range is above the Coulomb barrier for many nuclei, the reaction products are very complex. Systems $E\frac{dE}{dx}$ with semiconductor detectors were successfully used in the identification of heavy charged particles yielded (protons, deuterons, tritons, ^{3}He and alpha particles) [W 11], [W 13], [G 16]. Figure 5.27 shows a spectrum given by a thin detector of 80 μ with barrier layer on Si and Na I scintillator [G 16]. The particles were yielded in (p, p'), (p, d) and (p, t) reactions induced by 22 MeV protons. The relative amplitude of every mass peak is about 1, 2, 3, 12 and 16 for p, d, t, ^{3}He and alpha particles. Criterions were stated for an optimum resolution of a dE/dx and E detector system [H 3]. The product $E\frac{dE}{dx}$ has a real meaning only if a small part from the energy E is absorbed in the thin detector. Using an empirical relation

$$R = a\,E^{1.73}, \tag{5.7}$$

where the constant a depends on the particle type, it was shown [G 22] that the thickness d of the thin detector absorbing an energy ΔE, is related to the particle energy through the relation

$$\frac{d}{a} = (E + \Delta E)^{1.73} - E^{1.73}. \tag{5.8}$$

With an identifying assembly which uses this relation (the operations are made electronically with logarithmic elements) [G 22], the ^{3}He and ^{4}He could be separated within a large energy range, whithout requiring a very thin detector, a difficult operation which could be hardly made with previous methods.

5.7. NEUTRON DETECTORS

Lacking electric charge, the neutrons do not produce ionizations during their penetration through matter. An indirect method is used for their detection, that is charged particles emitted as a result of nuclear reactions induced by neutrons are detected.

The simplest process of interaction of low energy neutrons with matter as a result of which charged particles are produced is the neutron diffusion on protons. When a neutron beam passes through a hydrogen rich susbtance, part of neutrons knock the hydrogen nuclei (protons), yielding them a fraction E_p of their energy which may vary between 0 and 100%, depending on the collision angle. In the case of a monoenergetic neutron beam with energy E_n, the emerging protons have a continuous energy spectrum from zero to E_n. The energy of recoiled protons varies function of the emergency angle φ (in the laboratory system) according to the law

$$E_p = E_n \cos^2 \varphi. \tag{5.9}$$

Other nuclear reactions generating charged particles are also important in the neutron detection.

Table 5.1 lists the reactions, their value Q and the cross-section at certain neutron energies.

Light nuclei are preferred, due to their lower Coulomb barriers for charged particle emission. The neutron incident energy E_n transforms into kinetic energy of charged particles with additional emission of energy Q, so that by linearly detecting both charged particles (e. g., alpha particles and ^{3}He) and summing up the coincident pulses a signal proportional to $E_n + Q$ is obtained.

The semiconductor detector for neutrons consists of an *SD* for charged particles (usually with surface barrier layer), on the frontal surface of which is deposited a thin layer of hydrogen rich material or a converting film of a material containing one of the ^{10}B, ^{6}Li isotopes. The reaction products are therefore detected in a 2π — geometry. The materials making the neutron-charged particle conversion cannot be introduced inside the detector volume because the allowable degree of impurities is very low.

Unlike charged particle semiconductor detectors which detect each particle entering the sensitive volume (because it induces ionizations), in the case of neutron semiconductor detectors only a small fraction of the total neutron number passing through the converting films meets a nucleus; the neutron — charged particle nuclear reaction has a certain cross-section which varies with the neutron energy. It is reasonable therefore to define in the neutron detection a characteristic called detection efficiency (the ratio of the number of detected neutrons to total number of neutrons passing through the detector sensitive volume).

The exothermic reactions ^{6}Li (n, α) ^{3}He and ^{10}B (n, α) ^{7}Li are the most used from all the reactions, in the slow neutron detection.

The best detection efficiency can be obtained when the thickness of the converting layer is at least equal to the range of the most penetrating reaction product. For instance in the ^{6}Li (n, α) ^{3}He reaction, the penetrating particle is the triton, and in the ^{10}B (n, α) ^{7}Li reaction the alpha particle has a longer range.

The amplitude spectrum of reaction products yielded by slow neutrons will be continuous, from zero to 2.73 MeV or 1.47 MeV for the reactions ^{6}Li (n, α) ^{3}H and ^{10}B (n, α) ^{7}Li, respectively.

For detecting the neutrons in non-spectrometric conditions, it is enough to record the integral number of pulses exceeding a certain discrimination threshold. When this discrimination threshold is fixed at $\frac{1}{4}$ of the spectrum maximum energy, 8.5—153 pulses are obtained with a ^{6}Li converter on *SD* with 25—450 mm^2 surface for 1000 slow

Table 5.1

Cross-sections and energy releases of some nuclear reactions used in neutron detection

Nuclear reaction	Q(MeV)	Cross-section in barns (10^{-24} cm^2) Thermal	1 eV	10 eV	100 eV	1 keV	10 keV	100 keV	1 MeV	10 MeV
$^1H(n, n)\,^1H$	0	35	22	21	21	20	18	13	4.2	0.94
$^3He\,(n, p)\,^3H$	0.76	5400	810	260	80	26	8.0	2.2	0.8	0.2
$^{10}B(n, \alpha)\,^7Li$	2.8	4010	640	200	65	22	6.3	1.9	0.2	—
$^6Li\,(n, \alpha)\,^3H$	4.8	945	140	45	14	4.3	1.4	0.7	0.3	0.04
^{239}Pu (fission)	200 MeV	746	35	(30)*	(30)	(6)	2.5	1.8	1.2	2.4
^{235}U (fission)	200 MeV	582	60	(30)	(20)	(8)	4	1.8	1.2	1.8
^{237}Np (fission)	200 MeV	—	—	—	—	—	—	0.005	1.2	2.5
^{238}U (fission)	200 MeV	—	—	—	—	—	—	—	0.018	1.0
^{232}Th (fission)	200 MeV	—	—	—	—	—	—	—	0.001	0.13

* The values in brackets are averaged values over large ranges of the cross-section variation.

neutrons/cm^2. When using ^{10}B, there are 10—180 pulses obtained in the same conditions. The reaction products are stopped inside the depletion layer, in both cases, if the semiconductor detectors are made of N-type Si with $\rho = 300\ \Omega$ cm and the working bias voltage is of 50 V. It should be noticed that neutron semiconductor detectors have been achieved even since 1959 [B 1].

From Table 5.1 it can be observed that fissioning ^{239}Pu and ^{235}U converters, yielding fission fragments of high energy, but with short ranges in Si can be also used in the slow neutron detection. The neutron induced fission of ^{239}Pu and ^{235}U has a cross-section greater than 1 barn in the neutron energy range 0—10 MeV.

Fission fragments can be also obtained with fast neutrons from ^{237}Np, ^{238}U and ^{232}Th converters (the threshold is $E_n \sim 1$ MeV). All these elements have an alpha activity too, and therefore they give a continuous background even in the neutron absence. Since the fragment energy is much higher than the alpha particle energy, the alpha background can be easily cancelled through an amplitude discrimination of the semiconductor detector pulses.

Fast neutrons are detected with a good efficiency by using the (n, p) elastic diffusion process. All the recoiled protons from a layer of hydrogen rich material are diffused forward. Since the maximum energy of a continuous proton spectrum is equal to the incident neutron energy, in order to obtain signals of maximum amplitude, it is recommendable to use *SD* with barrier layer on *N*-type Si with $\rho = 3200$ Ω cm and 100 V applied voltage at which the depletion layer has a thickness about the order of 6 MeV proton range.

Another detection method for fast neutrons consists in their previous thermalization in paraffin, e.g. the resulting thermal neutrons can be efficiently detected with *SD* and ^{6}Li F or^{10} B convertor. Practically, the semiconductor detectors for thermal neutrons are surrounded with paraffin.

Figure 5.28 shows the construction of an ORTEC neutron semiconductor detector. It is, essentially, a charged — particle *SD* equipped with a converting layer and its protection lid. The pieces used in the *SD* mounting contain a negligible quantity of hydrogen.

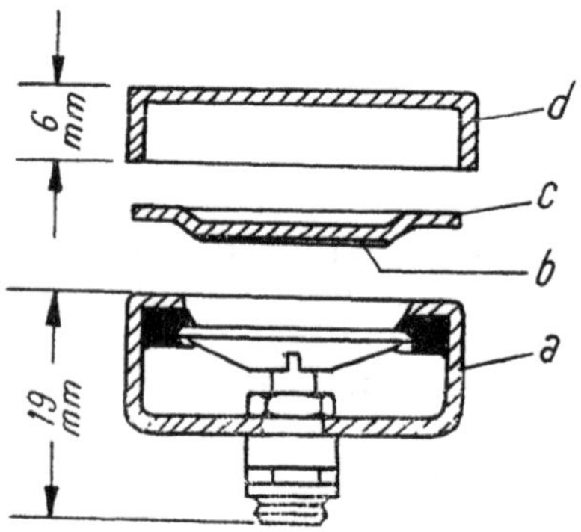

Fig. 5.28 — Cutaway view of neutron *SD*; *a* — barrier layer *SD*; *b* — neutron-charged particle converter; *c* — Al support ($\neq$ 0.5 mm); *d* = lid.

The methods used in the non-spectrometric detection with slight modifications, can be also used in the fast neutron spectrometry. In this case, two *SD* slightly spaced are used with a converting layer between them (e.g. ^{6}LiF) in a "sandwich" geometry. The fast neutrons with E_n energy induce ^{6}Li (n, α) ^{3}H reactions inside the ^{6}LiF layer yielding the whole energy to the alpha particle and triton. Both semiconductor detectors detect the reaction products and by summation of the coincidence pulses, we obtain a pulse with amplitude proportional to $E_n + Q$ (4.78 MeV) [L 14], [L 15], [B 36].

The high Q of the reaction provides a clear separation from the background : the peak given by slow neutrons is placed at an amplitude corresponding to 4.78 MeV energy. The construction of the spectrometric detector is shown in Fig. 5.29. A *SD* with barrier layer on *N*-type Si with $\rho = 3200\ \Omega$ cm and a working voltage about 150 V was used.

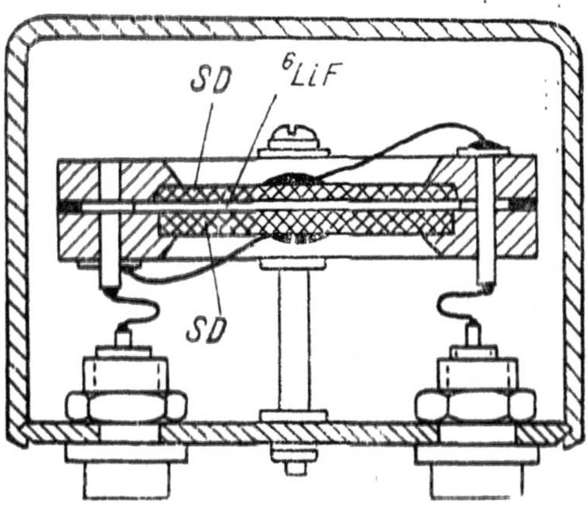

Fig. 5.29 — Cross-section view of a neutron detection spectrometric system with two *SD*.

Neutrons with energies higher than 7—8 MeV induce nuclear reactions on the Si isotopes. The background introduced by these reactions can be partially cleared up if working in coincidence. In addition a similar *SD* sandwich without ^{6}LiF, may be used to estimate the background and to subtract it from the useful spectrum.

The useful spectrum and the background can be simultaneously measured with a double detector of special construction [D 6], [D 3] (Fig. 5.30). Two semicircular barrier-layer semiconductor detectors were built (evaporating gold through a mask) on an *N*-type Si disk with $\rho = 1500\ \Omega$ cm and 2 cm diameter. The spacing between the two detectors was of 1 mm. A polyethylene film of 0.5—1 mg/cm^2 was laid on the surface of one semiconductor detector. The ensemble was enclosed in order to protect it against light. The (n, p) elastic diffusion was the process used, the polyethylene representing the hydrogen rich material.

The main "background" sources in the case of this detector type are:

— The recoiled protons coming from the water, or the hydrogen absorbed in the box walls. Their contribution can be reduced by the gold lining of the box.

— At low energies, the gamma-rays give a background from the summing up of pulses with low amplitudes but, generally, very numerous. This effect reduces when working with small load resistances (50 k Ω) and minimum biased voltages.

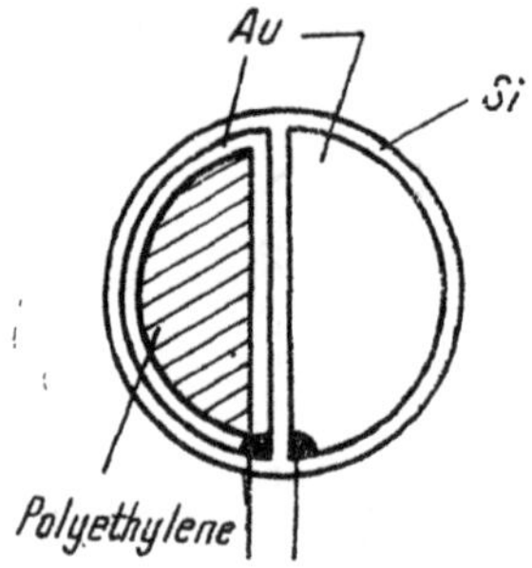

Fig. 5.30 – Double *SD* used in the neutron spectrometry.

— For $E_n > 3$ MeV there are prevailing reactions:

^{28}Si (n, p); $Q = -3.87$ MeV, isotopic abundance 92%;

^{28}Si (n, α); $Q = -2.66$ MeV;

^{29}Si (n, α); $Q = -0.021$ MeV; isotopic abundance 4.7%.

For $E_n = 6$—8 MeV the cross-section of the first two processes amounts to 0.4 b. This fact makes difficult the background problem beyond 7 MeV.

Figure 5.31 is a plot of the 4.2 MeV neutron spectrum given by the *SD* shown in Fig. 5.30. The thickness of the polyethylene layer was 1 mg/cm^2.

The energy resolution of the spectrometer with a neutron semiconductor detector is limited by the thickness of the converting layer used.

The smaller the thickness, the better the energy resolution, but the efficiency decreases exponentially.

A resolution, *FWHM*, of 300—400 keV [L 14] was obtained with a ^{6}LiF layer of 150 μg/cm^2. The detection efficiency was then of about 10^{-3} for slow neutron and 10^{-6}—10^{-7} for fast neutrons.

The variation with E_n of the product between resolution and efficiency was studied at a given converter thickness (1μ) [B 36]. An analysis of different methods used in the spectrometric detection of neutrons [D 7] with *SD* leads to the following conclusions.

The variation of the cross-section function of the neutron energy represents an important fact în this comparison. For (n, α) reactions

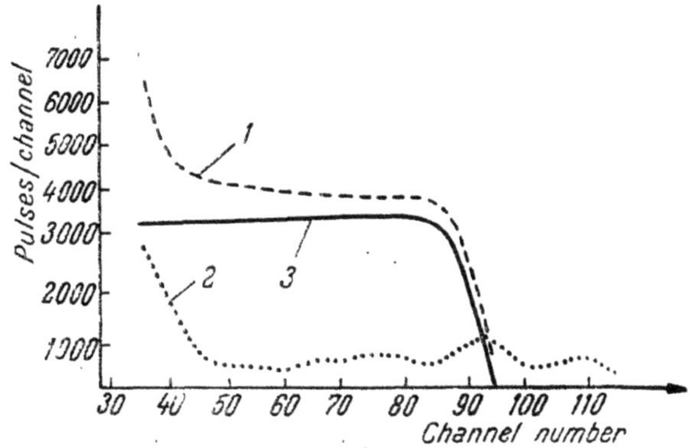

Fig. 5.31 — Spectra of 4.2 MeV monoenergetic neutrons obtained with the *SD* shown in Fig. 5.30. 1) — spectrum observed; 2) — background; 3) spectrum after background subtraction.

the cross-section variation leads to an efficiency decrease by three orders of magnitude from slow towards fast neutrons. The efficiency curves from Fig. 5.32 [D 5] were obtained within the energy range 0.1 — 10 MeV.

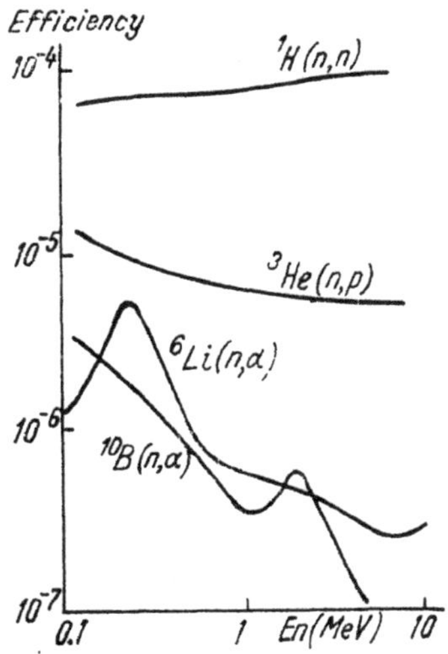

Fig. 5.32 — Detection efficiency versus neutron energy for different converter types [D 5].

Assuming that the energy dispersion is due to the heavier (hence slower) product resulting from the reaction, the material thicknesses (d) and the product $n\sigma$ were estimated [D 3] for 50 keV energy resolution (limited by the converter finite thickness) for neutron energies of 0.5, 1 and 5 MeV. The results are given in Table 5.2.

It is obvious from Fig. 5.32 as well as from Table 5.2 that a *SD* with recoiled protons has an efficiency with an order of magnitude higher than that of the ^3He detector which, in its turn, is by an order of magnitude better than all the others, due not only to its efficiency value but also to the lack of resonances as in the case of ^6Li and ^{10}B. In addition, the ^{10}B (n, α) reaction is not adequate for spectrometry because of competitive reactions, e.g., ^{10}B $(n, t2\alpha)$ with $Q = 0.33$ MeV [D 3].

Although the recoiled proton method has a high efficiency, the advantages are doubtful. The spectrum must be differentiated in order to find the incident neutron spectrum, which imposes severe conditions as concerns the precision of the initial spectrum. A typical neutron spectrum for fission requires an increase with a factor of 40 in the integral spectrum statistics. Then, practically, the efficiency of the recoiled proton spectrum is reduced below that of the ^3He detector.

Table 5.2

Converter thicknesses x_0 (in μ) necessary to obtain a 50 keV energy spread and the resulting product $n\sigma$ [D 3]

E_n	0.5 MeV		1 MeV		5 MeV	
Reaction	x_0	$n\sigma$	x_0	$n\sigma$	x_0	$n\sigma$
^6Li (n, α)	37	4.4	41	1.9	96	1.4
^3He (n, p)	35	63	38	61	70	49
^{10}B (n, α)	22	7.8	21	4	20	1.2
^1H (n, n)	100	530	160	600	625	860

It seems that ^3He is the most adequate conversion medium [D 3]. The quantity of ^3He can be slightly changed by varying the gas pressure. The *SD* with ^{3}He converter has several advantages in comparison with the ^{3}He proportional counter. The semiconductor detector has a lower background and does not require a tritium poor gas, since it works in coincidence. Such ^{3}He semiconductor detectors were constructed, [D 37] made of two identical *SD* with 1.3 cm useful diameter and 1.5 mm spacing between them. The *SD* were enclosed inside a steel weldtight box. A 60 μg/cm^2 ^{3}He layer exists between the two *SD* at a 3 atmospheres pressure. A 100 keV resolution with 10^{-5} efficiency is to be expected for fast neutrons. Difficulties also exist as concerns the background for $E_n > 8$ MeV.

The neutron semiconductor detectors are very useful, particularly when the flight-time method, very suited for working at a pulsed accelerator, is not applicable [D 12]. As an example of their use, in Figs. 5.33, 5.34, 5.35 some spectra obtained by various *SD* at different neutron energies are given.

Spectra such as those plotted in Fig. 5.33 and 5.34 were obtained with a ^{6}LiF semiconductor detector for slow neutrons [B 36] and for 2.95 MeV neutrons from the *D* (*d*, *n*) ^{3}He reaction, respectively.

The background in Fig. 5.34 *a* is assigned to 14.7 MeV neutrons yielded in the *T* (*d*, *n*) ^{4}He reaction due to a little quantity of tritium contained by the deuterium target.

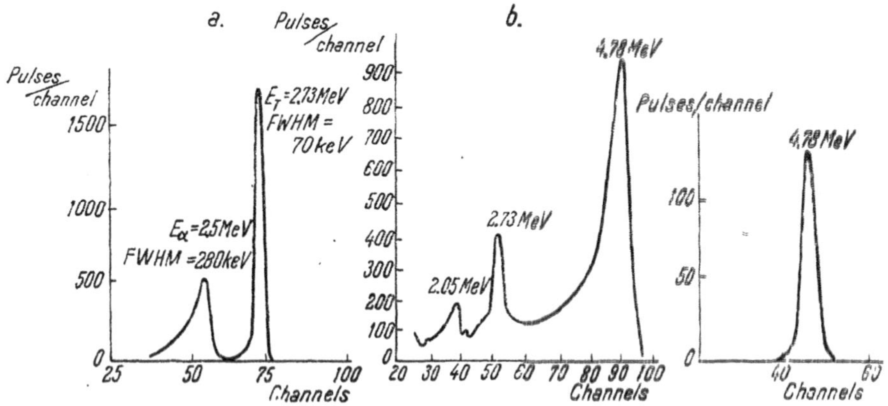

Fig. 5.33 – Thermal neutron spectra obtained with *SD* [B 36] *a*) a single *SD* covered with a ^{6}Li F layer of 1 μ, *b*) without coincidence, *c*) with coincidence.

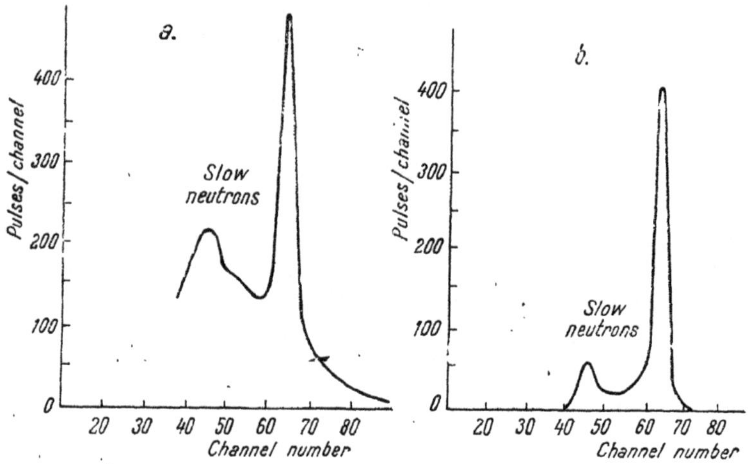

Fig. 5.34 — **2.95** MeV neutron spectrum recorded with a ^{6}Li F semiconductor detector [**L 14**], *a*) without coincidence; *b*) in coincidence.

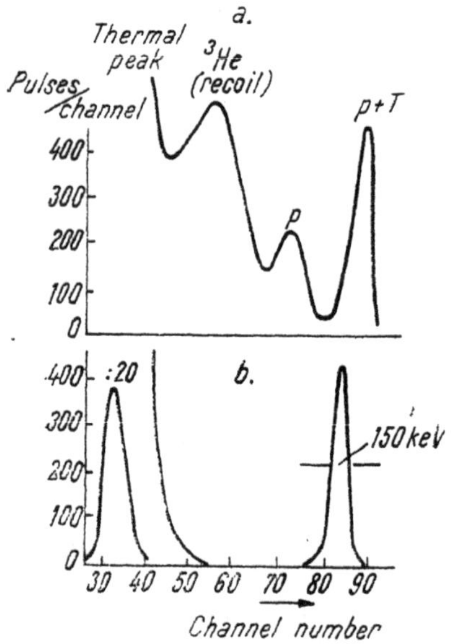

Fig. 5.35 — Spectrum of 3.2 MeV neutrons recorded with a ^{3}He converter *SD* [D 12], *a*) without coincidence; *b*) with coincidence.

Fig. 5.35 shows the 3.2 MeV neutron spectrum given by a *SD* with a ^{3}He converter [D 12].

Semiconductor detectors with ^{6}Li were used in measurements of reactor neutron flux distributions [L 1] and the study of some nuclear reactions at the reactor [L 2], [A 3].

5.8. SPECIAL DETECTORS

The study of detectors based on barrier-layer semiconductor detectors cannot be concluded without considering some types of special construction which should illustrate again the adaptability of junction semiconducting devices to problems raised by the experimental physics. We must also mention detectors with internal gain providing pulses of high amplitude.

5.8.1. *Multiple detectors.* In experiments with polarized charged-particle beams we measure the asymmetry introduced in a second particle-diffusion by an "analyser", the polarizing properties of which are well known (usually He or C).

The detectors are symmetrically placed on the left and right hand of this analyser, in the plane of the first reaction (or diffusion).

For such reactions, a triple detector [D 6] (Fig. 5.36 *a*) was built on the same support. If this ensemble is placed at 1—2 cm

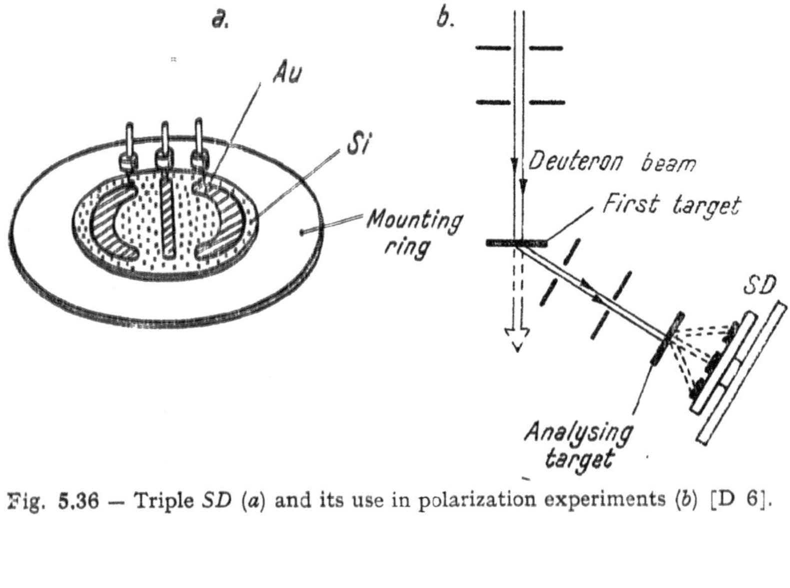

Fig. 5.36 — Triple *SD* (*a*) and its use in polarization experiments (*b*) [D 6].

distance from the carbon analyser (Fig. 5.36 *b*) both lateral detectors receive particles scattered at 40—60°, while the central *SD* works as a monitor. Any asymmetry introduced by the *SD* can be checked up by rotating it with 180°.

In polarization experiments the problem of the neutron and gamma-ray background is important due to the poor statistics. Using the *SD* described above a comparatively low background was obtained [D 12].

This triple detector is one of the simplest multiple *SD*. In experiments with magnetic spectrometers the spatial resolution is very important since the particle position after the magnetic analysis determines its momentum. A multiple detector of the form shown in Fig. 5.37 is used in such experiments [D 12]. Evaporating gold through a mask of thin wires, on a Si wafer, a 0.1 mm spacing of two neighbouring detectors was achieved without significant interaction between them. The coupling due to surface resistance can be weakened by reduction of the detector load resistance.

A similar linear system consisting of several (20—100) simple semiconductor detectors was used for the same purpose [J 1], [J 6]. Information can be obtained from each detector by means of an intricate system with individual preamplifier [P 2] or an ingenious method [B 19], [B 20] with each *SD* mounted as a capacity on an artificial delay-line, the identification of the detector from which the given pulse comes being made by the time spacing between the pulses reaching the two ends of the line. Valentin [V 1] described a method according to which the counting rate of each component of the 10× 10 *SD* system is visualized on an oscilloscope screen. The charge from every detector is stored by an integrator condenser discharging at every 100 μs and modulating the intensity of the spot which has a

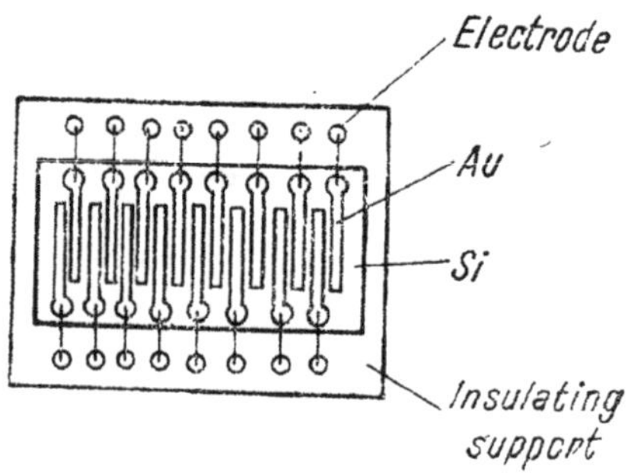

Fig. 5.37 — A multiple *SD* for the magnetic spectrometer [D 12].

certain position on the screen at that instant (corresponding to the *SD* position inside the system).

Recently [T 11] an indicator of the beam profile for a high energy accelerator was built with the aid of a 32 *SD* — system.

5.8.2. *Position-sensitive detectors.* Though, rather recently described in literature [N 2], [L 16], [L 17], [K 19], such detector types are already available under a commercial form. They allow a simultaneous estimation of both energy and position of charged particles. In this way angular distributions can be measured within a large angular range without resorting to mosaic systems which have a complex associated electronics. It is also possible that position-sensitive detectors should take the place of photographic plates now used in mass spectrometry or of other similar device [B 35].

The technology of this detector is that of surface barriers, but it is different from these ones, since, instead of one massive ohmic contact, it has two thin contacts at the surface ends.

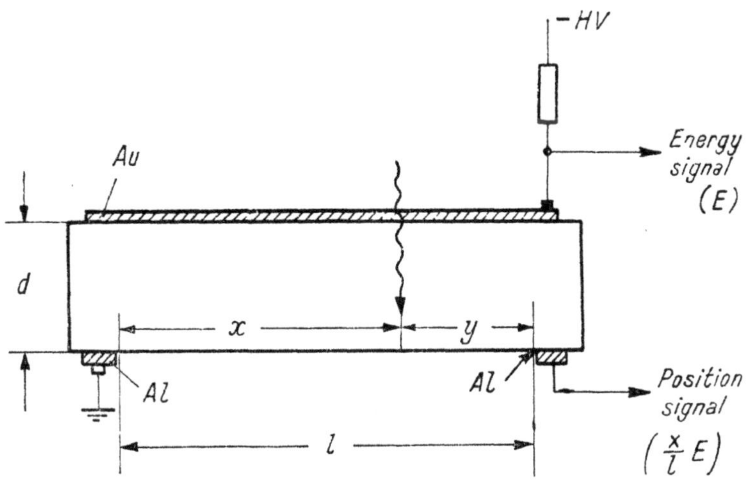

Fig. 5.38 — Cross-section view of a position sensitive *SD*.

In order to understand the working principles of this detector type, the Si parallelepipedic bar from which it is made (Fig. 5.38) is represented as a uniform resistance connected to small impedances at both ends.

The signal coming from a particle of energy E, penetrating the SD in a plane at a distance x from the zero potential electrode, will have an amplitude proportional to $\frac{x}{x+y} E = \frac{x}{l} E$, so that the position pulse amplitude is the greater, the closer to the position electrode are the particle ionizations. If this signal is divided by the coincident signal proportional with E, a position pulse is obtained, whatever be the energy.

These detectors require construction materials of high homogeneity. Such detectors with 35×4 mm² useful surfaces and depletion layer thicknesses up to 600 μ or even larger could be constructed. Thicknesses up to $2-3$ mm were obtained by using the compensation technique with drifted Li [L 16]. The energy spectra of 5.5 MeV alpha particles given by these detectors had a (2.5+3.3) % *FWHM* at room temperature and (1.2–1.9) % when a liquid nitrogen cooling

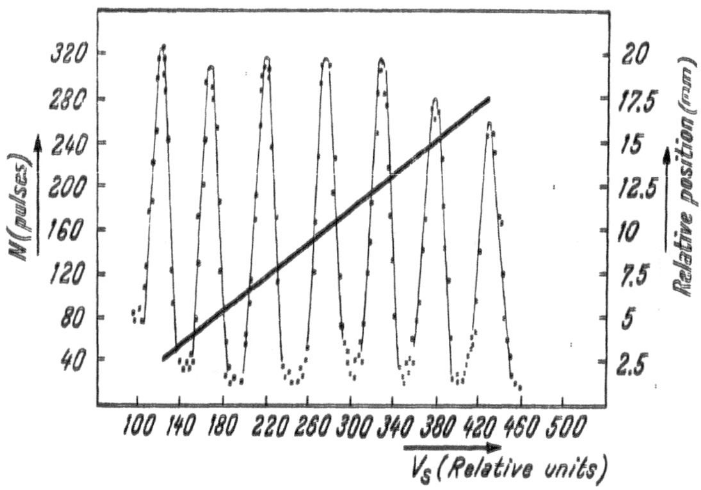

Fig. 5.39 — Alpha-particle position spectrum recorded with a positionsensitive detector [L 16] covered with a mask.

was used. It was estimated that a particle position could be determined with errors below 0.3 mm. Fig. 5.39 is a plot of the position spectrum of ^{241}Am alpha particles incident to the detector surface covered with a mask which has 7 slits of 1 mm with 2.5 mm spacings between them.

This complex device can be improved and there exists the tendency [N 2] to attach to it another contact in order to obtain a two-dimensional measurement of the position. Kalbitzer *et al.* [K 1] realized such an "$X - Y$ position sensitive detector". If a thin layer for dE/dx measurements is added on the frontal face, a single detector could then indicate both the energy and the position of the particle as well as its nature. Of course, these indications can be easily obtained with a very complex multiparametric (4 parameters) equipment.

The development of the "checker board counter" which is a dE/dx detector subdivided in a large number of sensitive fields was reported by Hofker *et al.* [H 14].

5.8.3. *Internal multiplication detector.* For low energy particle or quanta detection or for time applications it is suitable to use internal multiplication devices which permit to obtain large amplitude pulses. These internal multiplication semiconductor detectors are the solid state analogues of proportional and Geiger-Müller detectors exactly as the *SD* without internal multiplication are the analogues of the ionization chambers.

Similarly, it follows that it is possible to get a fast internal multiplication inside the *SD* by applying such a high reverse biased voltage that the electric field created in the depletion region should be strong enough to enable an avalanche multiplication phenomenon. The effect was observed [C 9], [M 29] for alpha particles inside a junction with thin depletion layer. An internal multiplication of about 18 was obtained, but the strong electric field raised great difficulties and the conditions of preserving and reproducing the multiplication were hardly achieved.

Better conditions for achieving a strong field were recently [H 18], [H 21] obtained by keeping the reverse current and the electric noise of the *SD* below reasonable limits (0.1 μ A and 20 keV, respectively). With Si as bulk material, the junction was made by Ga diffusion at 75 μ depth. In order to reduce the surface electric field at the points where the avalanche breakdown could happen, the device shape was that of a truncated cone with a 10° angle between the base and the generatrix. The total thickness of the device was 230 μ, such that for strong fields enabling the avalanche multiplication, the depletion layer was 150 μ thick, giving rise to a 3—5 pF capacity. The useful diameter was limited by a collimator at 2.3 mm.

Internal multiplications of pulses given by monoenergetic electrons (50 keV — 1 MeV) were obtained with this junction for maximum internal fields higher than 150 V/cm which generated the avalanche multiplication, without the junction breakdown.

The electron source was ^{106}Ru placed inside a magnetic spectrometer. The maximum multiplication was obtained for electrons of lower energy absorbed inside the surface region with strong field; for 50 keV electrons, the multiplication factor was about 50. The multiplication factor variation with voltage applied on the detector for electrons of 150, 200, 500 keV and 1 MeV is shown in Fig. 5.40. At 1890 V voltage, the curve *1* (E_c = 150 keV) tends to 2.3 MeV, *2* to 1.9 MeV, *3* to 1.1 MeV and *4* to 850 keV. In the analysis of the curves given in Fig. 5.40 it is necessary to take into account the detector "window" and the fact that the detector thickness permits the total stopping only of electrons with energies lower than 200 keV. The pulse rise times were below 20 ns (the minimum rise-time that could be observed on the synchroscope used).

The signal to noise ratio for detectors with internal multiplication was calculated by Haitz and Smits [H 2].

In the case of *SD* with barrier layer on Si used in fission fragment detection, an anomalous multiplication was noticed, i.e., it happened

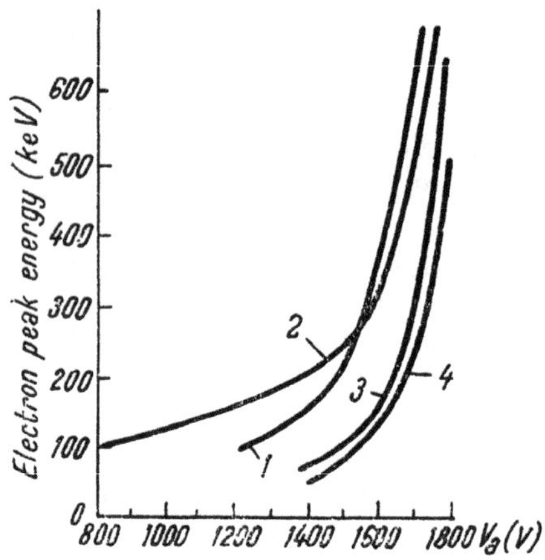

Fig. 5.40 — Variation of the electron peak position with the *SD* bias, at different energies [H 18]. 1) E_e = 150 keV; 2) E_e=200 keV; 3) E_e=500 keV (second peak); 4) E_e = 1 MeV (the peak of minimum ionization).

at electric field intensities much below that required by an avalanche multiplication. They tried to explain this phenomenon admiting that a tunnelling injection happens through the surface oxide layer [W 7].

The internal multiplication can be also obtained with transistor-type devices [W 17], [W 19]. Typical values of amplification coefficients of about 200—300 for 30—40 V collector voltages were obtained with a *N.P.N.* transistor made of a bulk material (*P*-type Si) of $\rho = 20{,}000\ \Omega$ cm having a base thickness of 250 μ.

Unfortunately, the problems raised up to now by the construction of *SD* with internal multiplication have not been yet solved up to their serial production stage, though in laboratories a series of successes as those mentioned above were scored.

5.9. *NIP* DETECTORS

The energy measurements for penetrating radiations require a large enough detection sensitive volume. In the case of $P-N$ junctions a thick depletion layer is obtained with bulk materials of high resistivity and with high reverse voltage. The preparation of high resistivity silicon is a difficult technical problem and the reverse biased voltage which can be admitted before excessive leakage currents are born is usually about hundreds of volts for carefully built devices and depends significantly on the surface conditions.

A method of getting thick sensitive regions, easily brought into being since it does not require a bulk material with high resistivity, is that consisting in the compensation of excess acceptors inside a region of the *P*-type bulk material ($\rho = 10-100\ \Omega$ cm) by means of lithium interstitial impurities. Using this method developed by Pell [P4], the resistivity of the compensated region can be increased up to $\simeq 10^2\ \Omega$ cm and its thickness exceeds 10 mm.

The name of *NIP* is due to the presence of two junctions inside the device : a N^+I junction at the frontal surface and an *IP* junction towards the ohmic contact. In terms of this quotation, *I* has the meaning of intrinsic (the resistivity of the compensated material is close to the intrinsic value).

At a given value of an increasing reverse voltage applied on *SD* the depletion layer extends all over the "*I*" region and a subsequent voltage increase does not practically produce a variation of the depletion layer thickness (hence of the *SD* capacity), but only an intensification of the electric field collecting charges yielded by ionization.

In the technology of junction devices meant for power rectifiers, the advantages of *PIN* diodes as concerns the breakdown voltage increase were well known, but their manufacturing required expensive material of high resistivity which could be hardly obtained. However, such devices were also used as *SD* [K 12].

Pell's technique [P 4] proceeds from a *P*-type bulk material (Si or Ge) with comparatively low resistivity, which is compensated by means of Li^+ donor ions, yielding a 0.03 MeV activation energy in Si. Li is used because in the case of other impurity types, the diffusion hides the ion drift effect [E 5]. Since Li is an interstitial impurity, its "diffusion" constant is 10^7 times greater [M 21] than that of substitution impurities (e.g. B or P).

At the beginning, the acceptor is uniformly spread all over the bulk material volume (*P*-type Si) at a level of N_A acceptors/cm³ (Fig. 5.41 *a*). The Li diffused at the surface (usually at temperatures of 350–400°C) will be distributed in depth according to the low [P 4]

$$N_D = N_0 \operatorname{erfc}\left(\frac{x}{2\sqrt{D_0 t_0}}\right), \tag{5.10}$$

where

D_0 — diffusion constant (at the temperature of the first diffusion), and

t_0 — diffusion duration.

In this way a N^+P ($N_0 \gg N_A$) diffused junction, linearly graded is obtained, the plane of the stoichiometric junction being placed at a distance C from the surface (Fig. 5.41 *a*).

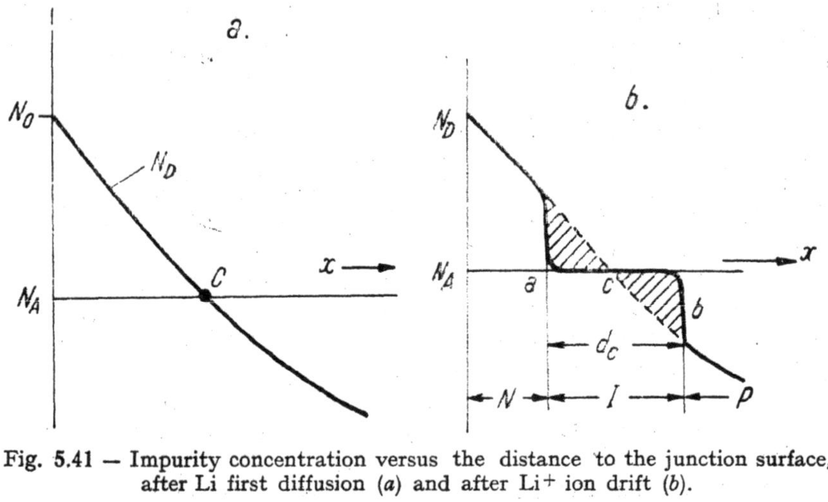

Fig. 5.41 — Impurity concentration versus the distance to the junction surface, after Li first diffusion (*a*) and after Li^+ ion drift (*b*).

If a junction formed in this way is supplied (in a second stage) with reverse voltage high enough that the Li^+ ion drift velocity exceeds that of the diffusion, a displacement of Li^+ ions takes place, from the Li^+ rich region towards the Li^+ poor region of the plane $x = c$ where an electric field exists.

This ion displacement has a resonable velocity when the temperature is high enough to yield a considerable mobility of Li ions. Since Li has a great chemical affinity for oxygen, it is very important that the bulk material used should have a low oxygen content ($<10^{15}$ atoms/cm^3 [M 21]) and thus the ion mobility keeps low at high temperature too. Because of the ion displacement around the plane $x = c$, the ion concentration becomes uniform at a level $N_D = N_A$. Indeed, at $x < c$ N_D cannot be less than N_A, because the excess acceptors would change the field configuration in such a way that the equalization $N_D = N_A$ would arise. For the same reason, a supercompensation ($N_D > N_A$) cannot take place in the region $x > c$.

Therefore, around the plane $x = c$ (between $x = a$ and $x = b$) an intrinsic region (Fig. 5.41 — b) appears, the thickness of which increases during the drift. It was demonstrated [P 4] that at $V_a =$ const and $T =$ const, the thickness d varies in terms of $\sqrt{t}$, where t is the drift duration

$$d = \sqrt{2\,\mu V_a t}. \tag{5.11}$$

On the basis of this relation a nomograph [B 28] was drawn, allowing an estimation of the time required for producing a given thickness at a certain reverse voltage (V_a) and temperature (T). For example, $d = 2$ mm in Si is obtained after 2 days drift at $T = 150°$ C and $V_a = 100$ V.

As far back as in 1961, *SD* [E 5] with thickness $d = 5$ mm but comparatively thick ($\sim 100\ \mu$) surface dead layers were constructed with this method. If phosphorus is diffused before Li, up to a depth of 1—2 μ we can obtain window thicknesses below 7 μ [B 28], since the quantity of diffused Li can be limited exactly at the value required by compensation.

A window thickness below 1 μ [W 6] could be also achieved with a method consisting in an extension of the compensated region up to 2 μ distance from the thin metalic contact ($< 0.5\ \mu$) on the *P*-type surface of the device.

PIN silicon semiconductor detectors of special form (with dimensions $1 \times 1 \times 60$ mm) were made [D 20] in order to be used in the focal plane of a magnetic spectrometer.

The Si compensated by the Li drift method can be used in surface-barrier *SD* construction [D 10]. A *P*-type frontal surface offers also the advantage of improved fast responses due to the difference between electron and hole mobility. A window below 0.05 μ [D 10] can be obtained in this case [D 4].

It was demonstrated [B 28] that when the reverse current of the device (I) is determined by the current generated in the depletion layer, the rate of realizing a given compensated thickness (d), is directly proportional to the power dissipated inside the junction ($P = V_a I$). If the drift takes place at constant temperature, the thickness of the compensated region obtained in a given time interval, is proportional to $\sqrt[3]{\mu P}$ [M 44]. When a fast compensation is desired, it is better to work with dissipated powers as high as possible in the junction. Automatic control systems were constructed which permit the use of dissipated powers higher than 50 W. The drift temperature (for Si) and the power dissipated in the junction are usually kept at constant values: the former at 100–200°C by placing the junction in an oil bath and the latter by means of an automatic control system [M 44].

Since the junction breakdown takes place at current densities higher than 50 mA/cm², it is necessary to stabilize not only the dissipated power but also the current [D 3] during the drift.

If the junction is placed in an oil bath, the temperature in its central region is higher than at the ends and a curved compensation profile is obtained [D 7]. A uniformly compensated layer can be achieved by keeping only two opposite faces of the crystal at high temperature.

A recently built, automatic equipment [G 23] does not require watching during the drift. When the compensated region reaches the distance 0.75 mm from the ohmic contact, an instrument shows the thickness of the non-compensated region and when this one decreases below a predetermined value, the temperature is automatically lowered at the room temperature and the drift stops.

If Li reaches the surface, it gathers there giving rise to a *N.I.N.* structure. When a voltage is applied on the device, one of the $N - I$ junctions would have a direct bias and the other a reverse one and the current injection produced by the direct biased junction would much increase the *SD* noise [D 7]. The sudden increase of the reverse current inside the device could be used as an indication of total compensation.

A long-term stability of the device characteristics is an imporsant problem of *NIP*-type *SD* built with the Li compensation technique which has not been yet utterly solved. Because of its high mobility, the Li^+ can also migrate at room temperature if the detector is kept for a long time without external voltage applied. For this reason, such *SD* are to be kept with reverse voltage applied on them [E 5].

A considerable improvement of the long-term stability of characteristics can be obtained if the bulk material used (*P*-type Si) is doped with Ga and not with B [N 3].

The instability of *PIN* detector characteristics is due not only to some "internal" causes, but to surface effects too, as for all *SD*. Recent studies [L 12] emphasized the existence of some *N*-type conduction layers at the surface, which extended from the Li rich region to the *P*-type layer on the opposite face. Therefore, strong electric field regions exist at surface in the neighbourhood of the $I - P$ frontier. Measurements showed that the largest part of the surface current was generated in this region with strong electric field. A model was suggested for the surface current generation [L 12]. The stabilization of the characteristics at surface was rather successfully achieved using the passivation technique with Si oxide.

Ambient storage effects and mounting problems of very large volume Ge (Li) detectors were studied in detail by G. Armantrout [A 13].

When the surface leakage currents are made minimum, the device static characteristics $I_D - V_A$ can be used for obtaining some information about the noise [M 48]. In this case a quantitative correlation can be established between these characteristics and the noise value.

The compensation with Li is not the only method to realize a large volume of material with resistivity close to the intrinsic value. A 200,000 Ω cm resistivity can be obtained in the case of a *P*-type Si with $\rho \sim 1000$ Ω cm, after an irradiation with slow neutrons [M 33], [V 5] and a "recovering" thermal treatment which ameliorates damages introduced by irradiation. Now, the compensation is due to the (n, γ) nuclear reaction on the ^{30}Si isotope.

$$^{30}\text{Si}\,(n, \gamma)\,^{31}\text{Si} \xrightarrow[2{,}6\text{ hours}]{\beta^-} {}^{31}\text{P (stable)}. \tag{5.12}$$

The reaction cross-section is 0.11 ± 0.01 barn and the isotopic contents of the natural element in ^{30}Si is 3.05%. A two mm thick *SD* which gave at 500 V applied voltage an equivalent noise $FWHM < 30$ keV could be made with this method [M 33].

The method cannot be applied in the case of Ge [V 5] since then both acceptor (^{71}Ga) and donor (^{75}As) are obtained, which compensate each other

$$^{70}\mathrm{Ge}(n, \gamma)\ ^{71}\mathrm{Ge} \xrightarrow[11\ \text{days}]{\beta^+} {}^{71}\mathrm{Ga} \tag{5.13}$$

$$^{74}\mathrm{Ge}\ (n, \gamma)\ ^{75}\mathrm{Ge} \xrightarrow[28\ \text{min}]{\beta^-} {}^{75}\mathrm{As}$$

For the *N*-type Ge, the compensation was achieved for thicknesses up to 2.7 mm [R 7] by introducing low levels in the energy gap in a bombardment with electrons resulting from irradiation at a ^{60}Co gamma-source of high intensity (600 Curie). A 0.5 mm thick *SD* made with this technique had a $FWHM = 9$ keV resolution for the ^{181}Hf line of 482 keV. The detector was working at the liquid nitrogen temperature and with 500 V applied voltage.

Thicknesses of the depletion layer of about 2—6 mm can be rather easily obtained in *NIP* detectors with Si (and with special precautions, even up to 10 mm). Such Si and Ge semiconductor detectors were made at the Institute for Atomic Physics.

In order to reduce the surface currents, guard-ring detectors were made from Ge (Li) [T 6].

Such *SD* can be used for detecting weakly ionizing particles (protons with energies higher than 10 MeV alpha particles higher than 40 MeV, electrons beyond 1 MeV a.o.) with resolutions $FWHM \approx 50$—60 keV [D 7].

Such detectors were used in the spectrometry of 40 MeV [P 3] and 60 MeV protons [B 16].

NIP-type *SD* gave also good results in the relativist particle detection (e.g. π-mesons with momentum 0.6—1.5 GeV/c) [K 13], [R 5].

We mention that "telescope" systems [T 10], i.e. a pile of *NIP* or even surface-barrier detectors, can be used in the spectrometry of penetrating particles. E.g., a barrier-layer detector of *N*-type Si with $\rho = 4000\ \Omega$ cm was built for such applications [D 12].

At 430 V, the depletion layer extended over the whole volume (thickness = 0.6 mm). The detector surface was 1.7 cm^2.

PIN detectors can be also used in the gamma-ray spectrometry.

Because of the low capacity, constant under bias voltage variation, the technique of negative feedback voltage amplification in preamplifier can be used in the case of *NIP* semiconductor detectors; it does not require a charge sensitive preamplifier [Z 2].

In thick *SD*, the transit-time of the charge carriers, determining the pulse front duration is quite long because

$$t_t = \frac{d^2}{\mu V_a}, \tag{5.14}$$

which leads to difficulties for time measurements. Likewise, these *SD* are not fitted for gamma-ray detection because the photoelectric peak is very small in comparison with the Compton-electron background due to the small *Z*-number of the Si (14). From this point of view Ge with its $Z = 32$ is better. Other materials with great *Z*-numbers, besides Si and Ge (e.g., Cd, Te) are considered by J. W. Mayer in [M 25].

Because the ratio of photoeffect cross-section to Compton effect cross-section is proportional to Z^4, an improvement of the factor $\left(\frac{32}{14}\right)^4 = 27$ is obtained in Ge in comparison with Si. That is in the case of a single absorption. Actually, due to the multiple processes (Compton diffusion followed by photoelectric effect), the advantage provided by Ge is more important. In order to decrease the influence of Compton-electron background, the detector volume ought to be larger so that by multiple photoelectric processes the whole energy of gamma-rays should be absorbed. Useful volumes up to 54 cm³ could be achieved.

In this respect the co-axial detectors [M 3], [T 7], [M 5], [M 4], [E 13] have large possibilities. Armantrout [A 14] resorts to a special technology for decreasing the capacity of plane *SD*.

It is to be noticed that at energies higher than 1.2 MeV, the effect of pair creation is also favoured in materials with large *Z*-numbers, its cross-section being proportional to Z^2.

In nuclear spectroscopy it is recommendable to use the photoelectric effects, since the energy of electrons yielded in this process is uniquely determined by the energy of incident radiations (in the case of Compton diffusion, a continuous spectrum of electron energies is obtained for a monoenergetic gamma-ray, and as concerns the pair creation, the positron annihilation leads to the appearance of new lines in the detector spectrum).

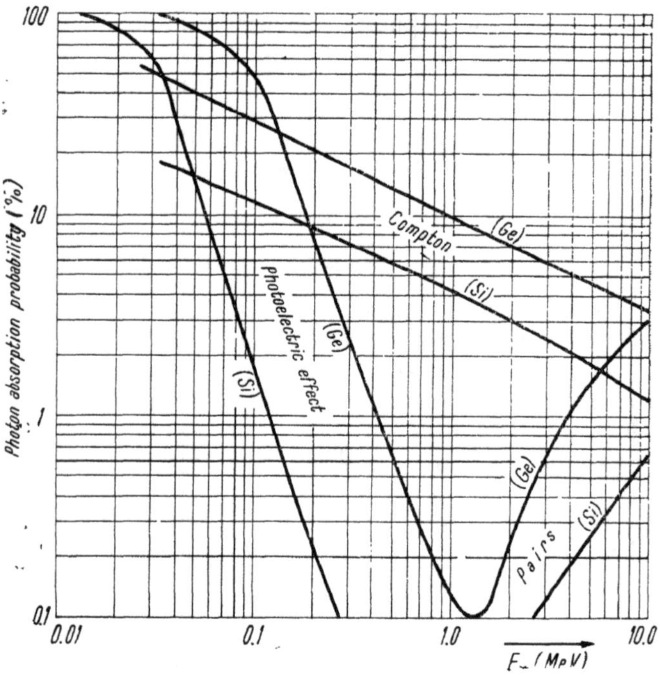

Fig. 5.42 — Variation of gamma-quanta absorption probability with the energy for Si and Ge semiconductor detectors of 3 mm thickness [G 21].

The variation of gamma-quanta absorption probability with the energy for Si and Ge semiconductor detectors with a thickness of 3 mm is shown by the curve in Fig. 5.42 [G 21], from which there results the advantage of using Ge, though Ge devices cannot work without being cooled at the temperature of liquid nitrogen. (Since Ge has an energy gap of 0.75 eV in comparison with 1.1 eV of the Si, the thermally generated current at room temperature is much higher for the former than for the latter).

After the Li drift too, the *SD* has to be kept at low temperature. Special Dewars are built for this purpose [K 16]. The technical difficulties of making *NIP*-type *SD* with Ge by the compensation method with Li are very important due to the strong condition of low oxygen admixture. However, very good results were obtained [H 6], [T 2], [T 3] and they are given in Table 5.3 [T 3].

At gamma-ray energies below 1 MeV the photopeak is used for spectroscopy. Two spectra of a ^{60}Co source given by a NaI (Tl) scintilator and *PIN*-type *SD* with Ge [T 3] are compared in Fig. 5.43. The resolution improvement by an order of magnitude is evident. A comparison with other devices used in the gamma-ray spectroscopy

Table 5.3

Efficiency and resolution (at $T=78°K$) of a *PIN*-type Ge detector with $\varnothing=19$ mm and $d=3,5$ mm [T3]

E_γ (keV)	Peak intrinsic efficiency % (± 20 %)	Measured resolution *FWHM* (±10 %) (keV)
122	55	3.3
661	0.7	4.3
1333	0.16	5.0
3918	~0.35*	7.8*
7639	~0.2*	9.8*

* Double-escape peak.

is given in Fig. 5.44 [H. 15]. At higher resolutions, the Ge (Li) semiconductor detector competes with many magnetic spectrometers as concerns the resolution. In applications, however, the detection efficiency is also of interest. In Fig. 5.45 the efficiencies [H 15] of different types of gamma spectrometers are compared. The experimental efficiency curve compared with calculations relying on the photoelectric process cross-section is drawn in Fig. 5.46 [E 12]. The data refer to a Ge *PIN* detector which has $d = 3.5$ mm thickness and 0.88 cm^3 volume. When the volume increases, the detection efficiency increases, more significantly towards higher energies. In [H 17], [K 15], [W 3] methods of numerical computing of the Ge (Li) *SD* efficiency for gamma-rays are reported. Efficiency measurements at energies within 500 and 1500 keV were made by Freeman and Jenkin [F 14].

Computer programs developed for the analysis of gamma-ray complex spectra obtained with Ge (Li) detectors are described by R. L. Heath [H 9].

Pair creation is the useful effect in gamma-ray detection at energies higher than 1.5 MeV. When a photon of energy $E_\gamma > 2\ m_0c^2$ ($m_0c^2 = 511$ keV) creates an $e^+ - e^-$ pair in Ge, the pair kinetic energy is all absorbed, but due to the small detector volume, the annihilation photons "escape" almost always.

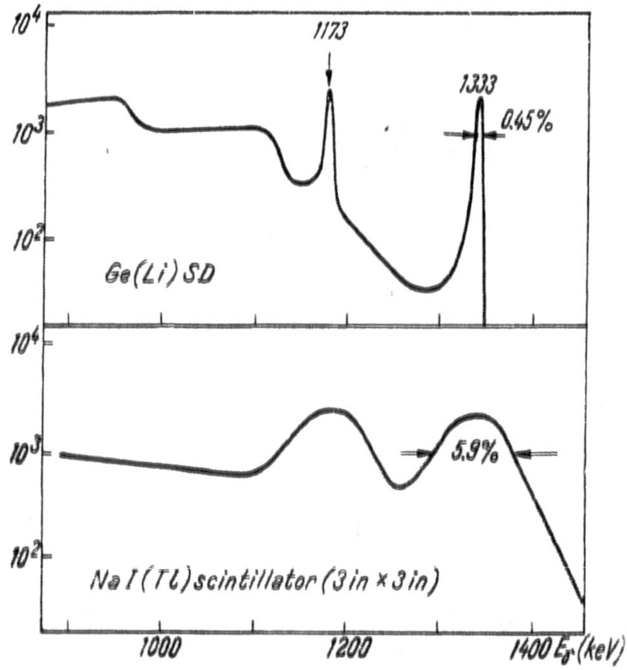

Fig. 5.43 — A comparison between ^{60}Co gamma spectrum recorded with *SD* (∅18 and $d=8$ mm) and NaI scintillator [T 3].

Therefore, most pair creation events give a "double escape peak" at the energy $E_\gamma - 2\ m_0c^2$. The probability that only one gamma-ray of 511 keV escapes is very small; in this case a peak of "single escapes" appears at $E_\gamma - m_0c^2$.

From Fig. 5.47 [T 3] we observe, in the case of the 2.7 MeV gamma-radiation yielded by a ^{24}Na source, a photopeak at 2754 keV, a peak of single escapes at 2243 keV and a peak of "double escapes" at 1732 keV. We see that the intensity of the double escape peak is by an order of magnitude above the Compton background and higher than the photoelectric peak. Lead absorbers can be used in order to separate the double escape peak from the photopeaks of low energy.

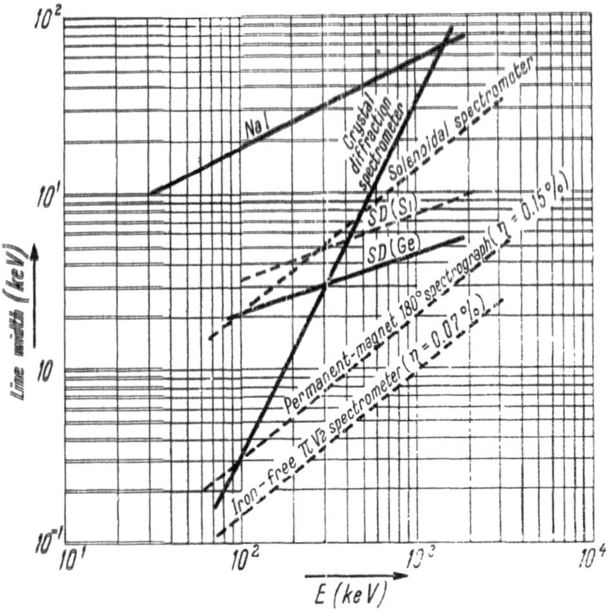

Fig. 5.44 — Comparison of energy resolutions of some nuclear spectrometers [H 15].

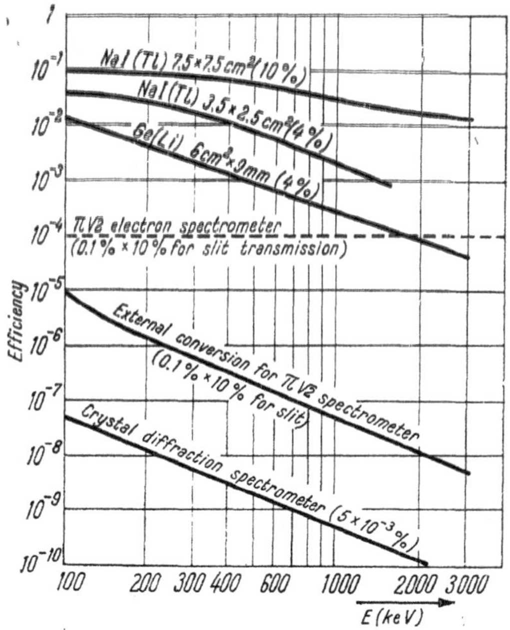

Fig. 5.45 — Comparison of practical efficiencies for some nuclear spectrometers [H 15]. The solid angle considered is given in brackets.

It was experimentally observed that there is no increase of doulbe-escape peak efficiency in Ge as it was to be expected from the theoretical cross-section. The efficiency can even decrease at higher energies (beyond 4 MeV). This fact can be explained by electrons and positrons being created with high energies so that they are able to go out from the detector volume, before losing the whole energy.

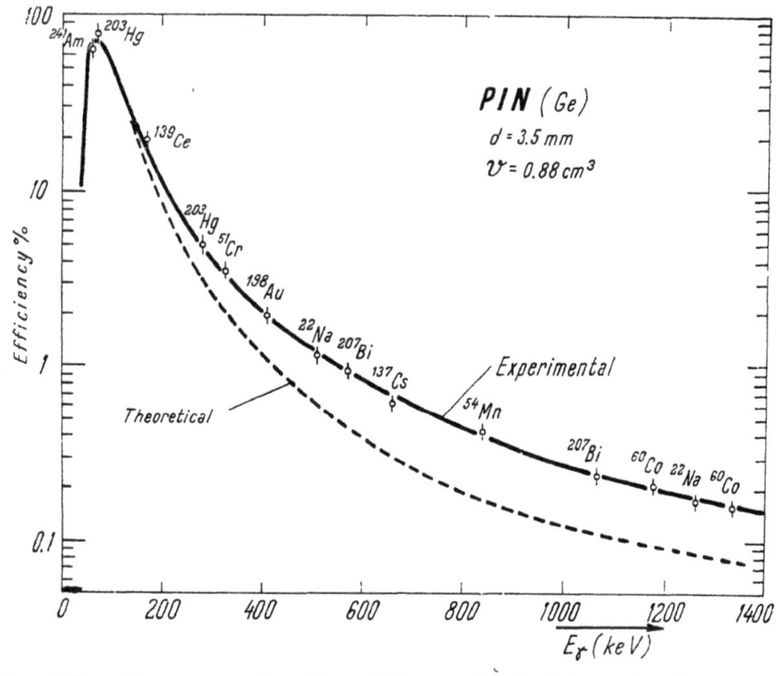

Fig. 5.46 — Gamma ray detection efficiency of a Ge(Li) semiconductor detector with $d = 3.5$ mm and volume 0.88 cm^3 [E 12].

The bremsstrahlung accompanying the electron slowing down can "escape" from the crystal. Both effects give pulses of lower amplitudes which enter the continuous background under the double escape peak. In larger volumes the effect has to be weaker.

The Compton diffusion background on wich the double escape peak is superposed can be much lowered by using a spectrometer with three crystals [T 3] : two large scintillators (5×3 cm^2) placed on both sides of the *SD* detect the annihilation quanta, "escaped" from the *SD*, in coincidence with the event of pair creation which occurs in the central *PIN* detector with Ge of high resolution.

As seen from Fig. 5.47 *b*, the ratio of double escape peak intensity of 2754 keV gamma-radiation (^{22}Na source) to the back ground

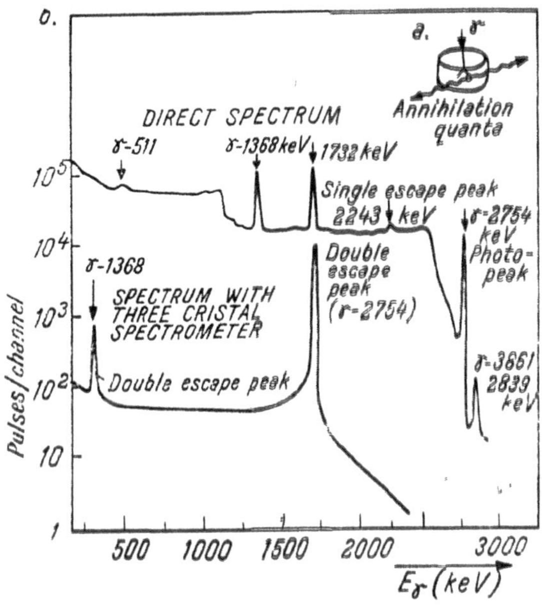

Fig. 5.47 — ^{24}Na spectrum obtained with a three-crystal spectrometer [T 3].

under it becomes 150/1 in the case of the spectrum controlled by the coincidence signal in comparison with 7/1 as it was before ; in addition, the spectrum was simpler.

For gamma-rays of medium energies (400—1500 keV) a Compton spectrometer [H 13] was built with silicon *PIN*-type *SD*, which gave a $FWHM = 18-20$ keV resolution and a 0.1% intrinsic efficiency. Another method was also used in the gamma-ray detection i.e., a thick Au converter of 4 mg/cm^2 was placed before a *NIP*-type

SD and the photoelectrons extracted from it were detected. The system converter — *PIN*-type *SD* with Si had an energy resolution $FWHM = 20$ keV [S 7].

Obviously, the restricted volume of this book could include only part of *SD* applications. In spite of the technical difficulties, not yet completely solved, the results obtained demonstrate the outstanding possibilities of semiconductor detectors as compared with other detector types (mainly, for the low energy range) and justify the assertion that their use on a large scale in the experimental nuclear physics will provide not only possibilities for more accurate measurements, but the obtaining of new qualitative results.

CHAPTER VI

AMPLIFICATION OF SEMICONDUCTOR DETECTOR ELECTRIC PULSES

The detector is in fact a radiation-electric signal transducer. In order to obtain information on the detected particle beam it is necessary to process the signals supplied by the detector. This function is fulfilled by the *SD* associated electronic equipment. The sensitivity of the measurement methods used in experimental nuclear physics and the accuracy of results depend on the parameters of these instruments. The requirements imposed on such an equipment are determined by the detector type and the specific character of the experiment in which it is used.

The excellent intrinsic performances of *SD* as concerns the energy and time resolution, make particularly attractive their use for the ionizing particle spectroscopy. The *SD* have begun to replace the old detector types (ionization chambers, scintillators, a.o.) and their successful application is due to the small energy dispersion given by the statistics of the charge pulse formation inside the solid state. For this reason the mean energy required for a charge-carrier pair creation is by an order of magnitude lower in Si and Ge than in gases and by two orders of magnitude lower than the mean energy spent for extracting a photoelectron from the photocathode of a scintillator associated photomultiplier. For example [D 25] a particle with 100 keV energy will yield about 100—330 photoelectrons in a scintillation detector, 3 000 electron-ion pairs in the ionization chamber and 30 000 electron-hole pairs in a *SD*. In the detection of monoenergetic particles of 100 keV there results hence standard deviations of (8.5—10) % for the scintillator, 1.8 % for the ionization chamber and 0.6 % for the *SD*.

The associated electronic equipment should have good performances such that to turn into account the *SD* possibilities. It is also necessary to choose correctly the detector and the whole spectrometric chain for the optimization of the system, in view of obtaining the maximum signal to noise ratio.

In the following, we shall refer particularly to problems concerning spectrometry equipment with *SD*, since these detectors are now used on a large scale in the amplitude analysis. The simple applications as monitors will not be mentioned since they do not impose specific conditions on the *SD* electronics. The apparatus used in time applications of the *SD* will not be presented because of difficulties still existing in this domain, although lately several successes have been recorded [K 5], [G 2], [W 1], [G 30], [E 14], [S 31], [B 10], [B 11], [B 12], [P 7], [M 38], [K 7], [Z 1].

Accordingly, our examination will concern the electronic devices which fulfil the main conditions of linearity, stability and energy resolution, required by the spectrometry applications of semiconductor detectors.

6.1. ELECTRIC CHARGE TO VOLTAGE PULSE CONVERSION

In spectrometry applications, the useful information given by *SD* is contained in the quantity of electric charge, the ionizing particle yields inside the detector volume. Indeed, it was checked up experimentally that the mean energy required for an electron-hole pair generation inside a solid is a constant for a certain material

$$\epsilon = \begin{cases} 3.6 \text{ eV in Si}, \\ 2.9 \text{ eV in Ge}, \end{cases} \tag{6.1}$$

ϵ being generally independent of the particle energy and type. It was shown in Section 3.2 that during an electron-hole pair collection, the charge that has to be taken into consideration is e and not $2e$ (e = electron charge). Since the number of pairs created by a particle losing all its energy E inside the sensitive volume of the detector is

$$N = \frac{E}{\epsilon}, \tag{6.2}$$

it follows that the charge of interest for us is

$$Q = \frac{E}{\epsilon} e, \tag{6.3}$$

i.e., there is a proportionality between charge and energy. By knowing the constants e and ϵ, and estimating the charge, we can determine the energy. In the case of 100% charge-collection efficiency in the detector, the charge Q is found again on the total capacity C_t, which is a sum of the SD capacity (C_d) and the preamplifier input capacity (C_{in}) which is considered to include also the capacity of the connections between detector and preamplifier

$$C_t = C_d + C_{in}. \tag{6.4}$$

The fast charge collection yields on the capacity C_t a voltage pulse with the amplitude

$$V_s = \frac{Q}{C_t}. \tag{6.5}$$

In the case of a $P - N$ junction, the current pulse shape is given by the relation (3.68)

$$i(t) = I_M \exp.\left(-\frac{t}{\tau_i}\right), \tag{6.6}$$

where $\tau_i = \varepsilon\rho = \varepsilon_0 \varepsilon_r \rho$, or taking into account that

$$\tau_i = \frac{d}{\mu \mathcal{E}_M} \text{ and } \mathcal{E}_M = \frac{2V}{d}, \tag{6.7}$$

then

$$\tau_i = \frac{d^2}{2\mu V}. \tag{6.8}$$

The electric charge Q is the current integral

$$Q = \int_0^\infty i(t)\,\mathrm{d}t = \int_0^\infty I_M \exp\left(-\frac{t}{\tau_i}\right)\mathrm{d}t = I_M \tau_i, \tag{6.9}$$

such, that the current pulse amplitude is

$$I_M = \frac{Q}{\tau_i} = \frac{e}{\varepsilon \epsilon \rho} E. \tag{6.10}$$

If, for instance, the particle energy is $E=1$ MeV, $N=2.85\times10^5$ pairs are yielded in Si, i.e., a charge $Q = 4.55\times10^{-14}$ C, and if $C_t = 100$ pF, the voltage pulse amplitude is 0.455 mV. For a resistivity $\rho = 1000\ \Omega$ cm there are a fall-time constant $\tau_i = 1.05$ ns and a current pulse amplitude $I_M = 43\ \mu$A.

The amplification of the voltage pulse in a usual voltage amplification circuit (Fig. 6.1) has the disadvantage that the signal amplitude depends on the voltage applied on the detector, since C_d depends on the bias voltage value. Signals given by particles with the same energy can have different amplitudes according to the detector bias voltage variation (incidental or wanted).

In fact, with the notations of Fig. 6.1 we have

$$C_{in} = C_s + C_g + C_a, \tag{6.11}$$

and if $C_d \sim C_{in}$ the capacity variation will affect the pulse amplitude. However, for *PIN* detectors the capacity variation above a certain threshold is insignificant; in addition, their capacity is usually rather small, so that the quantity $C_d + C_{in}$ can be considered to be a well-defined constant for a given detector. In this case, the amplifier with voltage negative-feedback (Fig. 6.1) may be used, for which the relations given below can be applied. In Fig. 6.1 C_d, R_d and R_s are the elements of the detector equivalent circuit, C_s is the stray capacity, R_D is the detector load resistance and R_G is the grid bias resistance.

The sensitivity (the ratio of output-voltage pulse amplitude V_o to input charge Q) is

$$A_q = \frac{V_o}{Q} = \frac{V_o}{V_{in}C_t} = \frac{\mathcal{A}}{C_d + C_{in}}, \tag{6.12}$$

with $\mathcal{A}$ the voltage amplification

$$\mathcal{A} = \frac{\mathcal{A}_0}{1 + \beta\mathcal{A}_0}, \tag{6.13}$$

where $\mathcal{A}_0 = \dfrac{V_{a1}}{V_{in}}$ and $\beta = \dfrac{R_1}{R_1 + R_f} =$ negative feedback coefficient; $1 + \beta\,\mathcal{A}_0 = F =$ feedback factor.

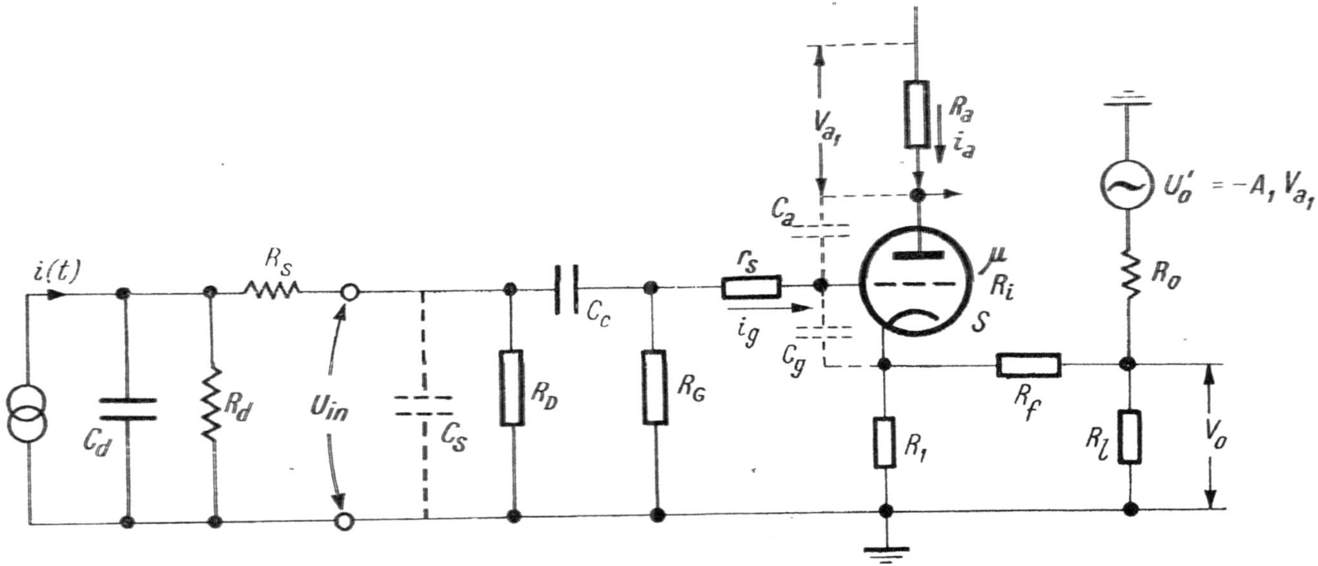

Fig. 6.1 — **The equivalent circuit diagram of the negative voltage feedback preamplifier** which has a *SD* at the input.

For $\beta \mathcal{A}_0 \gg 1$ and $R_f \gg R_1$ it results

$$A_q \simeq \frac{R_f}{R_1} \cdot \frac{1}{C_t}, \tag{6.14}$$

therefore

$$V_o = QA_q = \frac{Q}{C_t} \cdot \frac{R_f}{R_1}, \tag{6.15}$$

and using stable resistances, the spectrometer calibration for a given detector can be kept in time. However, in most cases $C_d \gg C_{in}$ and varies with the voltage, so that it is necessary to use a "charge sensitive preamplifier" (Fig. 6.2) which provides a great dynamical capacity

$$C_{in} = C_{dyn} \gg C_d, \tag{6.16}$$

with respect to which the variations of C_d are no longer of importance.

If $\mathcal{A}$ is the loop amplification and C_f is the feedback capacity, on writing that the charge Q is shared between C_t and C_f (Fig. 6.2 *b*) we obtain

$$Q = V_{in} C_t + (V_o - V_{in}) C_f, \tag{6.17}$$

and since $V_o = -\mathcal{A} V_{in}$, then

$$V_{in} = \frac{Q}{C_t + (1 + \mathcal{A}) C_f}, \tag{6.18}$$

and

$$A_q = \frac{V_o}{Q} = \frac{\mathcal{A}}{C_t + (1 + \mathcal{A}) C_f}. \tag{6.19}$$

In these relations

$$C_t = C_d + C_m + C_s + C_a + C_f. \tag{6.20}$$

The circuit given in Fig. 6.2 *c* is equivalent to that from Fig. 6.2 *b*.

Usually $\mathcal{A} \gg 1$ and is chosen together with C_f such that the inequality

$$C_{dyn} = \mathcal{A} C_f \gg C_t \tag{6.21}$$

be satisfied.

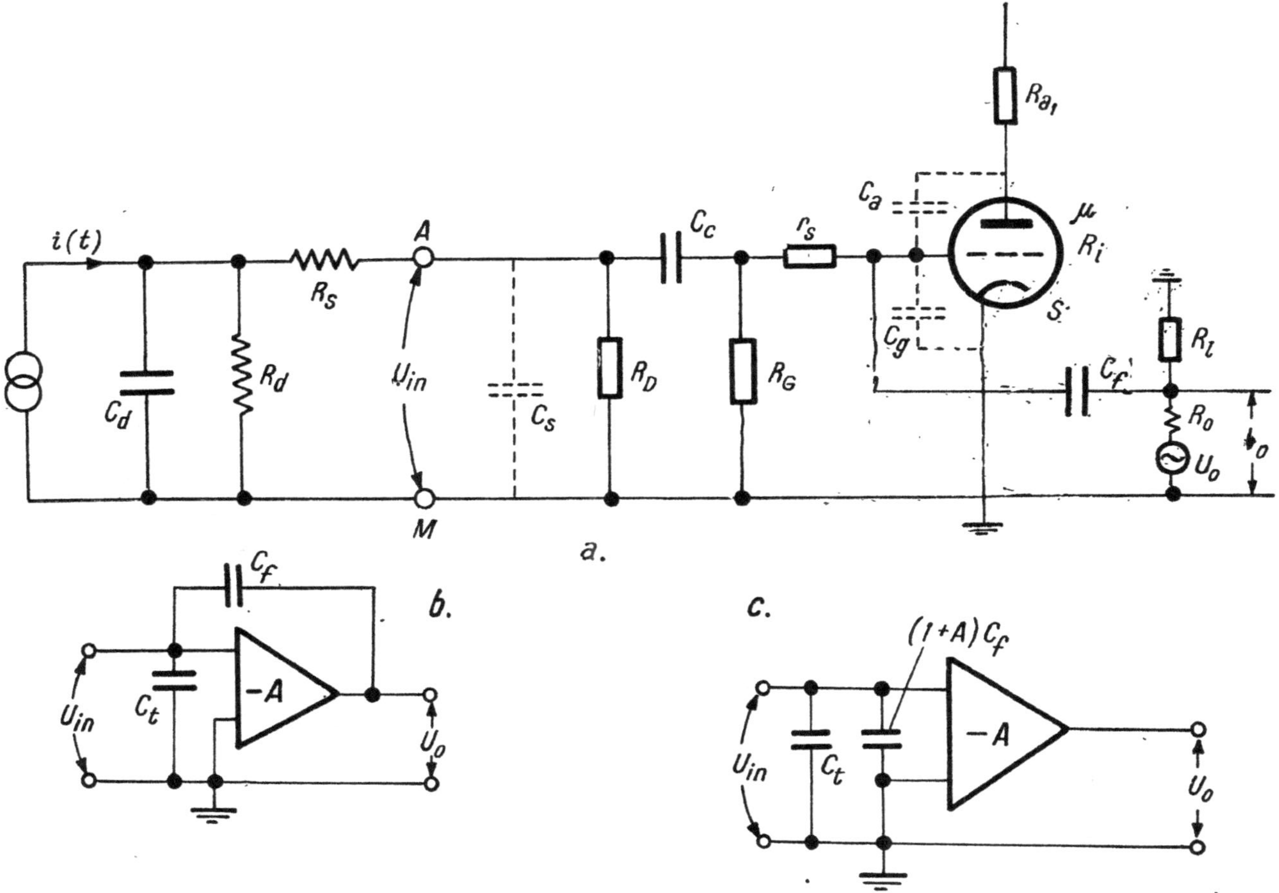

Fig. 6.2 — The equivalent circuit diagram of the charge sensitive preamplifier: *a* — with *SD* at input; *b* — simplified circuit diagram; *c* — the equivalent of the simplified circuit.

In fact C_{dyn} contains also the term $S\dfrac{R_i R_{a1}}{R_i + R_{a1}} C_a$ which may be neglected because the resistance R_{a1} is usually the input resistance of the circuit grounded grid configuration, of the order $\dfrac{1}{S}$.

In these circumstances the circuit sensitivity

$$A_q = \frac{V_0}{Q} = -\frac{1}{C_f}, \tag{6.22}$$

is independent of the detector capacity value C_d.

Without feedback (the feedback loop is interrupted) the sensitivity would be

$$A_q' = -\frac{V_0}{Q} = -\frac{V_0}{V_{in} C_t} = -\frac{\mathcal{A}}{C_t}. \tag{6.23}$$

Hence, the feedback factor is

$$F_q = \frac{A_q'}{A_q} = \frac{C_t + (1 + \mathcal{A})\, C_f}{C_t} \simeq \frac{\mathcal{A} C_f}{C_t}. \tag{6.24}$$

Without amplifier, the detector output voltage is $\dfrac{Q}{C_d}$ and with amplifier it is $-\dfrac{Q}{C_f}$; hence the voltage pulse amplitude increases proportionally to $\dfrac{C_d}{C_f}$.

From (6.22) and (6.21) it may be seen that for obtaining a better sensitivity it is necessary to take a small value of C_f and in such a case the greater is the total capacity C_t the higher is the amplification $\mathcal{A}$ required in order that relation (6.21) be satisfied. In a typical case, $\mathcal{A} = 1000$ and $C_f = 5$ pF. For a detector with $C_d = 100$ pF, the condition $\mathcal{A} C_f \gg C_d$ is satisfied and the sensitivity of the charge sensitive section of the preamplifier is $\dfrac{1}{5\,\text{pF}} = 0.2 \times 10^{22}$ V/C.

For a detector made of a given material (e.g., Si), the amplifier sensitivity may be given in practical units mV/MeV, considering relation (6.3). The transformation of one unit into another is made according to the formula

$$\frac{V_o}{E}\left(\frac{\text{mV}}{\text{MeV}}\right) = \frac{10^3}{10^{-6}} \cdot \frac{e}{\epsilon} \cdot \frac{V_o}{Q} = 0.45 \cdot 10^{-10} \frac{V_o}{Q}\left[\frac{V}{C}\right]; \qquad (6.25)$$

for silicon SD, and accordingly, in our case, 9 mV/MeV correspond to $C_p = 5$ pF as compared to 0.45 mV/MeV directly on the detector.

6.2. CHARGE-SENSITIVE-PREAMPLIFIER NOISE SPECIFICATION AND MEASUREMENT

In communications the amplifier noise is characterized by the noise factor or the noise equivalent voltage $V_{eff} = \sqrt{< v^2_{noise} >}$ at the input, useful quantities due to the resistive nature of signal sources. This unit of measure used in transmissions is less significant for pulse analysis. The "crest" value V_{cr} [W 2] would be more fitted, but it gives rise to difficulties since it is necessary to consider the probability of the existence of a ratio V_{cr}/V_{eff}. For example, for a Gaussian noise one obtains the data of Table 6.1.

Table 6.1

The probability to obtain a given V_{cr}/V_{eff} ratio in the case of a Gaussian noise

V_{cr}/V_{eff}	1.64	2.57	3.29	3.89	4.4
p(%)	10	1	0.1	0.01	0.001

The ratio of the time during which V_{cr}/V_{eff} is exceeded to the total time of observation, is denoted by p. Hence, assuming a ratio $V_{cr}/V_{eff} = 3$, the probability to exceed it is $< 1‰$. The noise amplitude distribution is not necessarily a Gaussian one.

In nucleonics, the signal sources are capacitive and the signal consists in electric charge pulses. It is then normal to specify the noise in terms of noise equivalent charge $\sqrt{< Q^2_{noise} >}$ or number of elementary noise equivalent charges $\sqrt{< N^2_{noise} >}$.

The efficient value of the noise equivalent charge $Q_{eff} \equiv \sqrt{< Q^2_{noise} >}$ is that charge quantity which applied at the amplifier input for a short time, as compared to the rise-time of its transient characteristics, yields at the output a voltage pulse with amplitude equal to the observed noise efficient voltage.

Since in nuclear physics, the measurement of the detected particle energy is of great interest, and the noise contributes to the worsening of the spectrometer energy resolution (as shown in Section 3.10), the noise is usually expressed in units of equivalent energy $\sqrt{< E^2_{noise} >} = E_{eff}$ (the energy lost by a particle which crossing the detector yields an electric charge equal to the noise equivalent charge). At last, the noise can be often expressed by the full width at half maximum *FWHM* (keV).

Since when making the equivalence in energy the material constant ϵ interferes, it is necessary to specify to which type of material (Si, Ge, a.s.o.) it refers, if the noise is expressed in terms of energy.

The transformation from one unit of measure to another can be easily made by means of Eqs. (3.178) and (6.3). The *FWHM* is usually measured in keV and thus we can estimate

$$E_{eff} = \frac{FWHM}{2.36} \text{ (keV)}, \tag{6.26}$$

$$N_{pairs} = \frac{FWHM \cdot 10^3}{2.36 \cdot \epsilon} \simeq 120\, FWHM \text{ (keV)}, \tag{6.27}$$

for Si *SD* and

$$Q_{eff}\,[\mathrm{C}] = Ne = 1.6 \times 10^{-19}\, N. \tag{6.28}$$

For example, at a $FWHM = 10$ keV it results $E_{eff} = 4.25$ keV, $N = 1200$ elementary charges and $Q_{eff} = 1.9 \times 10^{-16}$ C.

In a linear amplifier, such as those used in spectrometry, the most simple method of estimating the noise equivalent charge (Q_{eff}) [F 5] consists in measuring the amplitude (V_o) of the signal yielded when applying a known charge quantity Q_i at the input and a noise efficient voltage at the output V_n (e.g., with a thermal instrument); Q_{eff} is estimated as

$$Q_{eff} = \frac{V_n}{V_o} Q_i,$$

since according to relation (6.22) we can write $A_q = \frac{V_o}{Q_i} = \frac{V_n}{Q_{eff}}$.

Thus, for $V_o = V_n$ it results $Q_{eff} = Q_i$ (definition of noise equivalent charge). We notice that though the measurements are made at the output, it is not necessary to know the circuit amplification for determining the quantity Q_{eff}.

The pulse generator injects at the amplifier input a charge quantity Q_i which shared out among the three capacities gives at the point A a voltage

$$V_A = \frac{Q_i}{C_d + C_i}.$$

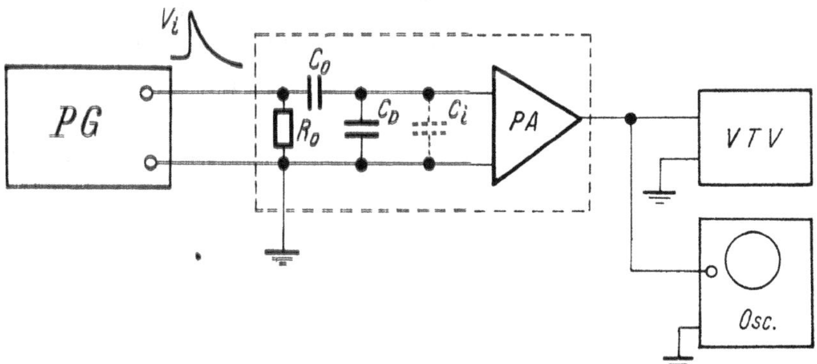

Fig. 6.3 — Experimental set-up for measurements of preamplifier noise equivalent charge. *PG* = pulse generator; *PA* = preamplifier; *VTV* = r.m.s. value vacuum-tube voltmeter, *Osc* = broadband cathode-ray oscilloscope.

This voltage can be written, however, as a direct result of the voltage capacitive division

$$V_A = V_i \frac{C_o}{C_o + C_d + C_i}.$$

Then

$$Q_i = V_i C_0, \tag{6.29}$$

if $C_0 \ll C_d + C_i$.

The capacity C_0 is usually taken of the order of 1 pF ; thus the condition $R_0 C_0 < t_c$ is fulfilled.

The output signal can be measured on a calibrated cathode-ray oscilloscope with a standard pulse generator.

The output noise level must be measured with a broad-band voltmeter for r.m.s. voltages (V_{eff}).

Attention should be drawn to the fact that the statement of some authors [F 5] concerning the possibility of measuring the noise with a voltmeter for mean value is not true, since only in a sinusoidal regime there exists a well-defined relation between V_{eff} and V_{mean}.

The measurement is made as follows : a sufficiently great charge is applied to the amplifier in order to surpass the noise and to permit an accurate estimation of the output pulse amplitude V_0 at the oscilloscope. Usually $Q_i = 10^{-14}$C (10 mV through 1 pF).

Then the pulse generator is put out (it is better to disconnect it, to avoid noise) and the noise level V_n is measured at output with an r.m.s. value voltmeter and $Q_0 = \frac{V_n}{V_0} Q_i$ is calculated.

The efficient value voltmeter may be replaced by an amplitude analyser.

Then we measure the pulse amplitude spectrum, which results when applying pulses with constant amplitudes at the input. The resulting Gaussian standard deviation represents the noise efficient voltage V_n

$$V_n = \frac{FWHM\ [\mathrm{V}]}{2.36}. \tag{6.30}$$

From equation (6.28) it follows then [D 1]

$$Q_{eff} = \frac{FWHM\ [\mathrm{V}]}{2.36} \cdot \frac{V_i C_0}{V_0}. \tag{6.31}$$

If the spectrometer is calibrated with a radioactive source of known energy, the *FWHM* noise value in keV is obtained directly on the analyser for a given capacity at the preamplifier input and for a certain time constant of the integration and differentiation circuit in the main amplifier.

6.3. AMPLIFIER-NOISE SOURCES

The detector is connected to the first vacuum-tube of the amplifier (terminals $A - M$) by means of a circuit, the diagram of which is drawn in Fig. 6.2 *a*.

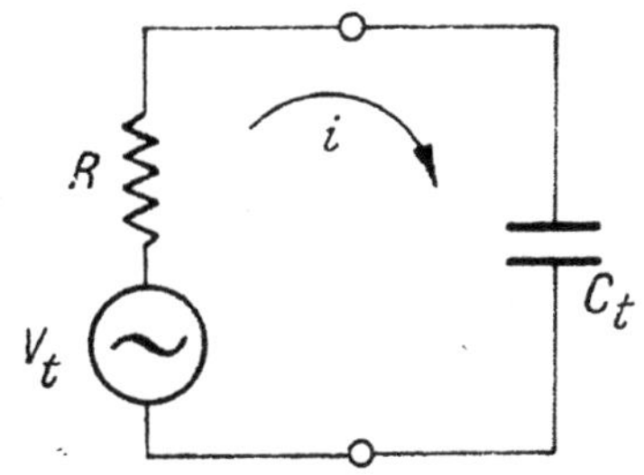

Fig. 6.4 — Simplified equivalent circuit configuration of the preamplifier input circuit, used in noise calculation.

The coupling capacity C_c is always chosen great enough such as to allow the neglecting of its reactance at frequencies within the spectrum of the pulse considered. We shall denote by R the equivalent resistance of the two resistances R_D and R_G connected in parallel. Then

$$R = \frac{R_D R_G}{R_D + R_G}. \tag{6.32}$$

If, in a first approach, we neglect the series resistance r_s we obtain a simple input equivalent circuit (Fig. 6.4). It is known [G 7] that the main noise types which appear in such circuits at the vacuum-tube inputs are : the resistance thermal noises, the current shot-noises and the flicker noises resulting from the fluctuations of cathode emission currents. The partition noises occur only in multi-electrode tubes (tetrodes, pentodes), which, for this reason, are not used in input circuits. If the first-tube amplification is high enough, the noise of the other vacuum-tubes may be neglected in the estimation of the signal to noise ratio.

The thermal noise of a resistance R and the shot noise of a current I are at their origin "white" noises, i.e., their spectral density is constant all over the frequency spectrum

$$< \delta V^2 > = 4kTR\delta f \tag{6.33}$$

and respectively

$$< \delta I^2 > = 2eIs\delta f, \tag{6.34}$$

where

k — Boltzmann's constant; T — absolute temperature; δf — an elementary frequency-interval; e — electron charge; s — a factor $(0 < s \leqslant 1)$ introduced in the case $I = I_g$ (the grid current) and including the correlations between electron movements, due to the space-charge effect [B 30].

But the input circuit $R - C_t$ (Fig. 6.4) alters this spectrum due to the presence of the capacitive element. For instance, the thermal noise voltage with a certain pulsation ω which appears at the C_t capacity terminals, is

$$V_{th,c} = \frac{i}{\omega C_t} = \frac{V_t}{\sqrt{1 + (\omega RC_t)^2}} \tag{6.35}$$

and consequently the spectral density of the noise appearing on the capacity C_t due to thermal agitation in the resistance R is

$$\frac{< \delta V^2_{th,c} >}{\delta f} = \frac{4kTR}{1 + (\omega RC_t)^2} \simeq \frac{4kT}{\omega^2 C_t^2 R^2}, \tag{6.36}$$

for

$$\omega \gg \frac{1}{RC_t}.$$

The mean square value of the noise voltage appearing on the condenser

$$< V^2_{th,c} > = 4kTR \int_0^\infty \frac{df}{1 + (\omega RC_t)^2} = \frac{kT}{C_t} \tag{6.37}$$

is independent of the resistance value R, since

$$\int_0^\infty \frac{df}{1 + (\omega C_t R)^2} = \frac{1}{2\pi} \int_0^\infty \frac{d\omega}{1 + (\omega C_t R)^2} = \frac{1}{2\pi} \frac{1}{RC_t} \frac{\pi}{2}. \tag{6.38}$$

The amplifier which transmits this signal has not, however, an infinite frequency band ; it amplifies only the frequency components within its pass band. It will be shown that in this case the resistance R enters the expression of the thermal noise voltage.

In a similar way, it can be shown that the noise-voltage spectral density of the vacuum-tube grid current varies according to the same law as the detector noise

$$\frac{< \delta V_{ge}^2 >}{\delta f} = \frac{2esI_g}{\omega^2 C_i^2}. \tag{6.39}$$

Here $I_g = | I_{g+} | + | I_{g-} |$, i.e., the positive and negative components of the grid current (Fig. 6.5) interfere, because even for a

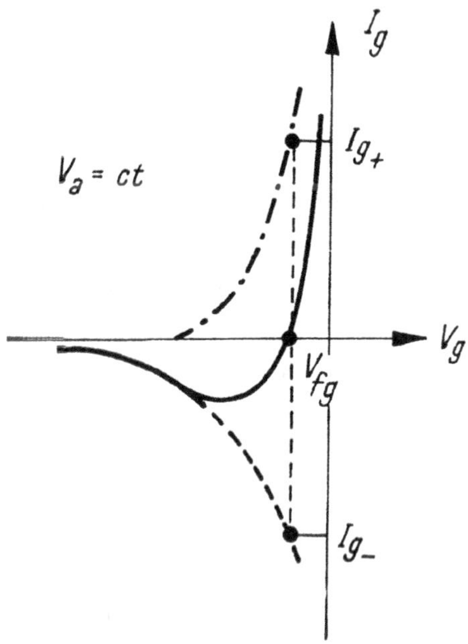

Fig. 6.5 — $I_g - V_g$ characteristic of a vacuum tube; I_{g+} is the electron component; I_{g-} — the ion component of the grid current.

zero grid current obtained for a "floating-grid" regime ($V_g = V_{fg}$ there is a noise due to random fluctuations of both components.

The positive component of the grid current is the grid-cathode diode current which varies approximately exponentially with the voltage; the negative component is formed especially of the electron current between grid and plate, released from the grid by plate emitted photons. This negative component can be approximately estimated [B 30] by the relation

$$| I_{g-} | = A_g I_a V_a^{3/2}, \tag{6.40}$$

where

V_a — anode voltage, I_a — anode current, $A_g \simeq 10^{-10}$, for most tubes in a floating grid regime $I_g = 2 \mid I_{g-} \mid$.

The shot noise of the tube can be taken into account by replacing the real tube (with noise) by an ideal noiseless tube which has a noise equivalent resistance R_{eq} in series with its grid.

The resistance R_{eq} being in series with the vacuum-tube grid, the noise spectral density is not affected by the input circuit and hence

$$\frac{< \delta V_{sh}^2 >}{\delta f} = 4kTR_{eq}. \tag{6.41}$$

Very often, the noise equivalent resistance R_{eq} is estimated with

$$R_{eq} = \frac{B}{S}, \tag{6.42}$$

where $B = 2.5$ and S is the vacuum-tube slope. Measuring [B 30] R_{eq} for the tubes 6922 (equivalent E88CC), E83F and EC-1000, $1 < B < 5$ was found; generally, this factor increased when the plate-current increased.

Finally, for the flicker noise one can write the relation

$$\frac{< \delta V_f^2 >}{\delta f} = \frac{A_f}{f}, \tag{6.43}$$

where A_f is a constant, the experimental value of which is $A_f = 10^{-13} V^2$ in the case of oxide cathodes.

Considering that different voltage fluctuations are uncorrelated, their total effect is obtained after a square summing up, so that the noise spectral density at the preamplifier input will be

$$\frac{<\delta V^2_{noise}>}{\delta f} = \frac{<\delta V^2_{sh}>}{\delta f} + \frac{<\delta V^2_{f}>}{\delta f} + \frac{<\delta V^2_{ge}>}{\delta f} + \frac{<\delta V^2_{th,c}>}{\delta f}, \tag{6.44}$$

or taking into account the relations obtained above and the fact that $\delta f = \frac{\delta\omega}{2\pi}$, it can be written as

$$\frac{<\delta V^2_{noise}>}{\delta\omega} = \frac{2\,BkT}{\pi S} + \frac{A_f}{\omega} + \frac{1}{\pi\omega^2 C_t^2}\left(esI_g + \frac{2\,kT}{R}\right), \tag{6.45}$$

or denoting the right-hand side by $N^2(\omega)$

$$\frac{<\delta V^2_{noise}>}{\delta\omega} = N^2\,(\omega). \tag{6.46}$$

From equation (6.45) we observe that while the shot noise is uniform all over the frequency spectrum, the flicker noise and particularly the grid-current and bias resistance noise prevail at low frequencies. This frequency dependence permits to measure each noise component apart (correspondingly choosing the pass band of the amplifier for measurement). All the measurements made showed that usually, the flicker noise can be neglected with respect to the other noise types. For estimating the tube noise, instead of using the parameters S, I_g and C_t which are difficult to measure and less significant for practice, other parameters [B 30] are suggested: R_s, R_p and C_t (Fig. 6.6 *b*). Equating the grid current noise with the thermal noise given by an equivalent resistance R_{gn}, the input circuit from Fig. 6.2 *a* used in the computation of noise, becomes the circuit from Fig. 6.6 *a* the simple equivalent of which is given in Fig. 6.6 *b*. The values of the equivalent circuit elements are calculated [B 30] by

$$C_t = C_d + C_f + C_{in} + C_s \tag{6.47}$$

$$R_s = R_{eq} + \left(\frac{C_d + C_f}{C_t}\right)^2 r_s + \left(\frac{C_d}{C_t}\right)^2 R_{sd}, \tag{6.48}$$

$$\frac{1}{R_p} = \frac{1}{R_{gn}} + \frac{1}{R_D} + \frac{1}{R_g} + \frac{1}{R_{pd}}, \tag{6.49}$$

and the noise in the resistances R_s and R_p is calculated with a constant voltage generator connected in series with R_s

$$< \delta V_{sn}^2 > = 4kTR_s \, \delta f, \tag{6.50}$$

and a constant current generator in parallel with R_p

$$< \delta i_{pn}^2 > = \frac{4\,kT}{R_p} \, \delta f. \tag{6.51}$$

The resistances R_s and R_p can be easily measured by means of noise generators (Fig. 6.7).

In order to measure R_s, the resistances in Fig. 6.7 *a* are connected in series with the first vacuum-tube grid, the feedback capacity C_{fb} is disconnected and the output noise is measured by using the optimum integration and differentiation time constants ($\tau \simeq 1$ μs) in the main amplifier. The noise equivalent resistance R_s is equal to the value of the series resistance introduced for which the output noise increases with a factor of $\sqrt{2}$ as compared to the short circuit condition at the input.

In order to measure R_p, the resistances in Fig. 6.7 *b* are connected in parallel with the input. R_p is the parallel resistance value which makes the noise to increase by a factor of $\sqrt{2}$ in comparison with open circuit condition. This measurement is made with the feedback capacity connected and with the maximum differentiation and integration time constant $\tau_{max} \gg 10$ μs.

The stray capacities have to be minimum, and the noise generator resistances must be separated from R_G by a coupling condenser in order to prevent the drift of the amplifier working point during the measurement.

6.4. EFFECTS OF AMPLIFIER SHAPING CIRCUITS ON NOISE SPECTRA

The noise at the amplification chain output depends on the spectral density of the input noise $N(\omega)$ and the variation of the amplifier voltage gain with the frequency $\mathcal{A} = \mathcal{A}(\omega)$. Usually, the main amplifier, which receives the signal from the preamplifier, includes the

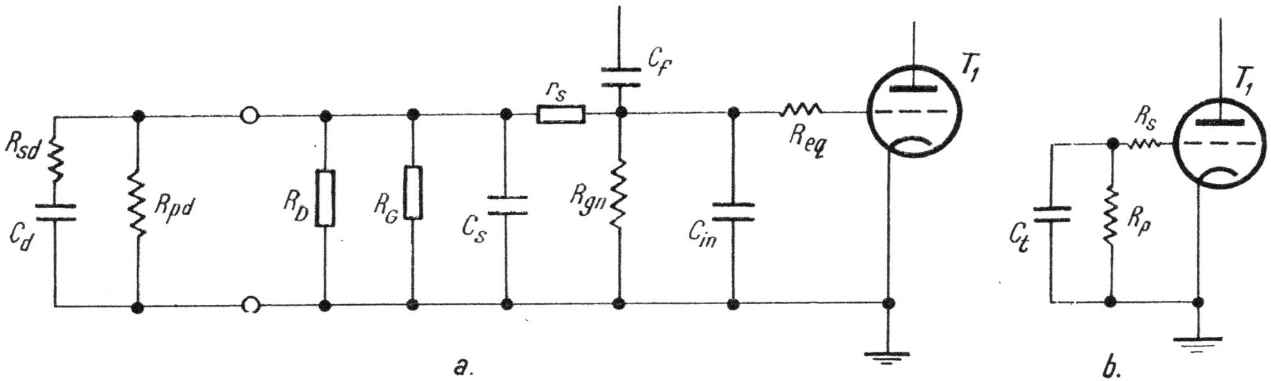

Fig. 6.6 — The equivalent circuit diagram of the preamplifier input circuit used in noise calculation; *a* — detailed diagram; *b* — simplified diagram.

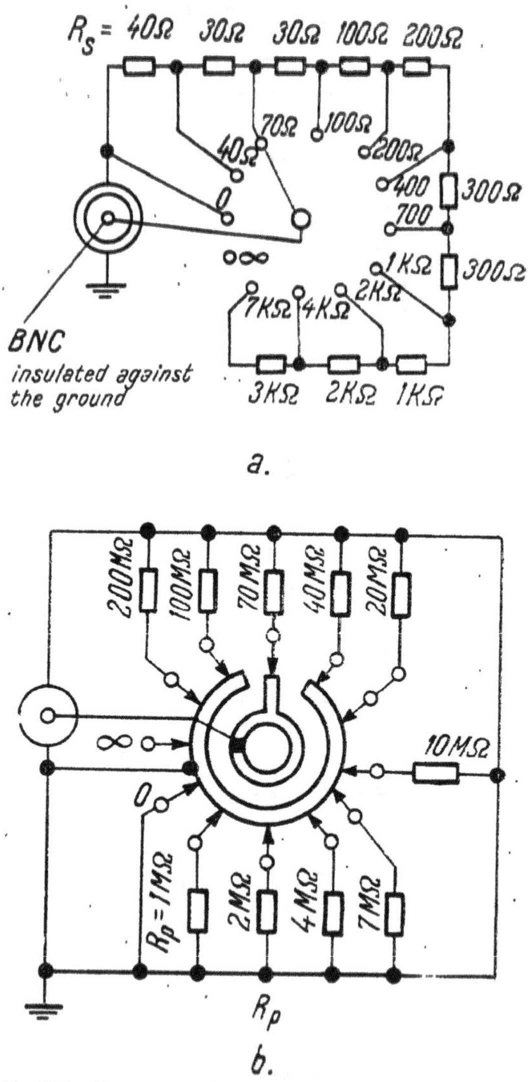

Fig. 6.7—Constant voltage (*a*) and constant current (*b*) noise generators [B 30).

shaping circuits limiting up and down the frequency spectrum transmitted.

The noise voltage mean square value at the amplifier output is computed according to the relation

$$< V^2_{o\,noise} > = \int_0^\infty N^2(\omega)\, \mathcal{A}^2(\omega)\, d\omega, \tag{6.52}$$

where $N(\omega)$ is determined from equations (6.45) and (6.46).

If f_c is the central frequency of the amplifier pass band (the frequency value for which the gain is maximum and has the value $\mathcal{A}(f_c) = \mathcal{A}_0$), an effective noise band of the amplifier can be defined as the frequency band within $\frac{f_c}{k}$ and kf_c of an amplifier with gain $\mathcal{A}_0 = \text{const.}$ and the same output noise power (assuming equal output and input impedances) as the real amplifier i.e.

$$\mathcal{A}_0^2 \int_{\frac{f_c}{k}}^{kf_c} N^2(\omega)\, d\omega = \int_0^\infty N^2(\omega) \mathcal{A}^2(\omega)\, d\omega. \tag{6.53}$$

The effective noise band will be

$$B_{noise} = \left(k - \frac{1}{k}\right) f_c. \tag{6.54}$$

Evidently, B_{noise} depends on the actual noise type $N(\omega)$ and the amplitude characteristic $\mathcal{A}(\omega)$ of the amplifier stage. The gain-frequency characteristic expression $\mathcal{A}(\omega)$ depends on the type and number of amplifier integration and differentiation circuits. In linear amplifiers used in nucleonics, the pass band is limited at the upper end by integration circuits RC and at the lower end by differentiation circuits RC or shaping lines DL (unloaded or in shortcircuit). We may use many combinations of integration and differentiation circuits. Six types of amplifiers are studied in [T 16]:

- $RC - RC$ (a differentiation circuit and an integration circuit);
- $RC - (RC)^2$ (a differentiation circuit and two identical integration circuits);

$(RC)^2 - (RC)$ (two identical differentiation circuits and an integration one);
$(DL) - (RC)$ (a delay line and an integration circuit);
$(DL) - (RC)^2$ (a delay line and two identical integration circuits);
$(DL)^2 - (RC)$ (two identical delay lines and an integration circuit).

If τ_d is the differentiation circuit time constant, τ_D the time delay of the delay line and τ_i the integration circuit time constant, the optimum (the lowest noise) of these amplifier types is achieved for $\lambda = 1$, 1.06, 1.38, 2.623, and zero, respectively, where $\lambda = \frac{\tau_d}{\tau_i}$ for $(RC)^j - (RC)^k$ amplifiers and $\lambda = \frac{2\tau_D}{\tau_i}$ for $(DL)^j - (RC)^k$ amplifiers. The paper [T 16] concludes that the $RC - RC$, $RC - (RC)^2$ and $DL - (RC)^2$ amplifiers are suited for minimum noise performances. The relations for shot noise and grid current calculation [T 17], are given în Table 6.2

Table 6.2

Calculation relations for the main noise sources in case of different shaping circuits

Shaping circuit	Shot noise V_{sh}^2	Grid current noise V_g^2
$(RC) - (RC)$	$\frac{\lambda}{\lambda+1} \cdot \frac{a}{\tau_i}$	$\frac{1}{2C_t^2} \cdot \frac{\lambda}{\lambda+1} \lambda b \tau_i$
$(RC) - (RC)^2$	$\frac{1}{2}\left(\frac{\lambda}{\lambda+1}\right)^2 \frac{a}{\tau_i}$	$\frac{1}{4C_t^2}\left(\frac{\lambda}{\lambda+1}\right)^2 (2\lambda+1) b\tau_i$
$(RC)^2 - (RC)$	$\frac{1}{2}\left(\frac{\lambda}{\lambda+1}\right)^2 \frac{2\lambda+1}{\lambda} \cdot \frac{1}{\tau_i}$	$\frac{1}{4C_t^2}\left(\frac{\lambda}{\lambda+1}\right)^2 \lambda b \tau_i$
$(DL) - (RC)$	$\frac{1}{2}[1-\exp(-\lambda)]\frac{a}{\tau_i}$	$\frac{1}{4C_t^2}[\lambda-(1-\exp(-\lambda))]\, b\tau_i$
$(DL)^2 - (RC)$	$\frac{1}{8}[3-\exp(-\lambda)][1-\exp(-\lambda)]\frac{a}{\tau_j}$	$\frac{1}{16C_t^2}[2\lambda-(3-\exp(-\lambda))]\, b\tau_i$

In this table $a = k\,TR_{eq}$ and $b = eI_g$

Complex filtering circuits can be also used for achieving the best performances. The analysis and the synthesis of such circuits were made by Blankenship and Newlin [B 31], [N 4], Blalock [B 24], Nygaard [N 6], Deighton [D. 14], a.o.

6.5. *RC—RC* AMPLIFIER SIGNAL TO NOISE RATIO

The most simple *SD* associated amplifier is, obviously, of the *RC*— — *RC* type, i.e., an amplifier where all differentiation circuits have time constants much higher than τ_d and all integration circuits have time constants much lower than τ_t, so that the pass band is determined only by one differentiation circuit and one integration circuit. The equivalent circuit diagram of such an amplifier is given in Fig. 6.8. On the left side of the input terminals AA' the simplified input circuit of the preamplifier with *SD* connected, is shown.

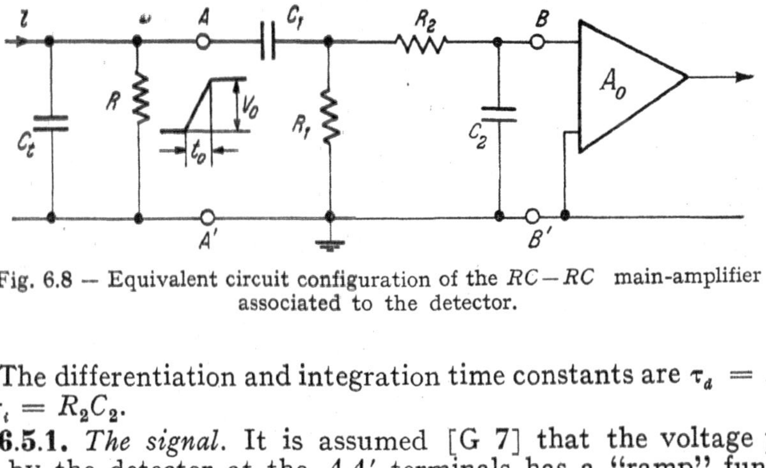

Fig. 6.8 — Equivalent circuit configuration of the *RC—RC* main-amplifier associated to the detector.

The differentiation and integration time constants are $\tau_d = R_1 C_1$ and $\tau_t = R_2 C_2$.

6.5.1. *The signal.* It is assumed [G 7] that the voltage pulse given by the detector at the AA' terminals has a "ramp" function shape *), with t_0 the pulse rise time and $V_0 = \dfrac{Q}{C_t}$ the amplitude, where $Q = N_\varepsilon$ is the total collected charge.

*) The "ramp" function approximation of the pulse with exponential decay is valid if $RC_t \gg \tau_d$.

Under these conditions, the signal at the $B - B'$ terminals has the expression [G 7]

$$V_{(0-t_0)} = \frac{V_0 \tau_d}{t_0}\left[1 - \exp\left(-\frac{t}{\tau_i}\right)\right] - $$
$$- \frac{V_0 \tau_d^2}{t_0(\tau_d - \tau_i)}\left[\exp\left(-\frac{t}{\tau_d}\right) - \exp\left(-\frac{t}{\tau_i}\right)\right], \qquad (6.55)$$

for $0 < t < t_0$ (front)

$$V_{(t_0-\infty)} = \frac{V_0 \tau_d^2}{t_0(\tau_d - \tau_i)}\left[\exp\left(\frac{t_0}{\tau_d}\right) - 1\right]\exp\left(-\frac{t}{\tau_d}\right) -$$
$$- \frac{V_0 \tau_d \tau_i}{t_0(\tau_d - \tau_i)}\left[\exp\left(\frac{t_0}{\tau_i}\right) - 1\right]\exp\left(-\frac{t}{\tau_i}\right), \qquad (6.56)$$

for $t_0 < t < \infty$ (decay).

The pulse amplitude is

$$V_M = \frac{V_0 \tau_d}{t_0} \cdot \frac{\left[\exp\left(\frac{t_0}{\tau_d}\right) - 1\right]^{\frac{\tau_i}{\tau_d - \tau_i}}}{\left[\exp\left(\frac{t_0}{\tau_i}\right) - 1\right]^{\frac{\tau_d}{\tau_d - \tau_i}}}. \qquad (6.57)$$

In the case often met with in practice when $t_0 \ll \tau_d, \tau_i$, by expanding the exponentials into series and retaining the first order terms, it results after simple transformations

$$V_M = V_0 \left(\frac{\tau_i}{\tau_d}\right)^{\frac{\tau_i}{\tau_i - \tau_d}} = V_0\, \alpha^{\frac{\alpha}{1-\alpha}}, \qquad (6.58)$$

where we have denoted $\alpha = \frac{\tau_i}{\tau_d}$. When $\alpha \to 1$ ($\tau_d = \tau_i$), we obtain *)

---

*) If we denote $X = \alpha^{\frac{\alpha}{1-\alpha}}$, then applying the logarithm $\ln X = \frac{\alpha \ln \alpha}{1-\alpha}$ and the L'Hôpital's rule, we obtain $\lim_{X \to 1} (\ln X) = -1$, hence $\lim_{\alpha \to 1} X = \frac{1}{e}$, where e is the radix of the Neperian logarithms.

$$V_M = \frac{V_0}{e} = \frac{V_0}{2.718}, \quad (6.59)$$

i.e., the signal is **2.7** times diminished.

6.5.2. *The noise.* The noise is estimated with the aid of equation (6.52). The square modulus of the circuit transfer function between the input terminals $A - A'$ and the output terminals $B - B'$ varies with the frequency

$$\mathcal{A}^2(\omega) = \frac{\omega^2 \tau_d^2}{(1 + \omega^2 \tau_d^2)(1 + \omega^2 \tau_i^2)}, \quad (6.60)$$

(if the two circuits are separated by intermediate stages), hence the noise at the $B - B'$ terminals will be

$$V^2_{noise} = \int_0^\infty N^2(\omega) \cdot \mathcal{A}^2(\omega)\, d\omega$$

$$= \int_0^\infty \left\{ \frac{2B\,kT}{\pi S} + \frac{A_f}{\omega} + \frac{1}{\pi \omega^2 C_i^2} \left[es\,(I_g + I_a) + \right.\right.$$

$$\left.\left. + \frac{2kT}{R} \right] \right\} \frac{\omega^2 \tau_d^2 d\omega}{(1 + \omega^2 \tau_d^2)(1 + \omega^2 \tau_i^2)}.$$

After performing all integrations, it results

$$V^2_{noise} = \frac{BkT}{S} \cdot \frac{\tau_d}{\tau_i(\tau_d + \tau_i)} + \frac{A_f\, \tau_d^2}{\tau_d^2 - \tau_i^2} \ln \frac{\tau_d}{\tau_i} +$$

$$+ \frac{1}{2C_i^2} \left[e\,(I_g + I_a) + \frac{2kT}{R} \right] \frac{\tau_d^2}{\tau_d + \tau_i}. \quad (6.61)$$

The mean square voltage of the shot noise at the $B - B'$ terminals is

$$V^2_{sh} = \frac{BkT}{S} \cdot \frac{\tau_d}{\tau_i(\tau_d + \tau_i)}, \quad (6.62)$$

and we observe that for a wide band amplifier (a short τ_i), this noise increases. In order to lower it, an input tube with high mutual conductance is required.

As distinguished from this noise, which is of importance in the high frequency range, the grid-current noise, the detector-current noise and the thermal noise of the grid resistance

$$V_g^2 = \frac{1}{2C_i^2}\left[e\,(I_g + I_d) + \frac{2kT}{R}\right]\frac{\tau_d^2}{\tau_d + \tau_i} \tag{6.63}$$

prevail, mainly in the low frequency region of the spectrum (a long τ_d). The detector load resistance and the grid resistance in parallel with it can be chosen sufficiently great such that the thermal noise given by their equivalent R be negligible.

The detector must work in conditions of low reverse current, and the input tube must have a low grid current (both its ionic and electronic components $I_g = |I_{g+}| + |I_{g-}|$).

Usually the detector current is higher by, at least, an order of magnitude than I_g, hence it is not necessary to take special precautions for reducing the grid current of the input tube.

However, very low currents can be obtained in the detector when it operates at the liquid nitrogen or helium temperature or when a guard-ring *SD* is used. In this case, the lowering of the grid current can bring important improvements of the signal to noise ratio.

For a given vacuum tube, the minimum grid current is obtained in working conditions of high negativation, low plate (or screen in the case of pentode) voltage and small dissipation. The tubes have to be selected (I_g is measured after a preliminary 2 hour ageing). Because we use resistances $R > 10$ M Ω in order that their thermal noise be negligible, the input tube usually works in "floating-grid" regime.

In the case of a vacuum tube with oxide cathode the flicker noise is

$$V_f^2 = \frac{A_f \cdot \tau_d^2}{\tau_d^2 - \tau_i^2} \ln \frac{\tau_d}{\tau_i}. \tag{6.64}$$

The constant $A_f = 10^{-13}$ was estimated experimentally. In order to separate this noise from the shot noise and the grid current noise [G 7] the input is put in shortcircuit and thus the thermal and grid current cancel out. The dependence of the noise on τ_d allows a separation between V_{sh} and V_f. At short τ_d, V_{sh} prevails and at long τ_d it is only V_f which practically remains.

6.5.3. *Equal differentiation and integration time constants.* The examination of the variation of the signal to noise ratio and of the resolution time $\left(\text{defined as } t_r = \dfrac{V_0\,\tau_d}{V_M}\right)$ vs τ_d/τ_i [G 7] leads to the conclusion that the maximum signal to noise ratio and the minimum resolution time can be obtained when $\tau_d = \tau_i = \tau$, i.e. when the integration and differentiation time constants are equal.

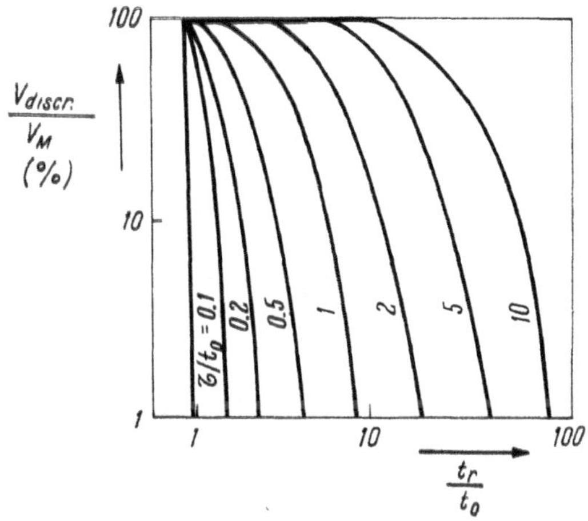

Fig. 6.9—Time resolution of the $RC-RC$ amplifier.

In such a case [G 7], the resolution time t_r is calculated as the pulse duration at a given level of discrimination. The result is given in Fig. 6.9.

The effective noise band width (B_{noise}) of this amplifier in terms of the central frequency f_c [S 3] is $0.785\,f_c$ for a shot noise (white noise), ($1/f$) $0.505\,f_c$ for a flicker noise and ($1/f^2$) $0.785\,f_c$ for the grid current noise, as compared with the band at the 3 dB of the amplifier, $B = 2f_c = \dfrac{1}{\pi\tau}$.

The principal noise sources in terms of mean square charges V^2_{noise}, C^2_i are given in Table 6.3 at room temperature $T = 290°\text{K}$

([H 5] and [F 5] corrected by [T 17]). The units used are : volt, seconds, farad, ampère, coulomb, ohm (MKSA system).

The relations given in this Table can be obtained from equation (6.61) where we consider $\tau_d = \tau_i = \tau$ and $kT = 4\times10^{-21}$ Joule. The input equivalent noise charge is computed as

$$Q_{eq} = 2.718\sqrt{V^2_{noise}\,C_t^2}\,, \tag{6.65}$$

Table 6.3

Main noise sources $(V^2_{noise}\;C_t^2)$

Source	Theoretical relationship	Noise at 290° K (C^2)	Tipical values: Parameters	Tipical values: Noise (C^2)
Shot noise	$\frac{kTR_{eq}}{2\tau}C_t^2$	$2\times10^{-21}\frac{R_{eq}C_t^2}{\tau}$	$S=10$mA/V	8×10^{-34}
Grid-current noise	$\frac{eI_g}{4}\tau$	$4\times10^{-20}I_g\tau$	$C_t=40$ pF	1.2×10^{-34}
Detector noise ("white" component)	$\frac{eI_d}{4}\tau$	$4\times10^{-20}I_d\tau$	$\tau=1\mu$s	40×10^{-34}
Thermal noise of the parallel resistance R	$\frac{kT}{2R}\tau$	$2\times10^{-21}\frac{\tau}{R}$	$i_g=3$ nA $i_d=0.1$ A	2×10^{-34}
Thermal noise of grid series and cathode-resistance	$\frac{kTR_s}{2\tau}C_t^2$	$2\times10^{-21}\frac{R_sC_t^2}{\tau}$	$R_g=10$MΩ $R_s=50\Omega$	1.6×10^{-34}
Flicker noise	$\frac{A_f\cdot\tau}{2}C_t^2$	$\frac{A_f\cdot\tau}{2}C_t^2$		0.8×10^{-34}

Total $V^2_{noise}\,C_t^2 = 53{,}6\times10^{-34}\,C^2$

and the number of elementary noise charges

$$N_e = 6.25\cdot10^{18}Q_{eq}\,. \tag{6.66}$$

Finally, for silicon for which $\in = 3.6$ eV, the energy dispersion produced by the noise is

$$FWHM\ [\text{keV}] = \frac{N_e}{120} = \frac{Q_{eq}}{120\ e} = 1.43 \times 10^{17} \times V_{noise}\, C_t \qquad (6.67)$$

or its square

$$(FWHM)^2 \simeq 2 \times 10^{34}\, (V_{noise}\, C_t)^2 \qquad (6.67')$$

For the example given in Table 6.3, we obtain a noise equivalent charge $Q_{eq} = 2.7 \sqrt{V^2_{noise}\ C^2_t} = 20 \times 10^{-17}$C. Hence $N_e = 1250$ electrons and the $FWHM_{\text{Si}} = 10$ keV.

The importance of low currents in the detector may be clearly seen. If $I_d > 1$ μA, all the other noises are negligible in comparison with the detector noise. In this case it would be advantageous to use small time constants (τ) for the amplifier. We notice also in this Table that the resistance thermal noise and the vacuum-tube flicker noise are, generally, negligible in comparison with the detector own noise and the vacuum-tube shot noise.

For a $(RC) - (RC)^2$ amplifier, the noise yielded by these two sources can be determined [B 30] from the nomograph given in Fig. 6.10, where the R_s, R_p and C_t, elements of the equivalent circuit drawn in Fig. 6.6 *b* appear.

In this case, the noise given by the resistance R_{sn} can be estimated in terms of the electron number, as :

$$N_{es} = 0.53\, \frac{R^{1/2}_{sn}\,(\Omega)\; C_t\,(pF)}{\tau^{1/2}\,(\mu s)},$$

and the noise of the resistance R_{pn} as

$$N_{ep} = 910\, \frac{\tau^{1/2}\,(\mu s)}{R^{1/2}_{pn}\,(M\Omega)}.$$

The grid current (in nA) equivalent to a resistance R_{pn}, is calculated from

$$I_{gt} = \frac{2kT}{e} \cdot \frac{1}{R_{pn}} = \frac{52}{R_{pn}\,(M\Omega)}.$$

At an anode voltage V_a and an anode current I_a, the grid current is approximately

$$I_{gt}\,(\text{nA}) = 2\times 10^{-4}\, I_a\,(\text{mA})\cdot V_a^{3/2}\,(\text{V}),$$

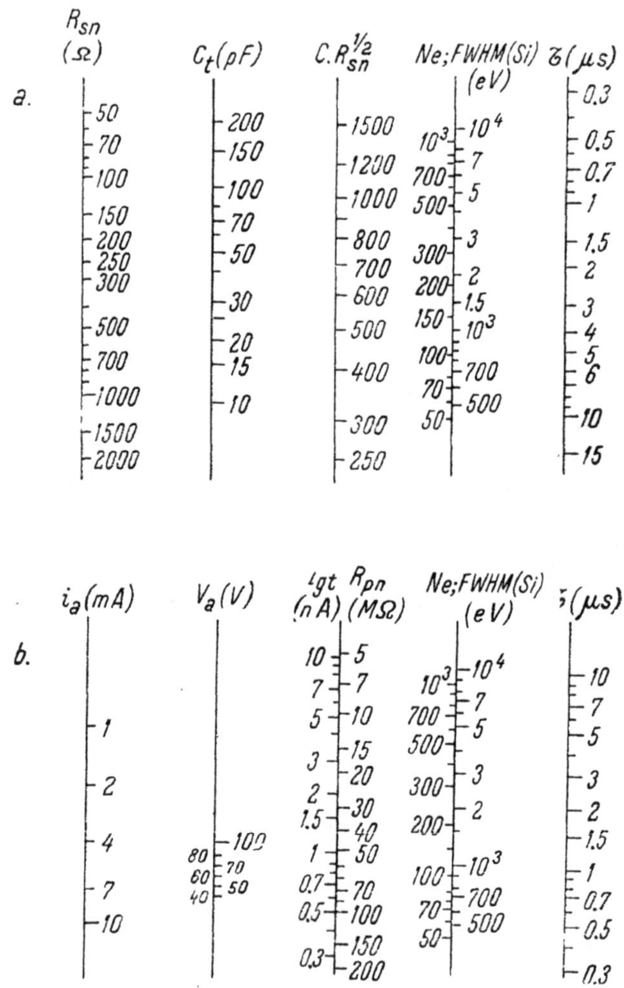

Fig. 6.10 — Nomographs used to calculate the plate-current type noise (*a*) and grid current type noise (*b*) [B 30].

but, in fact, if the resistance R_{pn} is measured as shown above, the estimation of I_{gt} is no longer necessary.

6.5.4. *Optimum time constant.* From Table 6.3 one may observe that for the $(RC) - (RC)$ amplifier, one of the main noise components is proportional to τ (shot noise) and another one is inversely proportional to $\tau = \tau_i = \tau_d$ (detector and grid current noise). Thus, the noise may be diminished by a variation of the time constant τ (Fig. 6.11).

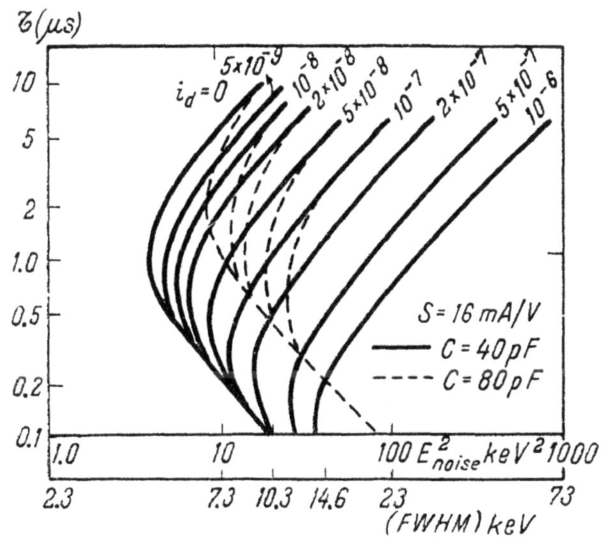

Fig.6.11 — The noise versus time constant of the $RC-RC$ amplifier [G 21].

The noise is minimum when the two noise sources become equal

$$\frac{kTR_{eq}}{2\tau_{opt}} = \frac{\tau_{opt}}{4C_t^2}\left[e\,(I_g + I_d) + \frac{2kT}{R}\right],$$

that is

$$\tau_{opt} = C_t\sqrt{\frac{2kTR_{eq}\cdot R}{e\,(I_g + I_d)\,R + 2kT}} \tag{6.68}$$

and

$$V_{noise,\,min} = \sqrt[4]{\frac{1}{2C_t^2}\,kTR_{eq}\left[e\,(I_g + I_d) + \frac{2kT}{R}\right]}, \tag{6.69}$$

because the flicker noise has been neglected.

In almost all cases, a sufficiently great resistance R, can be taken such that

$$\frac{2kT}{R} \ll \mathrm{e}\,(I_g + I_d), \tag{6.70}$$

and considering $R_{eq} = \dfrac{2.5}{S}$, we get

$$\tau_{opt} \simeq C_t\sqrt{\frac{5kT}{SeI_a}}, \tag{6.71}$$

$$V_{noise,min} \simeq \sqrt[4]{\frac{1.25\,kTeI_a}{C_t\,S}}, \tag{6.72}$$

where we set $I_a = I_g + I_d$.

We observe that while the voltage signal is inversely proportional to C_t, the noise voltage is inversely proportional to $\sqrt{C_t}$; though the noise voltage decreases when C_t increases, the signal to noise ratio worsens, since the signal at the $B - B'$ terminals in Fig. 6.8 is

$$V_s = \frac{Q}{2.718C_t} = \frac{e}{2.718\,\varepsilon}\,\frac{E}{C_t},$$

whatever be the value of τ for $\tau_d = \tau_i = \tau$.

For the $(RC) - (RC)^2$ amplifier, the optimum time constant [B 30] is given by the relation

$$\tau_{opt}\,[\mu s] = 5.8\cdot 10^{-4}\,C_t\,[\mathrm{pF}]\,\sqrt{R_s\,[\Omega]\,R_p\,[\mathrm{M}\Omega]}, \tag{6.73}$$

where

$$N_{noise,min} = 22\,\sqrt{2C_t\,[\mathrm{pF}]}\,\sqrt[4]{\frac{R_s[\Omega]}{R_p[\mathrm{M}\Omega]}}. \tag{6.74}$$

From this equation we deduce that the quantity $C_t^{-1/2}\left(\frac{R_s}{R_p}\right)^{-1/4}$ can be considered the figure of merit of the vacuum-tube. The greater the value of this quantity, the lower the noise. The ratio R_s/R_p is almost constant at the variation of the vacuum-tube anodic current up to a certain maximum value which depends on the vacuum-tube type.

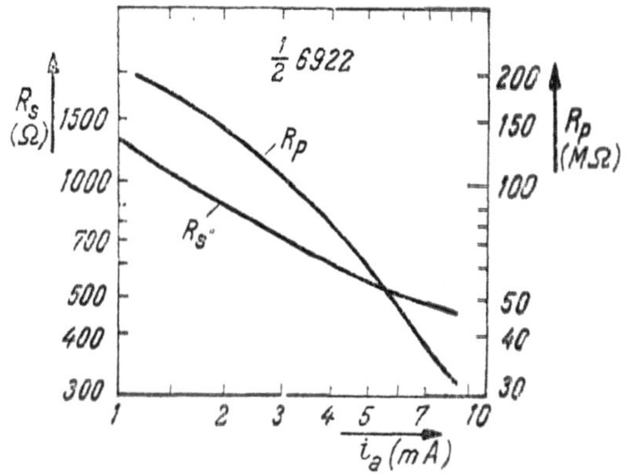

Fig. 6.12 — Variation of the series (R_s) and parallel (R_p) noise equivalent resistance with the current for the vacuum tube 6922 [B 30].

For a 6922 vacuum-tube (the equivalent of E88CC) this maximum current is about 1 mA, as one may see from Fig. 6.12 in which the variations of R_s, R_p, were plotted [B 30] in terms of I_a for half of the tube 6922 .In the C_t expression the vacuum-tube input capacity also appears. The vacuum-tube catalogues usually specify, the "cold" capacity (the filament unheated). The electron plasma surrounding the hot cathode decreases the apparent space between grid and cathode and an increase of 2—6 pF of the capacity is observed when applying the filament voltage (Table 6.4). It is hence important to measure the tube capacity when it is hot and at the chosen working point.

Considering [H 5] that the most important component of the detector current is the thermal current generated in the depletion region *)

$$I_d = 38 \frac{\sqrt{\rho V}}{\tau_0} \qquad (6.75)$$

the thickness of which is

$$d\,[\mu] = 0.22 \sqrt{\rho V}, \qquad (6.76)$$

and writing

$$C_t = C_{in} + C_d,$$

where

$$C_d \left[\frac{\text{pF}}{\text{cm}^2}\right] = \frac{3.3 \cdot 10^4}{\sqrt{\rho V}}, \qquad (6.77)$$

we obtain from equation (6.72)

$$Q^2_{noise\ min} = \left(\frac{\Sigma}{\tau_0 S}\right)^{1/2} [0.7\, C_{in} (\rho V)^{1/4} + 2.2 \cdot 10^4 \Sigma (\rho V)^{-1/4}] \qquad (6.78)$$

where Σ is the detector surface in cm^2 and τ_0 the minority carrier lifetime in μs.

Table 6.4

Variation of the input capacity with the anodic current, for some vacuum tubes

Vacuum tube	C_{in} (pF)		
	unheated	$I_a = 2$mA	$I_a = 7$mA
1/2 E88CC(6922)	4.7	6.6	6.7
EC-1000 (8254)	5.2	6.7	7.0
E83F (6689)	8.1	10.7	11.6
E810F (7788)	15.3	19.6	21.7

Observe from this relation that another optimization can be made, i.e. the bias voltage on the detector is varied up to the value

$$V_{opt} = \frac{10^9 \Sigma^2}{\rho C_{in}^2}, \qquad (6.79)$$

*) For a guard-ring detector, the approximation is rather good.

for which

$$E^2_{noise\ opt\ opt} = 250\,\Sigma\left(\frac{C_{in}}{\tau_0 S}\right)^{1/2} [\text{keV}^2], \tag{6.80}$$

for silicon detectors. Usually, however, the optimum voltage is very high and there exists the danger of breakdowns before it is reached, such that the value of the optimum voltage on the detector is estimated from an experimental curve i.e. the variation of the energy resolution vs voltage. Indeed, if $\Sigma = 0.8$ cm², $\rho = 1500\ \Omega$ cm, $\tau_0 = 500\ \mu$s ; $S = 16$ mA/V ; $C_{in} = 25$ pF, then $V_{opt} = 614$ V and we obtain $E_{noise} = 7.8$ keV. ($I_d = 72$ A ; $C_t = 35$ pF, $\tau_{opt} = 0.62\ \mu$s).

It should be mentioned, before closing this chapter, that at high counting rate [A 7] the usual technique of charge sensitive amplifiers does not give satisfactory results [B 38] due to pile-up. In such a case, the *SD* current mode of operation [P 15] is preferable.

CHAPTER VII

SEMICONDUCTOR DETECTOR ASSOCIATED ELECTRONICS

7.1. SPECTROMETERS WITH SEMICONDUCTOR DETECTORS

In addition to the detector and preamplifier, a spectrometer with semiconductor detectors consists of a main amplifier including the shaping circuits intended to give a maximum signal to noise ratio, an expander and an amplitude analyser. A multichannel amplitude analyser is generally used but it may be replaced by a single channel analyser. For calibration a pulse generator (usually with mercury relay) which injects a known electric charge quantity at the preamplifier input is also used. A block diagram of a spectrometer with semiconductor detector is shown in Fig. 7.1. The *SD* voltage supply is also represented.

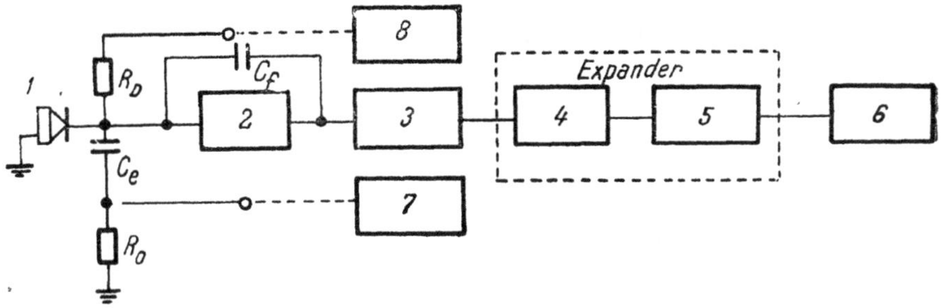

Fig. 7.1 — Block diagram of the spectrometer with *SD*: *1* — semiconductor detector; *2* — charge sensitive preamplifier; *3*—main amplifier; *4*—clipper; *5*—post-amplifier; *6* — amplitude analyser; *7* — pulse generator; *8* — *SD* voltage supply.

The system optimization (for maximum energy resolution) requires a variation of the detector bias voltage and of the amplifier shaping circuit time constants.

The energy resolution of the spectrometer is determined by the statistics of electron-hole pair creation and by the detector and amplifier electric noise. The energy dispersion (*FWHM*) is

$$\Delta E = \sqrt{\Delta E_{int}^2 + \Delta E_{el}^2}. \tag{7.1}$$

where

$$\Delta E_{int} = 2.36\sqrt{fE\epsilon}, \tag{7.2}$$

f being the "Fano factor" in *SD*, considered [B 30] to be less than 0.5.

$$\Delta E_{el} = \frac{2.36\times 2.718\,\epsilon}{1.59\cdot 10^{-19}} V_{noise}\,C_t, \tag{7.3}$$

and for a $(RC)-(RC)$ amplifier with an optimum time constant equation (6.72) gives

$$(V_{noise}\,C_t)_{min} = \sqrt{C_t}\,\sqrt[4]{\frac{1.25k\,T\,e\,I_a}{S}}. \tag{7.4}$$

These relations show that a good energy resolution can be obtained in the case of detectors with low reverse currents and capacities, and input vacuum tubes with a small $\frac{I_g}{S}$ ratio, or in other words, the figure of merit of the system detector-amplifying tube is $\frac{S}{(I_g + I_a)C_t^2}$. It is also necessary that the mean energy required by electron-hole pair creation (ϵ) be low. Accordingly, it is to be expected that at low temperatures germanium provides better results than silicon.

7.2. CHARGE SENSITIVE PREAMPLIFIERS

The preamplifier main parameters are the charge-sensitivity, dynamical capacity and the noise. When specifying the noise it is necessary to give the preamplifier input capacity value at which the measurement was performed, the type of the main amplifier shaping circuits and their time constants.

When a preamplifier meant to a given application is designed, it is of importance to know the *SD* principal parameters, i.e., the barrier capacity, the reverse current at the working voltage and the intrinsic noise, since for putting into evidence the detector characteristics, a preamplifier with corresponding parameters is required. In order to obtain the best performances, the circuit type used, the input vacuum-tube (or transistor) and its working regime, must be well chosen and the circuit lay-out carefully made, using efficient screens and ground connections suitably placed.

7.2.1. *Circuit types used in charge sensitive preamplifiers.* The first stage of the preamplifier is usually a cascode circuit which provides a pentode amplification and a triode noise. In this way the input element, vacuum-tube or transistor, noise is low and the contribution to noise of the other stages is negligible due to a high gain provided by the first stage.

The most simple charge sensitive preamplifier will be thus a cascode with capacitive negative feedback (Fig. 7.2) having cathode or White-follower at the output, able to provide a sufficiently small output impedance required when working on a matched *RF* cable.

In fig. 7.2 the same notations were used as in Fig. 6.1 for the components R_D, R_G, r_s, C_f and C_c, and for this reason they are no longer explained. At the "test" input (where the pulse generator is connected), the resistance R_0 equal to the characteristic impedance of the cable used and the standard capacity C_0 (usually of about 1 pF) through which a known electric charge quantity is injected at the preamplifier input are mounted.

The resistors r_s, r_1 and r_2 have small values (50—100 Ω) and are used for eliminating parasitic oscillations. The resistances R_3 and R_4 together with R_k ensure the static working point of the cathode follower T_3, and R_5, R_6 together with R_k determine the tube negativation. In order that the useful signal should not be divided in the ratio $\frac{R_4}{R_3 + R_4}$, the condenser C_3, the reactance of which at frequencies within the signal spectrum represents a shortcircuit in comparison with the resistance R_4, is also used.

By varying the grid bias voltage V_{g2} of the tube T_2, the plate current of the tube T_1 connected in series with the tube T_2, changes and the signal to noise ratio is thus modified.

Figure 7.3 shows the change of the noise curve shape function of capacity when the current I_a varies inside the tube T_1 [B 30] mounted in a preamplifier of the type drawn in Fig. 7.2. Within a precision of 10 % a good agreement with the noise estimated from the nomographs of Fig. 6.10 was found.

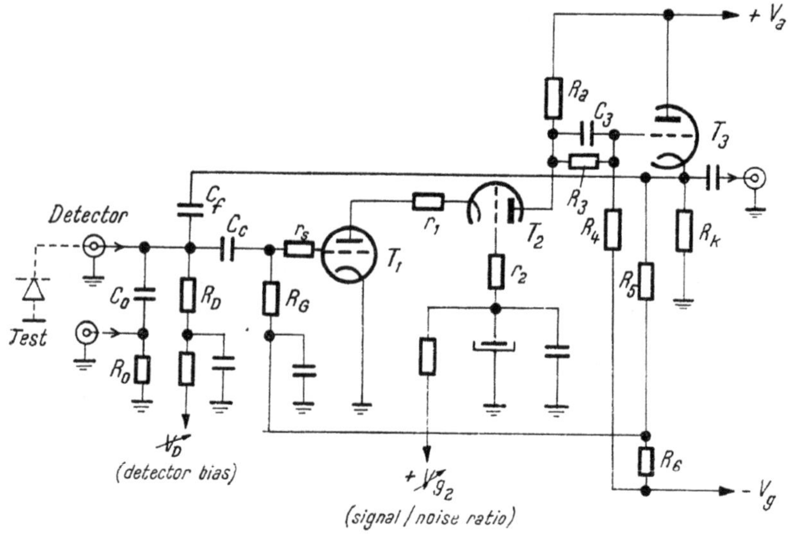

Fig. 7.2—Charge sensitive preamplifier with cascode having a cathode follower at the output.

In addition to the considerations concerning the noise, already studied in the previous chapters, the loop gain, the rise-time of the amplifier step-response and the output impedance are also of interest for the design calculation of this circuit. These quantities can be calculated using the equations given in Appendix III [P 9] for the amplifier without feedback ($\mathcal{A}_0$, t_{r0}, and R_{e0}). Since we are interested in the edge-response, the negative feedback coefficient is

$$\beta = \frac{C_f}{C_f + C_{in}} \simeq \frac{C_f}{C_{in}}, \tag{7.5}$$

and the feedback factor

$$F = 1 + \beta\,|\mathcal{A}|_0 = 1 + \frac{|\mathcal{A}_0|C_f}{C_{in}} \simeq \frac{C_{dyn}}{C_{in}}, \tag{7.6}$$

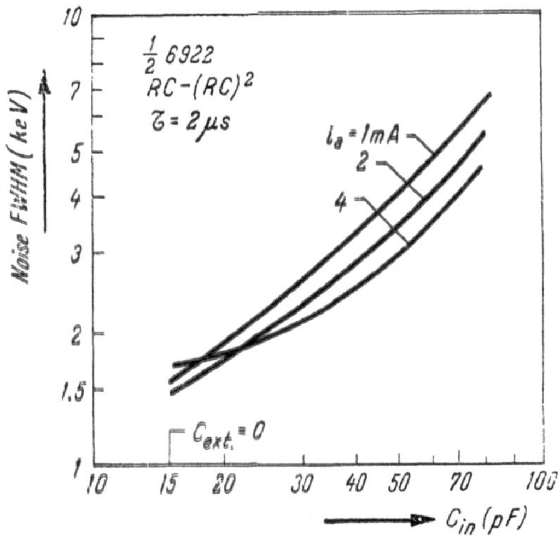

Fig. 7.3—Noise variation with the input capacity for a preamplifier with a 6922 vacuum tube at the input [B 30].

The transition to quantities of interest for the feedback amplifier can be easily made since we know that the rise-time of the negative feedback amplifier

$$t_r = \frac{t_{r0}}{F} \simeq C_{in}\frac{t_{r0}}{|\mathcal{A}_0|\,C_f}, \tag{7.7}$$

and its output impedance

$$R_e = \frac{R_{e0}}{F} \simeq C_{in}\frac{R_{e0}}{|\mathcal{A}_0|\,C_f}. \tag{7.8}$$

In all these equations the input capacity appears; when it increases, the rise-time of the transient characteristics and the output impedance increase too. Obviously, the output impedance refers here to the cascode and not to the cathode follower, for which it is known that $R_e \simeq \frac{1}{S}$, S being the vacuum-tube slope.

The dynamic capacity is

$$C_{dyn} \simeq |\mathcal{A}_0|\, C_f \tag{7.9}$$

and the charge-sensitivity $A_q \simeq \frac{1}{C_f}$, so that C_f can be determined from a compromise between the sensitivity (implying small C_f) and the dynamic capacity value (C_f great).

This circuit [B 30], [D 28] does not allow too high loop amplifications ($\mathcal{A}_0 < 300$) with reasonable rise time [B 44]; it is suited for detectors with low capacities (e.g., *PIN* type) which are sometimes used [G 17] with resistive negative feedback amplifiers.

It is known that the increase of the number of amplification stages seized with total negative feedback is not a convenient method of increasing the loop amplification since it easily leads to instabilities due to large phase differences introduced by every stage at high frequencies [D 11]. There exists a useful method [B 44] of increasing the dynamic capacity ($C_{dyn} \sim 2500$) which consists in the obtaining of a bootstrap load [W 10] at the cascode [P 9]. This circuit (Fig. 7.4) is one of the most used in charge-sensitive preamplifiers built since 1960 [C 5], [B 27] till now [G 20], [G 22], [D 1], [F 10], [G 12], [M 14], [P 12].

The bootstrapped load cascode amplifier is calculated in a way similar to the cascode, using the equations of App. III [P 9] and relations (7.7)—(7.9).

With the circuit given in Fig. 7.4 made at the Institute for Atomic Physics, we obtained [P 12] a feedback loop amplification $\mathcal{A}_0 = 600$ and a dynamic capacity $C_{dyn} \simeq 3000$ pF. The charge sensitivity of the preamplifier is 9 mV/MeV.

The variation of the preamplifier noise vs the value of the capacity connected to the input (without any optimization) is plotted in Fig. 7.5. For a zero external capacity, the noise is lower than 7 keV when the preamplifier is supplied by a 300 V stabilized power supply

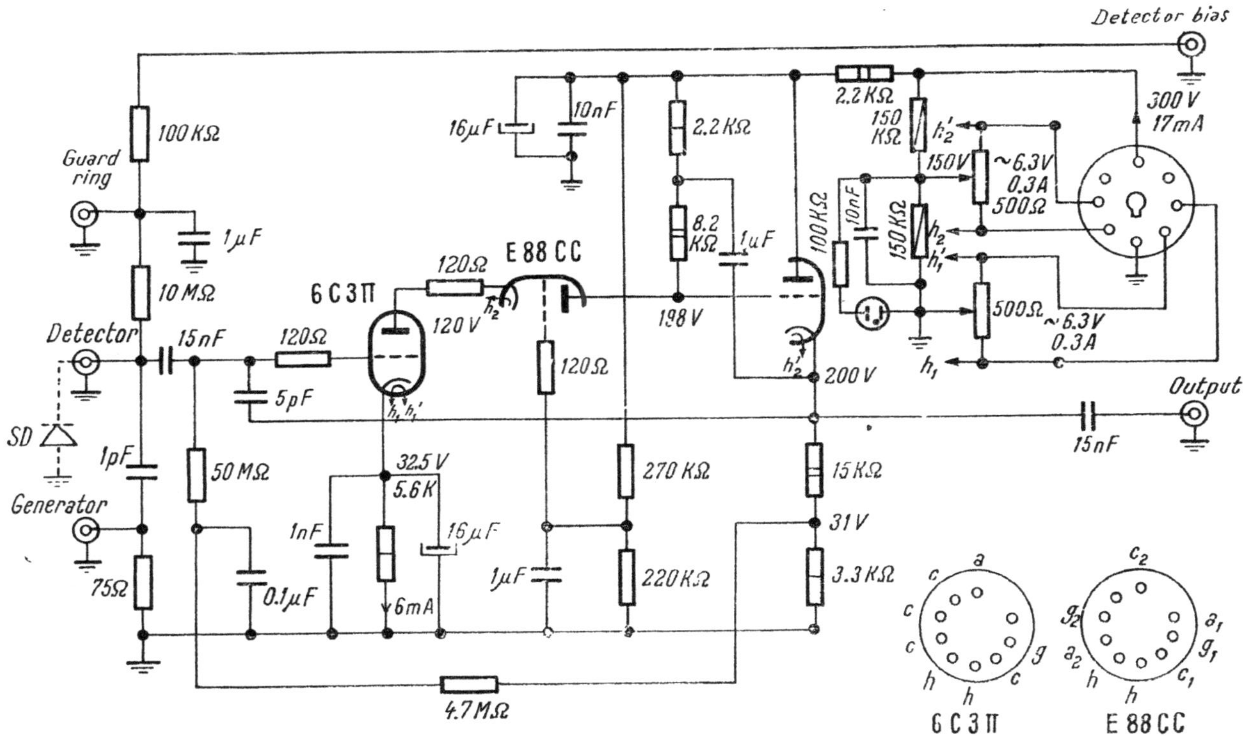

Fig. 7.4 — Charge sensitive preamplifier with bootstrapped plate load cascode.

with ripple less than 2 mV (top to top) and the vacuum-tube filaments are heated with alternative current.

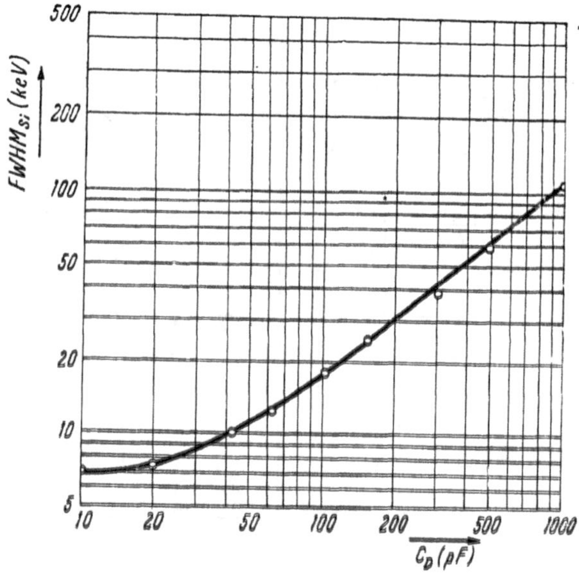

Fig. 7.5 — The preamplifier noise versus input capacity.

It is to be expected that this figure can be improved by using shaping circuits suitably chosen, and a tube filament supply with direct current.

Due to the precaution measures taken and to a filter against the hum effect, a low background was obtained, i.e. less than 2 pulses with amplitudes higher than 4 MeV after 8 hour working in the neighbourhood of important parasitic sources existing during the IAP-Cyclotron working.

Figure 7.6 shows the charge sensitive preamplifier. The dimensions of the copper box are $90 \times 105 \times 122$ mm^3.

With a recently built variant of this type of circuit [B 2] a loop amplification $\mathcal{A}_0 = 1200$ and $t_r \leqslant 0.3$ μs were obtained using a pentode (6J11P).

Another circuit [C 5], often used [T 3] is given in Fig. 7.7. After an input cascode (T_1 and T_2), the signal is amplified by an amplifying stage with cathodic coupling (T_3 and T_4) and the feedback loop closes. At the output there is a cathode-follower (T_5). The tube T_1, with high mutual conductance, is negatively biased by summing up the negative voltage from the resistance R_k which is great enough, and the positive voltage taken from the resistance R_2. The RC group is introduced to eliminate the oscillations at high frequencies. The anode of the tube T_3 and the grid of the tube T_4 are connected to earth through the capacities C_3 and C_4.

Fig. 7.6 — View of the charge sensitive preamplifier made at IAP.

A variant of this circuit was made by "Vakutronik" [G 1], [K 4] and is given in Fig. 7.8. Here, the bootstrapped plate load cascode (T_1, T_2) is followed by a White-follower (T_3, T_4) and a grounded-grid stage (T_5). A White-cathode-follower (T_6 and T_7) is used at the output. Though 7 tubes are used instead of 3, as in a bootstrapped

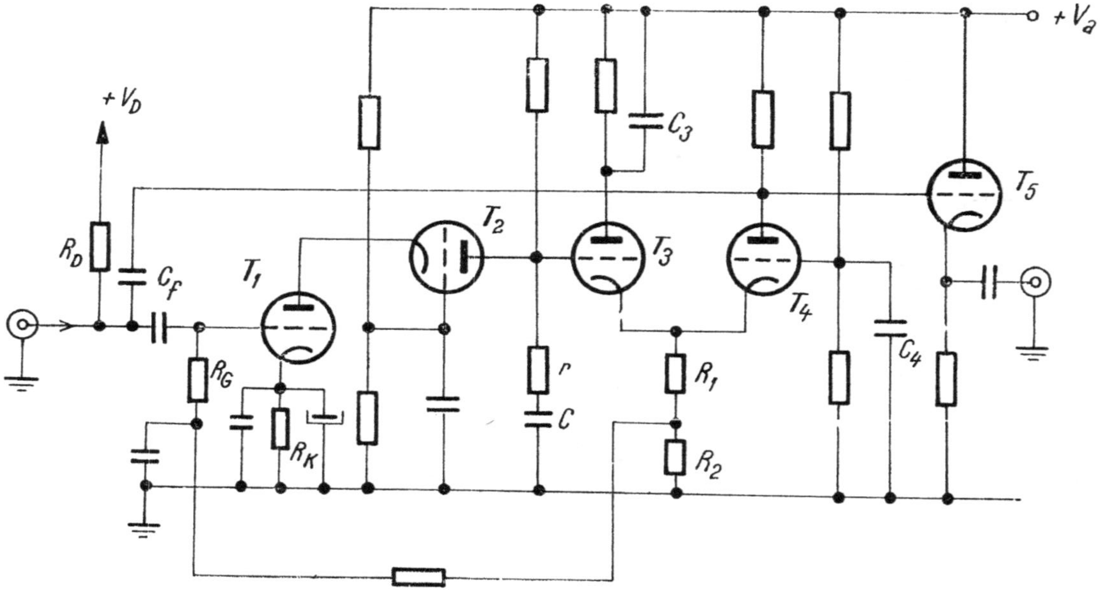

Fig. 7.7 — Charge sensitive preamplifier with cascode and differential amplifier.

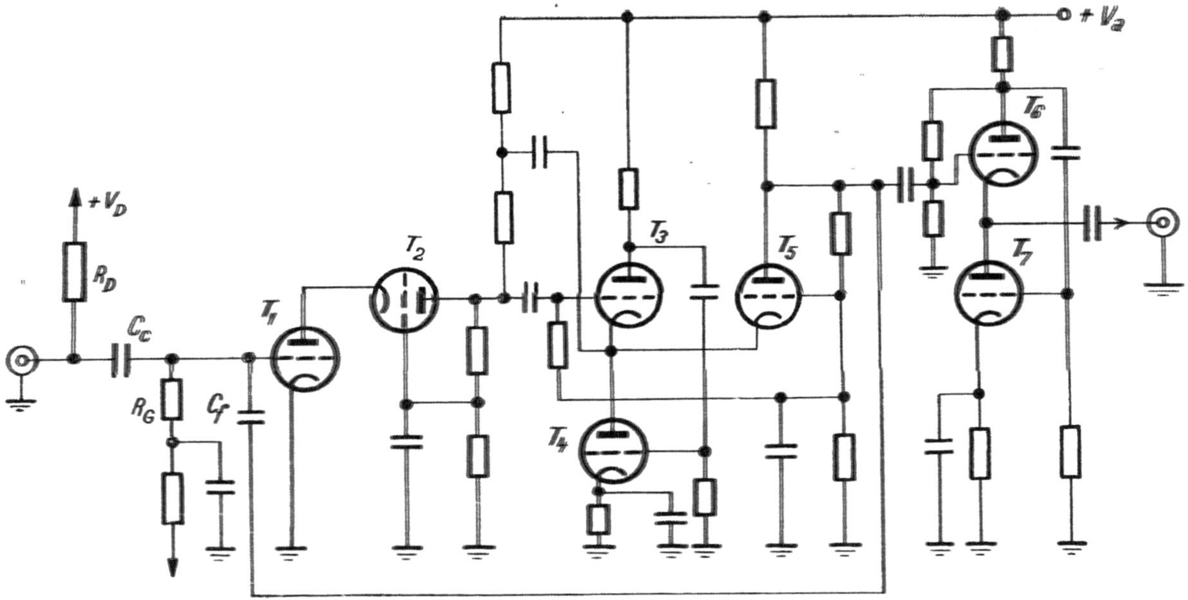

Fig. 7.8 — The Vakutronik-type charge sensitive preamplifier.

load cascode (Fig. 7.4), the performances of this circuit are not much improved in comparison with the other one.

The circuit given in Fig. 7.8 and reported in [H 1] provides the highest loop amplification.

Similarly to the bootstrap cascode, the cascode (T_1, T_2) of this circuit has a dynamical charge given by the tubes T_3 and T_4. It differs from the bootstrap cascode by the fact that the latter has a positive current feedback while the bootstrap cascode has a positive voltage feedback.

In addition, this configuration offers the possibility of inserting in the positive feedback loop a circuit able to modify the phase characteristic; under certain conditions, the gain band-width product is considerably increased.

A current I injected in the node A will flow through T_3, because the input impedance of this one is small

$$\frac{R_l + R_{i3}}{\mu_3 + 1} \leqslant R_f + \frac{R_{i4}}{\mu_4+1}. \tag{7.10}$$

This current will give rise, on the plate of the tube T_3, to a voltage IR_l which will be found again on the cathode of the tube T_4, yielding a voltage IR_f on the resistance R_f and if $R_f = R_l$, then at the point A the same current I returns, cancelling the initial current; an infinite impedance is obtained (finite-voltage with zero current). A more accurate analysis [H 1] shows that the impedance at the point A has the form

$$R = \frac{r}{1-g}, \tag{7.11}$$

where

$$r = \frac{R_l + R_{i3}}{\mu_3 + 1},$$

and

$$g = \frac{R_l}{R_f + \dfrac{R_f + R_{i4}}{\mu_4} + \dfrac{R_f + R_{i3}}{\mu_3}} \simeq \frac{R_l}{R_f}.$$

From these equations, we notice that it is possible that $g \simeq 1$ and thus $R \to \infty$.

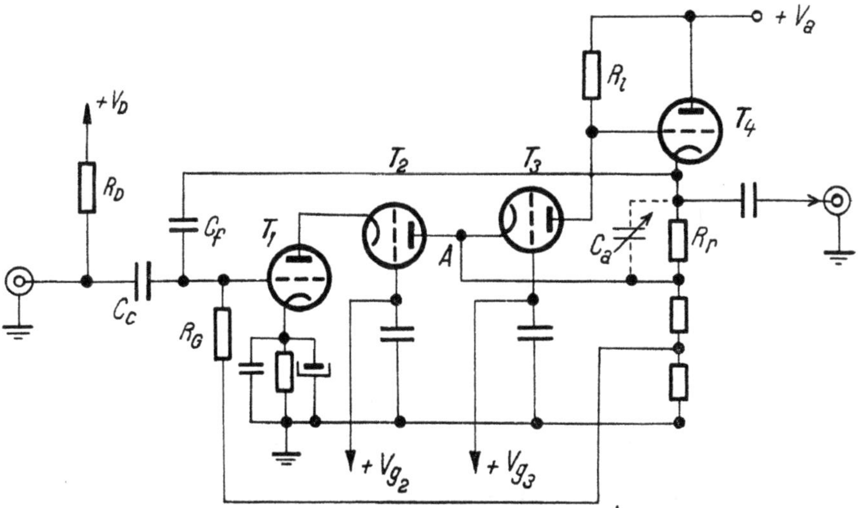

Fig. 7.9 — Charge sensitive preamplifier with current positive feedback circuit as cascode load.

The voltage amplification of this circuit is

$$\mathcal{A}_i \simeq \mu_1 \mu_2 \frac{\mu_3 R_l}{R_{i3} + R_r}. \tag{7.12}$$

The compensation at high frequency of the effect of parasite capacities in parallel with the anode T_3, can be achieved with a capacity C_a connected in parallel with R_r. Due to the high loop amplification, there is an important tendency to oscillations. The authors mention that their apparatus requires $\frac{C_{in}}{C_f} > 15$ in order that the preamplifier should not oscillate.

The loop gain achieved with vacuum tubes with $\mu_1 = 56$ and $\mu_2 = \mu_3 = 28$, and resistance $R_1 = 9\ \text{k}\Omega$ was 15 000—30 000, and the transient characteristic rise time varied from 20 to 100 ns for input capacities within 100 and 1000 pF (for every C_{in} value C_0 was modified such that overshoot should not exist at the output).

7.2.2. *Input vacuum tube and circuit.* The loop amplification required for obtaining a given dynamical capacity with convenient sensitivity is achieved by connecting the adequate tube to one of the circuits described above. In order to have a good sensitivity stability C_f must be a condenser with minimum temperature coefficient. The noise is not influenced by the circuit type, but only by the input tube and its parameters. Ordinarily, the triodes (6C3П, 6C15П, 417A, EC1000, 6H16Б), the Nuvistors for small volumes (RCA 7586, 8056), the double triodes 6H23П, 6H14П, E88CC (6922) or the pentodes 6Ж9П, E 810 F (7788), E 83 F connected as triodes are used.

In a preamplifier without external capacity connected at the input (the circuit proper capacity was 15 pF) a noise value of 1.5 keV (*FWHM* in Si) was obtained by means of a 6922 vacuum-tube [B 30], 2.2 keV with the tube 7788 (E 810 F) [B 27], 3 keV with 417 A and EC 1000 [B 27], [L 5]. These noise values are usually estimated at $\tau_d = \tau_i = 1 \mu s$ in the main $(RC) - (RC)$ amplifier. Electrometric tubes are not adequate [C 17] due to their low mutual conductances.

The performances mentioned above represent peak values, reached after a vacuum-tube selection. In the case of serial production (2000 unselected tubes) values were measured within 5 and 5.5 keV [G 25] in comparison with the theoretic value (4.3 keV). For small input capacities (12.5 pF), the theory cannot be well verified, due probably to the flicker noise which has not been accurately estimated. For great capacities (1000 pF), the calculations have been almost perfectly checked up.

The vacuum-tube type is chosen according to the detector capacity and current [G 22], [B 44]. High resolution experiments require a detector with small capacity (C_d) and low reverse current (I_d). In such a case, the vacuum tube must have a maximum figure of merit $\frac{S}{I_g C_{in}^2}$ (6 H 23 π, for instance).

Once the input tube is chosen, its working conditions must be examined. From the grid current standpoint the input tube can work in two regimes i.e. either with a high negative bias $|V_{g0}| > 1.61$ V which reduces the grid current to the ion component [C 17], or in a "floating-grid" regime when the ion component of the grid current equals the electronic component and a negative biased voltage settles down without being necessary to connect a resistance R_g to the grid [D 25]. This way of working is generally used, because in this case the resistance R_g can be as great as possible without affecting the negative bias of the tube. The influence of the preamplifier input resistance on the signal to noise ratio is studied in [K 2]. An $R_G \gg 50-1000\ \mathrm{M}\Omega$ is used in order that its thermal noise should be negligible. As an example [D 25], Fig. 7.10 gives the family of $I_a - V_g$ and $I_g - V_g$ characteristics of the vacuum-tube 6Ж9П connected as triode (the screen and the suppressor grid connected with the anode). On the characteristics $I_a - V_g$ are shown the static working-points at which $|I_{g-}| = |I_{g+}|$.

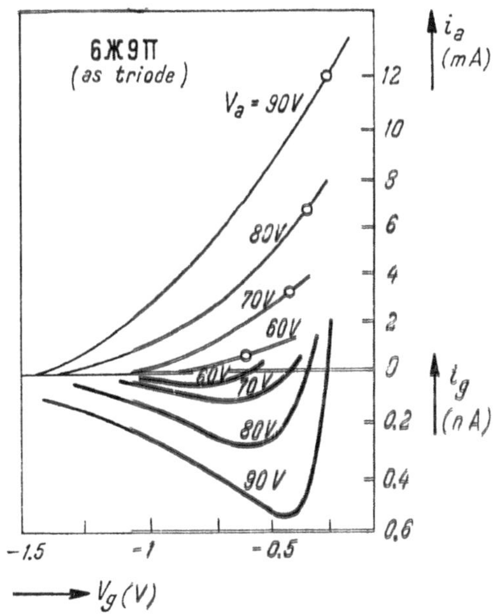

Fig. 7.10 — $I_a - V_g$ characteristic of the 6 ж 9 п vacuum tube connected as triode [D 25].

The static working-point of the vacuum tube (the anode current value when $I_g = 0$) is chosen by means of the characteristic $\dfrac{R_{sn}}{R_{pn}}$ obtained with the method described above [B 30], the tube figure of merit being $\left(\dfrac{R_{f\prime}}{R_{pn}}\right)^{-1/4} \cdot C_{in}^{-1/2}$.

The circuit must be carefully designed. For the input stages resistances with low noise and high stability (metallized or reeled) are chosen, and the filament voltage of the first tube is sometimes lowered [D 1].

When the vacuum-tube cathode is supplied at high potential against the earth, one of the filament electrode must be connected at a d.c. potential close to that of the cathode and not to earth. The anodic

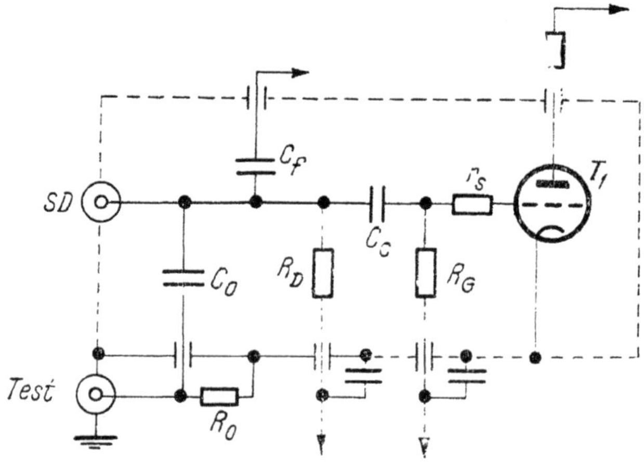

Fig. 7.11 – Screening of a charge sensitive preamplifier input [B 30].

supply voltage ($V_a = 300$ V) has a stabilization up to 1% and its hum reduces below 1.0 mV. The input tube circuit is carefully screened [B 30] as in Fig. 7.11 for instance, and the pieces are placed on the frame so that a minimum circuit capacity results. A great attention must be also paid to the obtaining of good earth connections [G 12].

7.3. MAIN AMPLIFIER

As in the case of scintillation detectors, the operation of *SD* requires nonoverloading linear amplifiers with a maximum gain of the order of 10 000.

In addition, for *SD* a higher versatility and a better stability in time are necessary.

In order to obtain the optimum signal to noise ratio, the amplifier is equipped with a differentiation circuit and one or two integration circuits (the $RC - RC$ or the $RC - RC^2$ type) with step-controlled time constants within 0.1 and 10 μs.

When choosing the place of the differentiation circuits inside the amplifier, account should be taken of the possibility of the appearance of overloadings due to the overlapping of fast succeeding pulsees.

The differentiation circuits are generally placed near the input [B 44], where the base-line shift does not exceed 0.1 V, and the integration circuits towards the output (lest the noise of the expander postamplifier should count).

At any rate it is not adequate to place the *RC* circuits between the stages included inside the negative feedback-loop.

In applications, where high counting rates are met with, the lower limit of the amplifier band width is determined by one or two shaping delay lines. In this case the output pulse is rectangular (a single line) or bipolar (two lines) and its duration, for the optimum signal to noise ratio, is given by the equation

$$2\tau_D = 0.8\ \tau_t\ . \tag{7.13}$$

A simplified block-diagram of a linear amplifier with a high versatility [G 12] is plotted in Fig. 7.12.

The commutator K_1 has several functions : in position 1 it achieves a direct coupling to the resistance $R_A = 1$ kΩ, "jumping" over the differentiation circuit ; in position 2, the variation of the capacities C_d allows a variation of the time differentiation constant within 0.2 and 5 μs ; in position 3 the delay line is coupled in short circuit L_1. The step by step control of the gain between 0 and 16 is achieved on the first two positions, modifying the plot position on the divider R_A and the resistance ratio $\frac{R_2}{R_1}$ so that the line matching condition should be accomplished $(R_g + R_1) \parallel R_2 = R_0$ on the position 3. For the first three positions the input impedance is ~ 1 kΩ and for position 4 it is $R = 100$ kΩ, here no attenuation being introduced.

The first amplification unit *AI* is a "triplet" with negative voltage feedback and a cascode as the first stage. By changing the degree of negative feedback with the aid of the potentiometer *P*, a fine control of the gain can be performed within the limits (0.1—1). The last stage has both anode and cathode load, and hence this first gain unit plays the part of an inverter ; the commutator K_2 selects the input pulses of negative or positive polarity. The commutator K_3 permits to introduce a second short-circuited delay line L_2.

A White-cathode follower *WCF* is followed by a two-position commutator K_4 : on the position 1 the signal is amplified by the

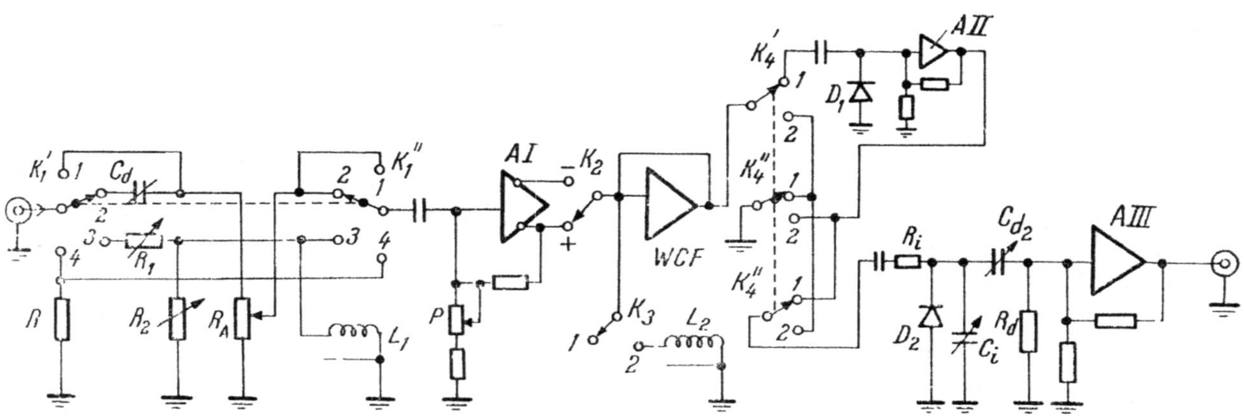

Fig. 7.12 — Block-diagram of a main amplifier.

triplet A II too, and on the position 2, used when the input pulse is high enough, the pulses go towards the final section A III. The diodes D_1 and D_2 have a limiting role in order to prevent overloadings due to pulses of high amplitude.

The integration circuit R_iC_i has a time constant that can be varied between 0.2 and 5 μs and the second differentiation circuit $C_{d2}R_d$ is used for higher time constants τ_{d2} (5 μs ÷ 5 ms). The last stage provides at the output a maximum linear amplitude of ~ 130V (a mono or bipolar pulse). For × 100 or × 1000 overloadings, the recovery time is 30 and 50 μs, respectively, and at a pulse-packet frequencies of $5 \cdot 10^4$/s (when working with a modulated beam) the base line shift is less than 0.5 %.

Fig. 7.13 — Front panel of the amplifier made at IAP.

Figure 7.13 gives a frontal panel picture of a linear amplifier and expander (see the next section) built at IAP according to the block scheme given in Fig. 7.12. The maximum gain it can provide is 20 × 25 × 30. The maximum output voltage amplitude is 100 V. The pulse given by the shaping line has a duration of 0.8 μs.

In some amplifying stages bootstrap dynamical loads are used (see appendix III) [P 10], which provide wide dynamical ranges and high gains with a small number of components.

For a fast recovery from overloading, Fairstein [F 7] suggests a triple differentiation. An excellent review paper concerned with pulse amplifiers used in nucleonics is that due to Fairstein and Hahn [F 6].

7.4. AMPLITUDE ANALYSER AND EXPANDER

The multi-channel amplitude analysers have been built in order to take advantage of all the *SD* possibilities in the amplitude analysis of some very wide energy ranges. For instance an amplitude analyser with 4096 channels is used in the study of energy spectra within the range 0—40 MeV [C 6]. In this manner in this energy range an accuracy of 10 keV/channel is obtained.

A drift of the system amplification, less than one channel, requires a stability of the amplifier gain and of the time amplitude converter gain and zero-point higher than 2.5×10^{-4}, a technical performance which can be hardly achieved. For this purpose an automatic amplitude—control system is used, which can compensate variations of $\pm 3\%$ and $\pm 1.5\%$ of the converter gain and zero-point respectively.

A generator of pulses with stable amplitude (10^{-4}) yields two reference pulses with repetition frequency of 50 Hz and amplitude ratio equal to eight. These pulses are alternatively applied to the preamplifier input and must be recorded in the 512-th and 4096-th cannel of the analyser, respectively. The reference pulse errors in "finding" the channels represent error signals for two automatic control systems of the converter gain (great pulse) and of its zero-point (small pulse).

Since this is a digital system, it does not lose the control even if a considerable number of reference pulses cannot be recorded because the pulses coming from the detector are processed by the converter.

This system is very complicated and expensive and in addition, there are very few cases when the whole energy range from zero to the maximum energy is considered.

In these circumstances the observation of a narrow spectrum region with corresponding accuracy can be made through the changing of the amplitude scale by translating and multiplicating it by a constant factor (Fig. 7.14). The expander inserted in the spectrometric chain between the main amplifier and amplitude-analyser makes this operation by clipping the pulse signals at a well determined stable threshold and by their subsequent amplification. For this purpose the expander contains a clipper and a "postamplifier".

From the example given in Fig. 7.14 one may observe the manner in which in the case of a 100 — channels analyser which uses an expander with cut-off threshold corresponding to the 67-th channel and a postamplification of 3, the peak *FWHM* extends over a number of channels three times greater ; thus it may be estimated with a better accuracy.

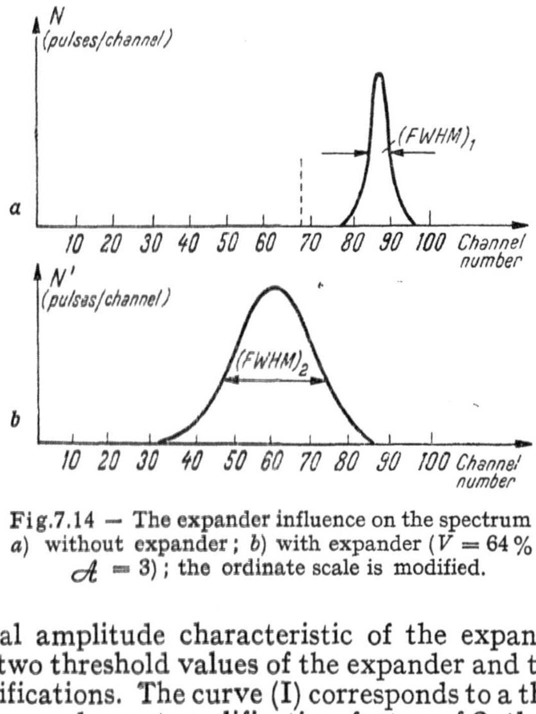

Fig.7.14 — The expander influence on the spectrum : *a*) without expander ; *b*) with expander ($V = 64\%$; $\mathcal{A} = 3$) ; the ordinate scale is modified.

The ideal amplitude characteristic of the expander is given in Fig. 7.15, for two threshold values of the expander and two corresponding postamplifications. The curve (I) corresponds to a threshold $V_{in1} = 0.5 \times (V_{in})_{max}$ and a postamplification factor of 2, the last stages of the postamplifier and main amplifier having the same dynamical range $(V_o)_{max} = (V_{in})_{max}$. If the cut-off threshold is placed at $V_{in} = 0.75\ (V_{in})_{max}$ as in the curve (2) then the necessary postamplification will be 4. The expansion factor is equal to the postamplifier gain.

If the spectrum interval is considered around 5 MeV energy and an analysis accuracy of 1 keV/channel is expected from a 256 — channels analyser, the maximum expansion factor necessary will be 20 for obtaining the equivalent $256\times20\simeq5000$ channels. The stability of the preamplifier sensitivity, of the main amplifier gain and of the expander cut-off threshold must be higher than $\frac{1}{5000}=2\times10^{-4}$. If the maximum linear amplitude of the signal at the postamplifier output is 100 V and the postamplification is 20, the signal corresponding to the maximum channel

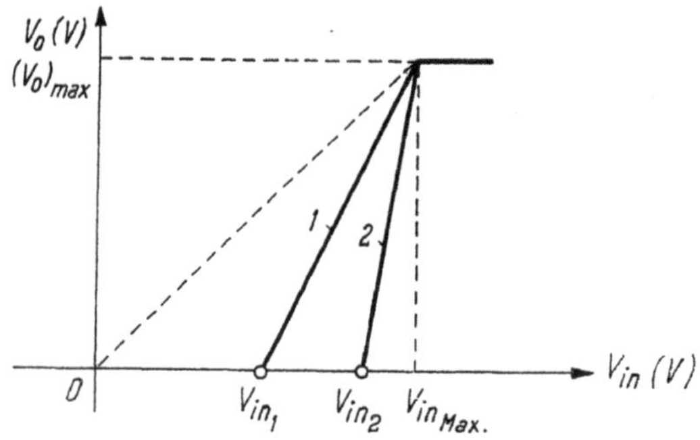

Fig. 7.15 — Ideal amplitude characteristics of an expander.

(256) will have at the postamplifier input $\frac{100}{20}=5\text{V}$, hence the channel width will be $\frac{5\text{ V}}{256}\simeq 20\text{ mV}$. The threshold must have a drift less than 20 mV, lest it should exceed the channel width. This value is very close to the limit that can be reached, if all possible precautions are taken in the design of the final amplifier and its power supply. The threshold potentiometer must be supplied from a gas-stabiliser tube or a regulated power supply using such a tube as reference source of voltage. The reference tubes must be protected by screens against air-currents; the same thing must be done for the respective resistances which are reeled [F 5].

Figure 7.16 shows the expander circuit diagram [G 20]. The serial type clipper is made of the diode D_1 and the resistance R_1.

The threshold is controlled with the potentiometer P. The diode D_2 together with the condensers C_1 and C_2 are meant to broaden the pulses, which have an exponential shape at the input and become very sharp after the base cut-off. By means of the feedback applied through the resistance R, a pulse with an almost constant plateau and a

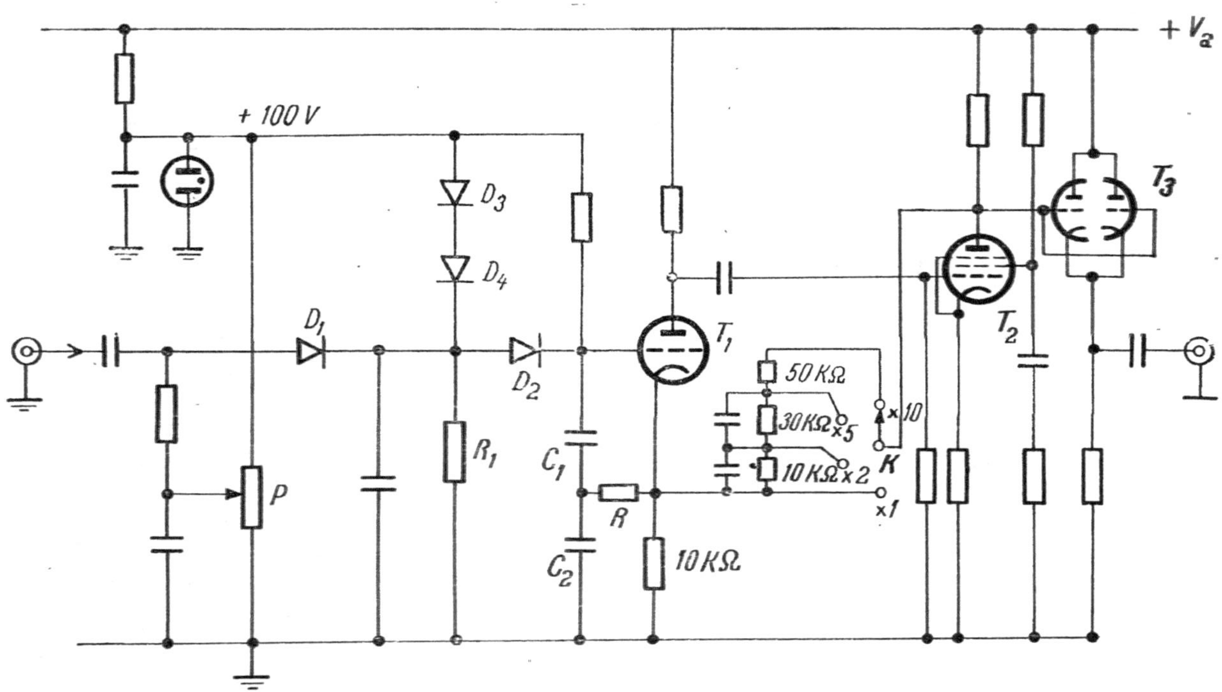

Fig. 7.16 — The circuit diagram of an expander.

duration representing a significant fraction of the total recovery time may be obtained. The diodes D_3 and D_4 compensate the effects of the voltage-drop charges on D_1 and D_2 when the voltage varies.

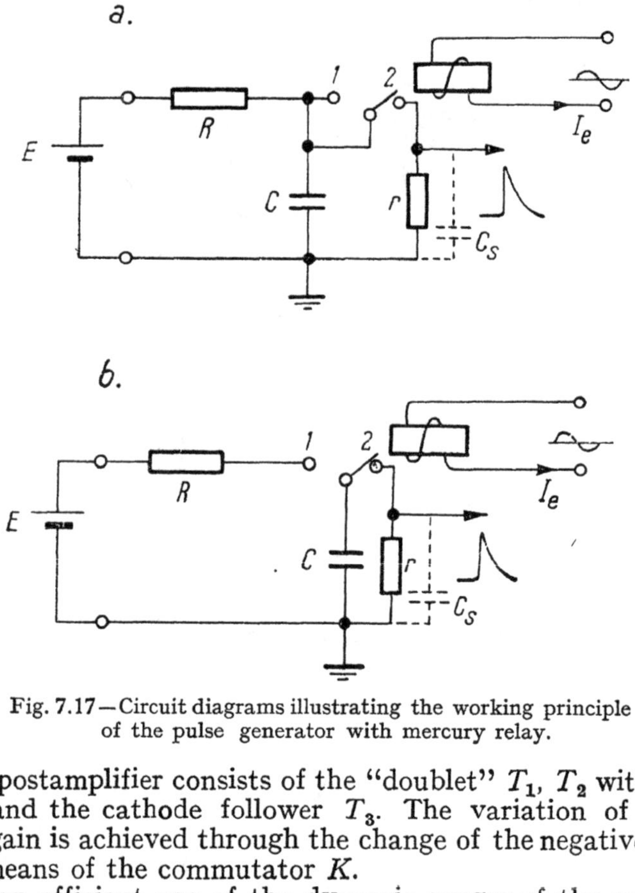

Fig. 7.17—Circuit diagrams illustrating the working principle of the pulse generator with mercury relay.

The postamplifier consists of the "doublet" T_1, T_2 with negative feedback and the cathode follower T_3. The variation of the postamplifier gain is achieved through the change of the negative feedback ratio by means of the commutator *K*.

For an efficient use of the dynamic ranges of the main amplifier final stage and postamplifier, a simultaneous control of both the threshold and the expansion factor is suggested [G 12]. For example, to a series of expansion factors 1, 2, 4, 8 there corresponds a series of thresholds (given in relative units) : 0, 0.5, 0.75, 0.875. Instead of the potentiometer *P*, a commutator is connected to achieve the given division.

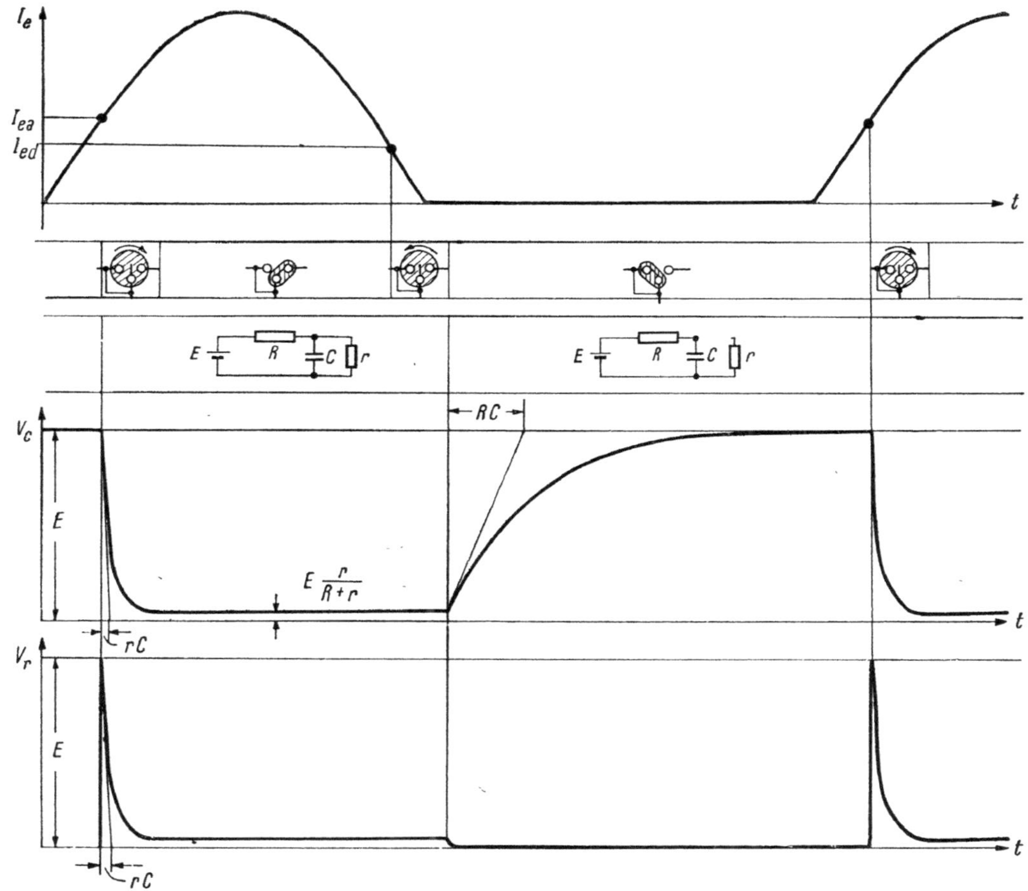

Fig. 7.18 *a* — Waveforms corresponding to the circuit configuration in Fig. 7.17 *a*.

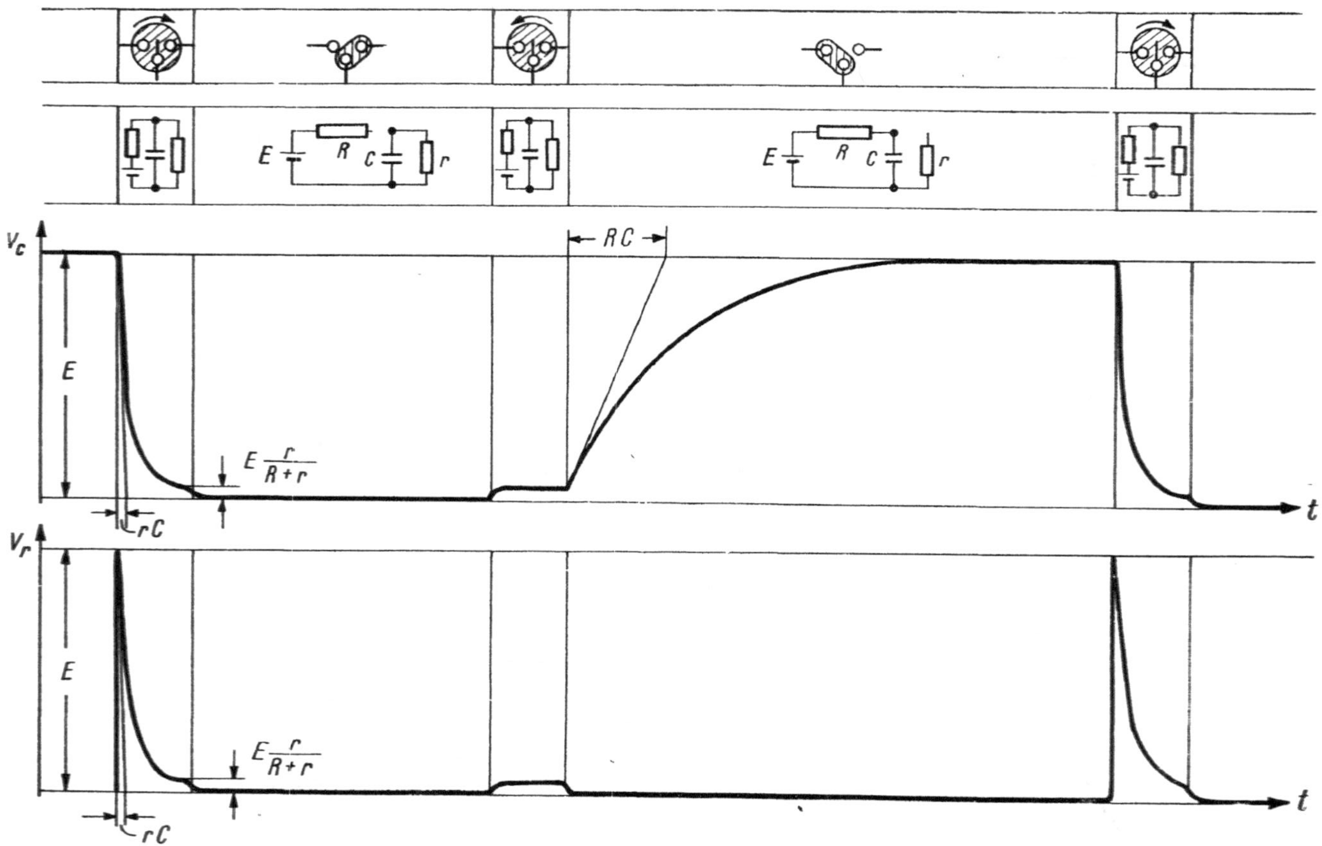

Fig. 7.18 *b* — Waveforms corresponding to the circuit configuration in Fig. 7.17 *b*.

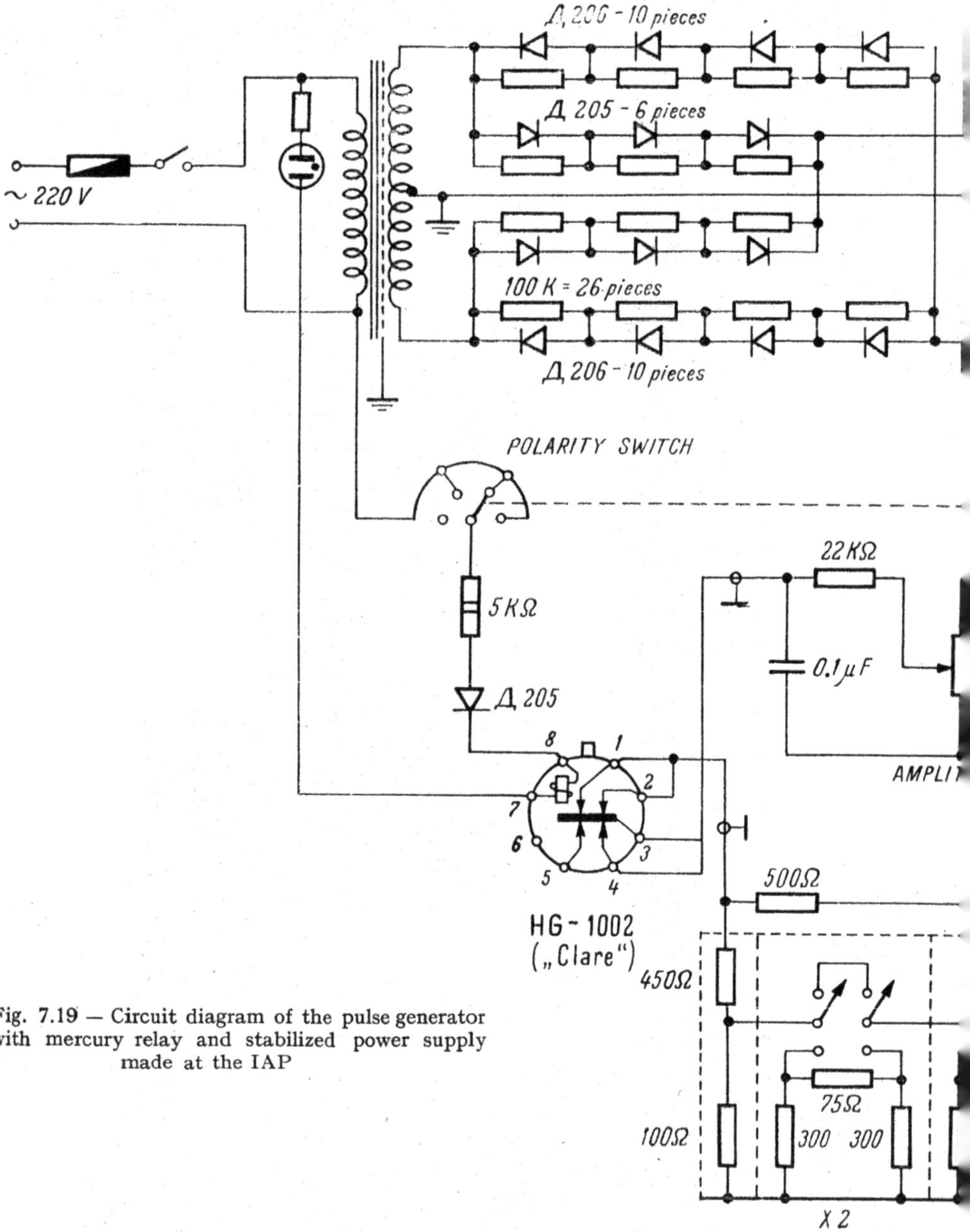

Fig. 7.19 — Circuit diagram of the pulse generator with mercury relay and stabilized power supply made at the IAP

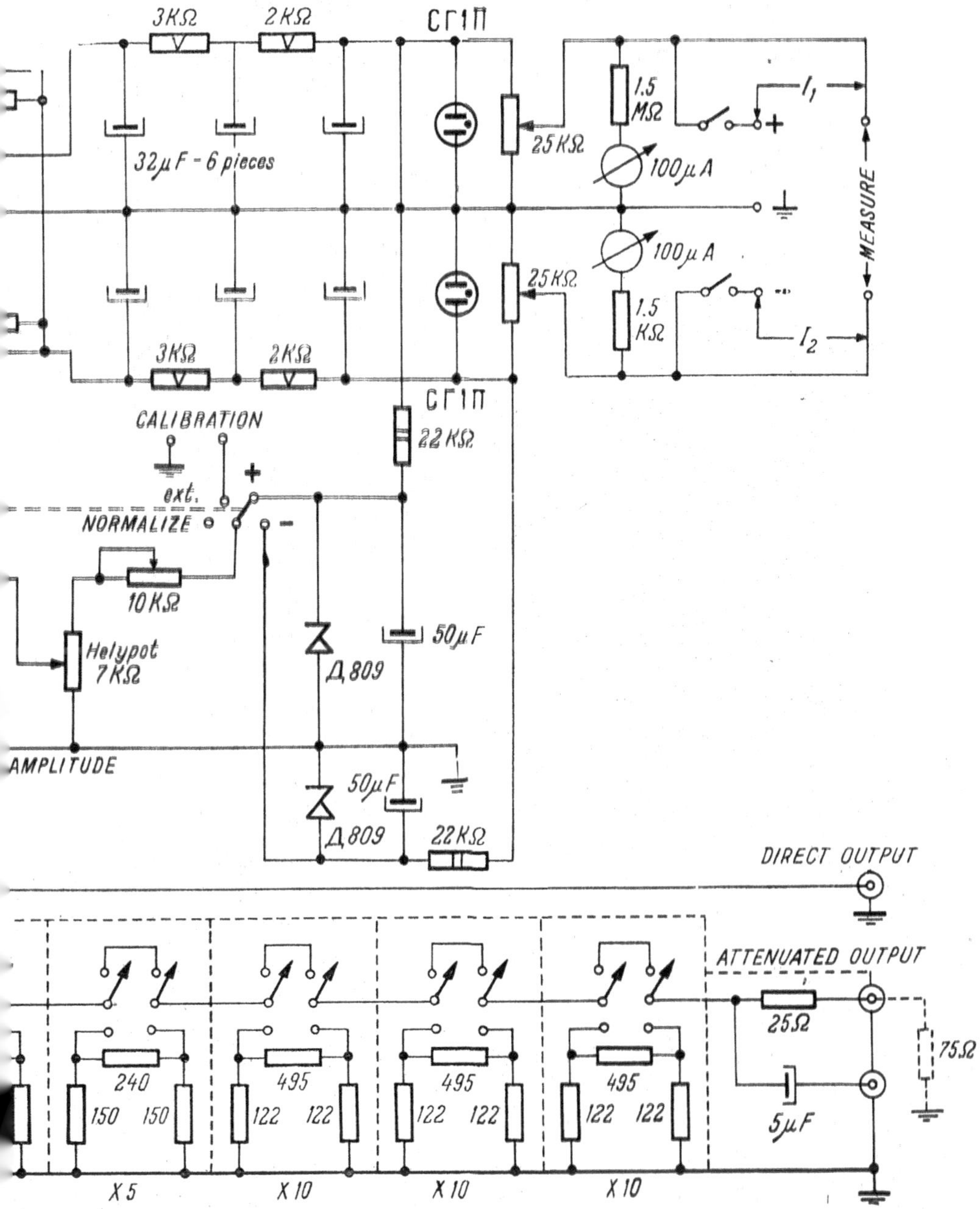
3KΩ
2KΩ
СГ1П
32μF - 6 pieces
25KΩ
1.5 MΩ
100μA
I_1
+
MEASURE
100μA
25KΩ
1.5 KΩ
I_2
3KΩ
2KΩ
СГ1П
CALIBRATION
22KΩ
ext.
+
NORMALIZE
-
10KΩ
Helypot 7KΩ
Д809
50μF
AMPLITUDE
50μF
Д809
22KΩ
DIRECT OUTPUT
ATTENUATED OUTPUT
25Ω
75Ω
5μF
240
150
150
495
122
122
495
122
122
495
122
122
X 5
X 10
X 10
X 10

7.5. HIGH AMPLITUDE STABILITY PULSE GENERATOR

The charge pulse yielded by a nuclear radiation detector with semiconducting devices (*SD*) can be quite well simulated by a voltage pulse with fast rise time (nanoseconds) and slow exponential decay (hundreds of microseconds), applied on the detector through a rather low coupling capacity.

Due to this fact, in nuclear-particle spectrometers equipped with *SD*, pulse generators with a rather simple structure are used on a large scale meant to make the energy calibration, to check up the linearity and the stability as well as to measure the energy resolution of the spectrometer.

Fig. 7.20 — Frontal panel of the pulser and power supply.

Usually, the calibration does not require an accuracy higher than 1% so that the pulse amplitude accuracy must not be still higher but for measuring resolutions of the order of 2 keV at 5 MeV, i.e., 0.4 ‰, pulse amplitude stability of about 10^{-4} is necessary (for a short time, during the spectrum measurement). If it is desired to measure the long term system stability, this stability has to be preserved during eight hours of working.

Pulses fulfilling all these conditions can be generated by means of a mercury relay. In order to simplify the generator construction, the pulse frequency usually chosen is that of the a.c. line.

Due to the fact that the pulse generator working requires a highly stabilized voltage with low ripple, its construction can be combined with that of a regulated power supply for the *SD* bias voltage, the more so as here the ripple must be also reduced and the *SD* currents are low (at most of the order of microamperes).

The working principle of the pulse generator with mercury relay is illustrated by the circuit diagram given in Fig. 7.17. There are used two variants (*a*) or (*b*) [A 16], [C 16]. In both cases, a highly stabilized power supply E charges a condenser C through the resistance R, as long as the commutator set in action by the relay is in the position (*1*). While the commutator is shut on the other position (*2*), the condenser C discharges on the resistance r and thus the voltage on the resistance r varies exponentially with the time constant rC from the value $E \dfrac{C}{C + C_s} \simeq E$ towards the value $E \dfrac{r}{R + r} \simeq 0$, because usually $C \gg C_s$ (stray capacity) and $r \ll R$.

By supplying the primary relay winding with simple rectified line current, the pulse frequency is 50 Hz and the voltages on the condenser and on the load resistance, in the case of the circuits given in Figs 7.17 *a* and 7.17 *b* have the variation in time drawn in Figs. 7.18 *a* and 7.18 *b*, respectively.

Due to the mercury relay, the pulse rise time is limited only by the contact resistance and the parasite capacity. As distinguished from the usual relay with dry contacts, where the contact resistance has a comparatively slow variation with the contact pressure, the mercury relay has a very fast variation of the resistance, which makes it suitable for the construction of nanosecond rise time pulse generators. The contact reed of these relays is grooved with several capillary channels towards a mercury tank and is plated with a metal (e.g. Ni), "wetted" by the mercury. In this way the whole contact surface is wetted by the mercury. This fact provides a series of advantages, besides the higher commutation rate i.e. the repeated contact due to reed vibrations disappears, the reliability is higher due to an increased corrosion resistance, the working duration is longer ($\sim$100 million operations).

When the reed passes from one contact to the other, there is a period (0.2 : 2 ms) indicated in Figs. 7.18 *a* and *b*), in which before the cutting of the former contact, the mercury provides a new one. This mode of operation "make-before-break" can become a disadvantage in certain applications; in our case it restricts the maximum pulse duration. As one may see in Fig. 7.18 *b* this makes a low amplitude $E\frac{r}{R+r}$ pulse to appear at the output; if the resistances r and R are chosen such that to satisfy the inequality $R \gg r$, the amplitude of this pulse is negligible. However, the resistance R cannot have too large values, because if we do not have $RC \ll 10$ ms, the capacity C does not charge up to the voltage E during the half-period of the line voltage and the pulse amplitude stability depends also on the stability of the resistance R and capacity C and not only on that of the voltage E. The capacity C must be great enough to achieve the desired decay time-constant (rC) and at any rate it cannot verify the inequality $C \gg C_s$. There exist relays which do not have the disadvantage mentioned above. If such relays are connected to the circuit given in Fig. 7.17 *b*, durations as long as possible can be obtained when using great capacities C since the resistance R should no longer satisfy the inequality $R \gg r$.

As a stabilized power supply (E), one can use according to the value and the stability expected for pulse amplitudes a transistor stabilized rectifier, a Zener diode, or a gas filled stabilization tube. In order to increase the voltage stability of the Zener diode, this one may be supplied by a constant current stabilizer and for rendering it insensitive to temperature variations, a simple thermostat made with a power transistor and a thermistor may be introduced.

Figure 7.19 gives the circuit diagram of the apparatus constructed at the Institute of Atomic Physics and Fig. 7.20 the photograph of its frontal panel.

The generator is constructed in compliance with the diagram shown in Fig. 7.17 *a*, where $r = 500\ \Omega$, $C = 0.1\ \mu$F and $R = 22$ kΩ. These values determine a pulse decay time constant $rC = 50\ \mu$s and a condenser charging time constant $RC = 2.2$ ms. The 50 μs decay constant is great enough when considering the usual 10 μs maximum time constants of the main amplifier differentiation circuits.

The generator can be used in three different modes. In two cases, positive or negative pulses are yielded at the output, using for this purpose two internal power supplies (E) i.e., Д — 809 type Zener diodes. The output pulse maximum amplitude is then 8 V. If a higher amplitude is necessary (up to 100 V) or the pulse amplitude is calibrated with a standard power supply, there is a third way of using it: it is the "external" way, when the external voltage supply of wanted polarity is connected to the "calibration" terminals.

A fine control of the pulse amplitude can be made by means of a ten-turn helicoidal potentiometer (built at IAP) and a step control (2.5—10) with screened π-cell attenuator.

The elements of this attenuator (Fig. 7.21) can be calculated with the equations

$$R_1 = R_0 \frac{1 - A^2}{2A}, \tag{7.14}$$

$$R_2 = R_0 \frac{1 + A}{1 - A} \tag{7.15}$$

where the impedance R_0 and the attenuation $A = \frac{v}{V}$ appear. The resistance values given in the scheme (Fig. 7.19) were obtained for $R_0 = 100\ \Omega$ and $A = \frac{1}{2};\ \frac{1}{5};\ \frac{1}{10}$ and achieved by connecting in series and in derivation different metallized resistors of standard values.

It is to be noticed that although the distance between the pulse generator and the preamplifier is short, it is necessary to work under matching conditions, due to the quite fast rise time of the pulse. In this respect the BNC connectors raise no difficulties in the case of nanoseconds rise times.

Using a normalization potentiometer, the helicoidal potentiometer may be calibrated in energy units for a given coupling capacity (C_i) at the charge sensitive preamplifier input

$$V\,(\mathrm{mV}) = \frac{45.5\,E\,(\mathrm{MeV})}{C_i\,(\mathrm{pF})}. \tag{7.16}$$

The mercury-relay excitation winding is supplied with a pulsed current obtained by monoalternance rectification from the 110 V winding of the line transformer primary circuit. The secondary circuit with median plug of this transformer is used to obtain high positive and negative voltages by rectifying both alternances. After a corresponding *RC* filtering, these voltages are used together with gas filled stabilization tubes СГ1П, both "inside" to yield standard voltages (E) for the pulse generator and outside to supply the *SD*.

The positive and negative voltage at the useful terminals is measured by two front pannel instruments. Wire-wound potentiometers of 25 kΩ permit a continuous variation of these voltages within 0 and 150 V.

An external sensitive microammeter can be easily connected to measure the reverse current of the semiconductor detector. Two of the pannel keys, I_1 and I_2 are used for this purpose. The external dimensions of the apparatus are 480 × 238 × 320 mm.

The measurements carried out with a 400 channel amplitude analyser of SA-40 type and a Ribet-Desjardins "241 A" oscilloscope have pointed out the following performances:

The pulse maximum amplitude (in the external working regime) is 8 V; the dynamical scale of the attenuator is 10000 : 1; the helicoidal potentiometer has 100 × 10 = 1000 divisions; its nonlinearity is less than 0.3%.

The pulse amplitude stability:

— for short pulses it is better than $\frac{1}{800}$;

— during 8 hour working it is better than 1% (though the Zener diodes were not in thermostatic conditions);

— for a line voltage variation between 170 and 250 V, the amplitude changes with less than 0.7%.

The pulser noise is lower than 0.02% of the pulse amplitude. The pulse shape is exponential: the pulse rise time is shorter than 10 ns; the decay time constant is 50 μs. The output pulse polarity is positive or negative. The voltage ripple of the 150 V rectifiers is smaller than 5 mV top to top (below 0.007 %). The apparatus power requirements are 35 VA, 220 V and 50 Hz.

It is worth mentioning that in addition to the nuclear radiation spectrometry with *SD*, the pulse generator may be also used in a series of other applications.

A high pulse stability may be also achieved with another type of vacuum-tube pulse generator [D 26]. For this purpose a differential amplifier with a current stabilizing cascode circuit connected to the cathode is used. Its short-term stability is 2×10^{-4}.

7.6. TRANSISTORIZED APPARATUS

Due to specific noises and to a comparatively high input current, the bipolar transistors have higher noises than the vacuum tubes, by about an order of magnitude. This conclusion was drawn after several measurements were carried out [F 18], [B 17], [B 44], [E 8], [E 9], [K 4], [L 7], [R 1] on different transistor types, such as П 403, П 411 A, OC872, 2N779, 2N930, 2N2484.

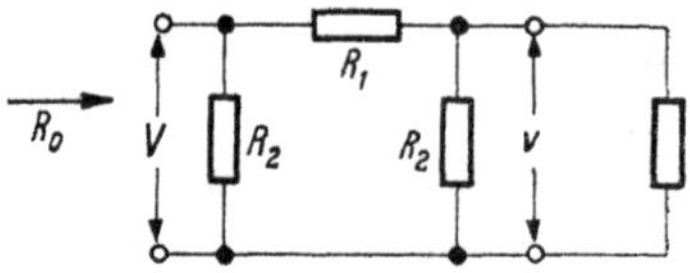

Fig. 7.21 — Circuit diagram of an attenuation π-cell with load impedance R_0.

For this reason, the transistor preamplifiers are considered well suited for high capacity detectors (higher than 1000 pF [O 12]) for which their noises are comparable with that of the vacuum tubes. There are also applications which require small dimensions [S1] and minimum electric power consumption. In transistor preamplifiers the same circuit types are used as in those with vacuum tubes [B 44]. For example Fig. 7.22 shows a preamplifier circuit diagram [C 5] in which one can easily observe, the transposition of the analogous vacuum-tube diagram (Fig. 7.4) (a bootstrapped plate load cascode).

At the present time there exists the field effect transistor which gives at small capacities of the *SD*, noises comparable or even smaller than those of the vacuum tubes.

A simple field effect transistor preamplifier circuit diagram [B 23] is given in Fig. 7.23.

Such transistors are mainly used in preamplifiers which work with small capacity *SD*, such as for instance the PIN-types.

The working principles of field effect transistors were formulated by Shockley [S 13] and their noise theory given by Van der Ziel [V 4]. At the present, many commercial types which may be used in *SD* preamplifiers exist [B 22], [B 23], [K 3], [R 1], [R 2], for instance FSP-401, 2N2500, 2N3113, 2N2606, 2N2607, 2N2609. High performances were obtained by Elad and Nakamura [E 3], [E 4] with 2N3823 silicon field effect transistors (0.7 keV) and with TI $\times$ 301 (0.3 keV) germanium transistors.

The field effect transistor working principles are based on a voltage control of the majority charge-carrier movement through a semiconducting channel. The channel conductance between source and drain electrode is modulated, by varying the channel cross-

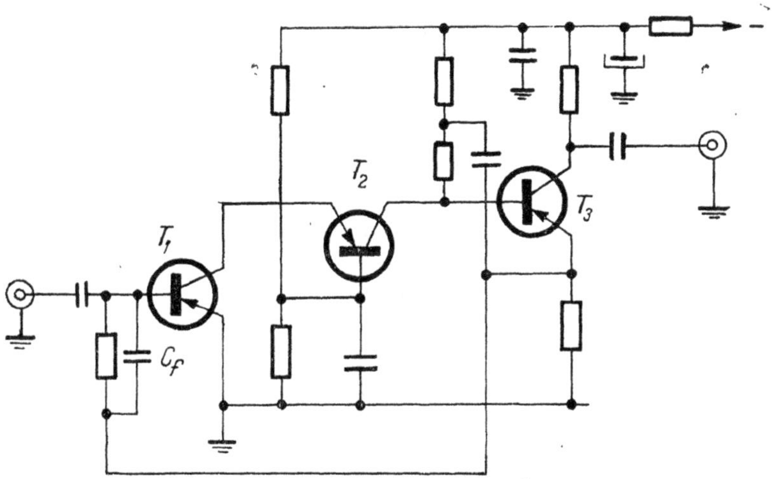

Fig. 7.22 — Charge sensitive preamplifier with bipolar transistors (bootstrapped plate load cascode circuit).

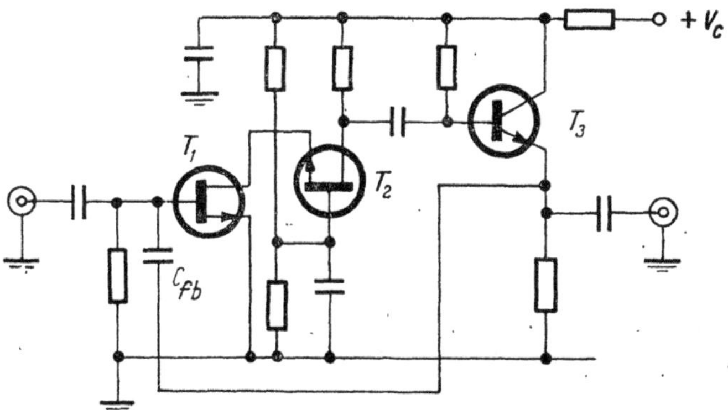

Fig. 7.23 — Charge sensitive preamplifier with field-effect transistors at the input (cascode circuit).

section as a result of the depletion region modification in the reverse biased gate-channel junction. By analogy with the vacuum tubes, the source plays the part of the cathode, the gate that of the grid and the drain that of the anode. The gate current is quite low ($10^{-2} \div 1$ nA) and the input capacity too ($1 \div 50$ pF), because the gate junction is inversely biased.

The characteristic $I_d - V_d$ is similar to that of a pentode.

The cut-off frequency and the transconductance are limited at the present [R 2], below 50 MHz and 5 mA/V, respectively, due to technological difficulties.

By applying the noise theory of charge sensitive amplifiers to field effect transistors, Radeka [R 1] comes to the conclusion that the main noise source is given in this case by the channel current fluctuations (the channel thermal noise) which may be equivalent to the noise of a resistance

$$R \simeq \frac{0.7}{g_m},$$

where g_m is the transistor transconductance (equal, in fact, to the direct-current channel resistance). It may be observed that the noise may be lower than in vacuum tubes $\left(\frac{2.5}{S}\right)$ which have $S = g_m$. As in the case of vacuum tubes, other noise sources are present here, which are given in Table 7.1 [R 1].

The noise is calculated for the most simple filter in the $(RC)-(RC)$ amplifier which gives a signal to noise ratio only 26% lower than the best possible theoretical signal to noise ratio [R 1]. The $FWHM$ noise value is calculated with the relation

$$FWHM = 2.36\ (\Sigma \overline{E}^2)^{1/2},$$

since the table gives the standard deviation (r.m.s. noise value). The following notations are used: R_n — noise equivalent resistance (k Ω); g_m — transconductance (mA/V); C_t — total input capacity without feedback (pF); $A_f = \frac{1}{f}$ (V²) noise figure; I_D — detector current (nA); $\tau = \tau_i = \tau_d$ = integration and differentiation time constant (μ s); R_g — input resistance (M Ω).

In the circuit diagram shown in Fig. 7.23, the second transistor should be also a field effect one, though there are transistors (e.g.

Table 7.1

Noise of field effect transistors in charge sensitive amplifiers

Noise source	Input equivalent mean square noise charge (C^2)	Input equivalent mean square noise charge (keV)^2 at T = 293°K in Si (3.6 eV/pair)
Channel thermal noise	$\overline{Q^2} = 4kTR\dfrac{C_t^2}{\tau} \simeq 2.8\,kT\cdot C^2\dfrac{1}{g_m\tau}$	$\overline{E^2} = 0.8\cdot10^{-2}\cdot\dfrac{R_nC_t^2}{\tau} \cong 0.56\cdot10^{-2}\cdot C^2\cdot\dfrac{1}{g_m\tau}$
Channel $1/f$ noise	$\overline{Q_f^2} = 4A_fC_t^2$	$\overline{E_f^2} = 2\cdot10^9 A_f C_t^2$
Shot noise of input current (gate or detector)	$\overline{Q_D^2} = 2e I_D \tau$	$\overline{E_D^2} = 0.16 I_D \tau$
Input resistance noise	$\overline{Q_R^2} = 4kT\dfrac{\tau}{R_g}$	$\overline{E_R^2} = \dfrac{8\tau}{R_g}$

Texas Instruments: 2N930 and 2N2586; National Semiconductors 2N2252) which do not add a significant noise. Table 7.2 lists the results of the measurements made on the transistors FSP 401 (Fairchild) and 2N2500 (Texas Instruments) connected in a bootstrapped plate load cascode. Blalock [B 25] built a broad band low-noise charge sensitive preamplifier with 2N3823 at the input.

Reference [2] describes the main parameters of some preamplifiers: 11 with transistors, 12 with vacuum tubes and 5 hybrids (a vacuum tube at the input, the remaining being transistors).

In [W 9] the construction of an integrated circuit charge sensitive preamplifier is reported.

The FSP401 transistor is well suited to *SD* with very small capacity, for whch it gives a *FWHM* < 2 keV for a large range of the time constant τ variation. The weak dependence on τ and the dependence on C_t, almost equal for low and high τ, point out that the $\dfrac{1}{f}$ noise prevails.

This transistor has a slope of the noise increase vs. input capacity of 0.5 keV/pF while the 2N2500 transistor has 0.14 keV/pF. For this reason the latter is well suited to higher capacities. Another advantage of Si field effect transistors consists in the possibility to be cooled at liquid nitrogen temperature (e.g., together with a Ge *NIP*-type semiconductor detector used in spectroscopy) when the values g_m and I_a are $\frac{2}{3}$ of the respective quantities at room temperature. A noise of 1.8 keV with a slope $\leqslant 0.06$ keV/pF was obtained with a preamplifier with 2N2500 cooled at 125° K.

In order to decrease the slope of the noise increase vs the detector capacity, several parallel field effect transistors are used at the preamplifier input [H 10], [S 22].

Some papers [A 12], [E 9], [G 26], [L 7] give the description of a complex, transistorised apparatus for *SD* (linear amplifiers and expanders, linear gates, integral and differential discriminators, fast-slow coincidence circuits, a.o.). The block-diagram of a linear transistorized amplifier [G 26] is drawn in Fig. 7.24.

The maximum gain of the amplifier is 1000 and the minimum rise-time of its transient characteristics is 80 ns. The apparatus has an input attenuator with a step by step control : 1 ; 2 ; 5 ; 10 ; 20 ; 50. There is also a coarse attenuation (1 ; 3 ; 9) at higher levels, lest the signal to noise ratio should be affected by the noise of the amplifier input stages. The fine gain is continuously controlled within the range 1—3, by varying the negative feedback factor of the first amplifying unit. An inverter and the differentiation circuits follow. An *RC* working mode may be chosen (time constants $\tau_d = 0.1$; 0.2 ; 0.5 ; 1 ; 2 ; 5 μs) or a *DL* shaping line (simple or double) ($2\,\tau_d = 0.25 - 1.5$ μs) and in this case an additional stage for the overshot and gain control is introduced.

The last three stages have nonblocking diode limiters when the signal exceeds a given level, the diode conducts, the negative feedback increases and the gain is much reduced over that level. Before the last but one stage, the integration time constant is changed ($\tau_i = 0.1$; 0.2 ; 0.5 ; 1 ; 2 ; 5 μs).

All the amplifying units, except the final one, are identical ; they include two complementary transistors placed so that the signal should drive them always into conduction. In the final stage, special precautions are taken, in order to obtain a linearity up to $\pm$ 8V amplitude, on an output impedance of about 1 Ω.

The transistor expander can be made on the same principles as similar devices with vacuum tubes (e.g. with differential amplifiers with threshold [E 9], [L 7]) or using other techniques. For instance [G 12] the difference can be made between the direct pulse

Table 7.2

The parameters of some field-effect transistors used in charge sensitive preamplifiers

Parameter / Transistor type	g_m (mA/V)	I_0 (mA) $V_{cs}=5V$ $V_{ps}=0.3V$	Saturation voltage (V)	I_p (nA) $V_{cs}=5V$ $V_{ps}=0.5V$	R_n (Ω)	A_f (V^2)	C_{ps} (pF) $V_{cs}=5V$ $V_{ps}=0V$	C_t (pF)	N_{noise} (e)	*FWHM* (keV) $=7\mu s$ $R_g=10^9\Omega$
FSP-401 *N* channel-Si	0.2	0.25	3	10^{-2}	3400	1.35×10^{-11}	2	4	200	1.65
2N2500 *P* channel-Si	1.5	1.5	2.6	5×10^{-2}	500	10^{-13}	26	29	400	3.3

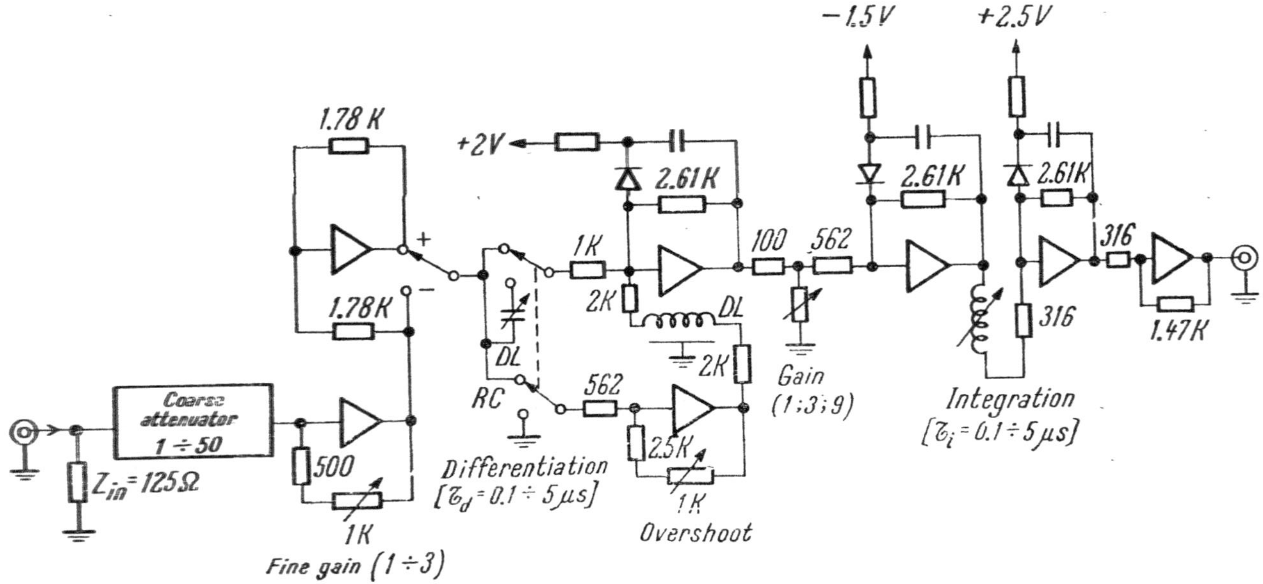

Fig. 7.24 — Block-diagram of a main transistorized amplifier [G 26].

from the linear amplifier output and the same pulse after having limited and reversed its polarity in the transistor T_1 (Fig. 7.25). The summing up is made in the operational amplifier SA.

Another interesting method [F 1] consists in a ballistic amplification of the sum of two electric charges injected at the preamplifier

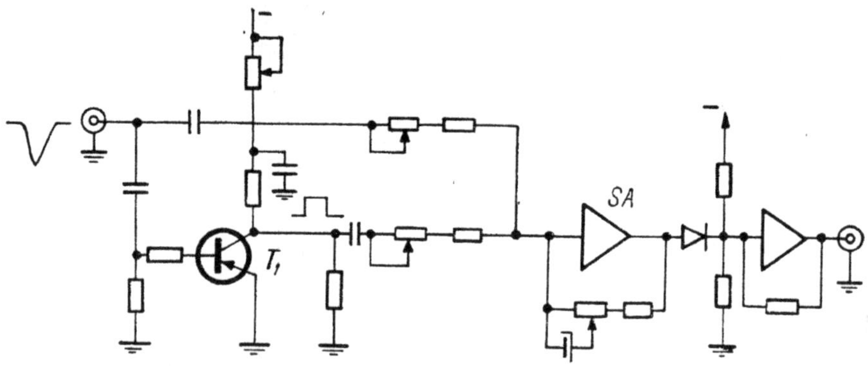

Fig. 7.25 — Circuit diagram of a transistorized expander (with summation at high signal level).

input (the charge given by the detector and an opposite charge given by the standard charge generator).

Since the charges are subtracted just on the detector, the whole amplifying chain (preamplifier, main amplifier) does not require a too high amplification stability. (If the multichannel analyser has 256 channels, a stability of $\frac{1}{256}$ is necessary). Of course, the requirement for $\frac{1}{5000}$ stability (for an expansion coefficient $= 20$) is still valid for the compensation charge.

The expander block diagram and the wave forms in three points are given in Fig. 7.26.

The work [F 1] studies effects due to compensation charge injection delay (t_i) from the charge given by the detector and the conditions in which these effects are reduced to minimum. The compensation system, including amplifier, monostable flip-flop and a standard step pulse generator realized with a normal cut-out transistor and a Zener diode is described. In order to provide the necessary stability $\left(\frac{1}{5000}\right)$, the Zener diode and this transistor are

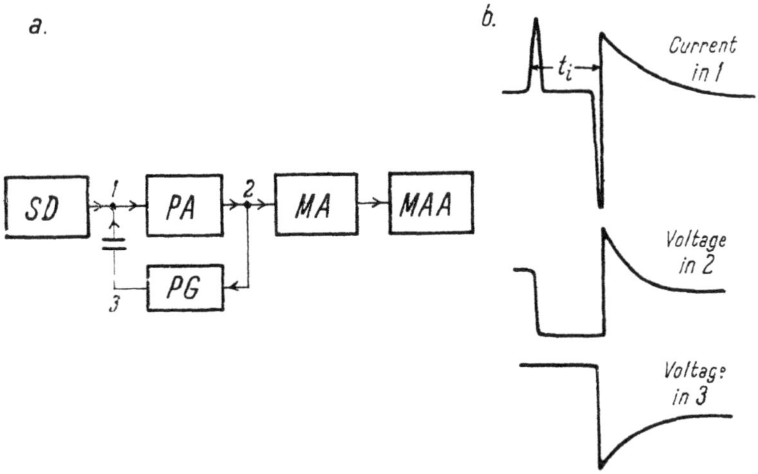

Fig. 7.26 — Block diagram of a transistorized expander (with summation at low signal level) (*a*) and the waveforms (*b*) [F 1].

set in a thermostat, made with a power transistor stabilizing the temperature up to ± 0.25°C.

We have briefly reported here some of the most important achievements in the transistorization of *SD* related equipment. New, interesting results are to be expected since this domain is intensely studied.

APPENDIX A I

BASIC PROPERTIES OF Si AND Ge

SILICON

Atomic structure. The silicon belongs to the IV-th group of the element periodic system. The silicon nucleus has 14 protons. The electron configuration consists of two shells, each of them filled with 2 electrons placed on the 3 *s* and 3 *p* orbitals. The number of electrons required to fill up the 3 *p* orbital (and at the same time the 3-rd shell) determines the silicon valency : 4.

Crystalline structure. The silicon has a face-centered cubic structure with a lattice constant (see Fig. A.I. 1 *b*) : $a = 5.43 \times 10^{-8}$ cm. The crystal structure is shown in Fig. A.I.1. The number written near every atom in Fig. A.I. 1 *b* represents its height against the cube base considered in the figure plane. The atom coupling is of a tetrahedric type : every atom is bounded to its four neighbouring atoms through covalent bonds.

Energy scheme. The electron effective mass in the conduction band is : $m_e^* = 0.25\ m_0$ (m_0 — mass of a free electron). The hole effective mass in the valency band is : $m_h^* = 0.33\ m_0$. The energy — gap : $E_g = 1.12$ eV. The ionization energies of donor and acceptor impurities are given in Table A.I.1.

Table A.I.1

Ionization energies of donor and acceptor impurities in Si

Donor impurities	$(E_c - E_d)$ (eV)	Acceptor impurities	$(E_a - E_v)$ (eV)
Au	0.85	B	0.045
P	0.045	Al	0.057
As	0.049	Ga	0.065
Sb	0.039	Zn	0.092 (0.3)

In Table A.I.1. E_d is the energy of donor levels, E_a — the energy of acceptor levels, E_c — the lower limit of the conduction band and E_v — the upper limit of the valency band.

a.

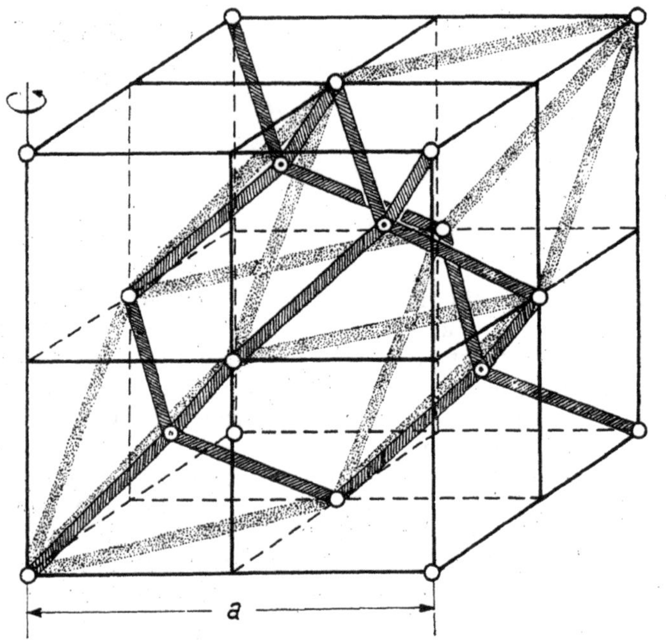

b.

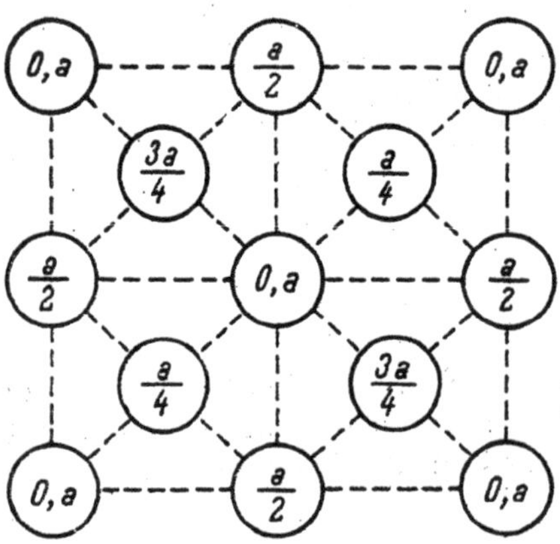

Fig. A.I.1 — Diamond-structure unit (face-centered cubic crystalline system) for Si and Ge; geometric (*a*) and analytic (*b*) representation.

Table A.I.2

Ionization energies of donor and acceptor impurities in Ge

Donor impurities	$(E_c - E_d)$ (eV)	Acceptor impurities	$(E_a - E_v)$ (eV)
P	0.01	B	0.01
As	0.01	Al	0.01
Sb	0.01	Ga	0.01
Li	0.01	In	0.01
Au	0.05	Ti	0.01

Table A.I.3

Physical constants of Si and Ge

Constant	Silicon	Germanium	Units
Atomic number (Z)	14	32	—
Atomic weight (A)	28.09	72.60	—
Mass numbers of stable isotopes	28 ; 29 ; 30	70 ;72 ;73 ;74 ;76	—
Lattice constant (a)	5.43	5.66	A
Number of atoms per volume unit	4.96×10^{22}	4.41×10^{22}	cm^{-3}
Density (δ)	2.33	5.33	$g.cm^{-3}$
Melting point	1420	936	°C
Linear coefficient of thermal dilatation	4.2×10^{-6}	6.1×10^{-6}	$(°C)^{-1}$
Thermal conductivity	0.20	0.14	$cal.cm^{-1} s^{-1} (°C)^{-1}$
Specific heat (0—100°C)	0.181	0.074	$cal.\ g^{-1}(°C)^{-1}$
Relative dielectric constant (ε_r)	12	16	—
Energy-gap (E_g)	1.11	0.75	eV
Intrinsic resistivity (ρ_i) $(T = 300°K)$	230 000	47	Ω cm
Intrinsic concentration of charge carriers (n_i) $(T = 300°K)$	2.5×10^{13}	6.8×10^{10}	cm^{-3}
Electron mobility (μ_e)	1350 $(T=300°K)$	3800$(T=300°K)$ 10000$(T=77°K)$	$cm^2.V^{-1}.s^{-1}$
Hole mobility (μ_h) $(T = 300°K)$	480 $(T=300°K)$	1800$(T=300°K)$ 15000$(T=77°K)$	$cm^2.V^{-1}.s^{-1}$
Electron diffusion constant (D_e) $(T = 300°K)$	35	92	$cm^2.s^{-1}$
Hole diffusion constant (D_h) $(T = 300°K)$	12,4	44	$cm^2.\ s^{-1}$

GERMANIUM

Atomic structure. Germanium belongs to the IV-th group of the element periodic system. The electron configuration has the first three shells filled up and 2 electrons on each $4s$ and $4p$ orbitals. The number of electrons necessary to fill up the $4p$ orbitals (and the 4-th shell) determines the germanium valency: 4.
Crystalline structure. Germanium has the same crystalline system as the silicon with a lattice constant somehow different : $a = 5.66 \times \times 10^{-8}$ cm.
Energy scheme. The effective mass of electrons in the conduction band : $m_e^* = 0.08\ m_0$ and $m_e^* = 1.58\ m_0$. The hole effective mass in the valency band : $m_h^* = 0.04\ m_0$ and $m_h^* = 0.4\ m_0$. The energy gap : $E_g = 0.75$ eV.
The ionization energies of donor and acceptor impurities are given in Table A.I.2.

Other constants concerning Si and Ge are given in Table A.I.3.

Table A.11

Main natural and artificial alpha sources

Z	Element	A	Name	Lifetime*)	Energy (MeV) and intensity of main lines (percents)
83	Bi	210	RaE	5.01 d	4.94 (5×10^{-5} %)
		211	AcC	2.16 min	6.617 (8.3 %) 6.273 (17 %)
		212	ThC	60.5 ,,	6.086 (10 %) 6.047 (24.6 %)
		213	Bi	47 ,,	5.86 ( ) (2 %)
		214	RaC	19.7 ,,	5.505 (0.018 %) 5.444 (0.022 %)
84	Po	210	Po	138.4 d	5.305
		211	AcC′	0.5 s	7.442
		212	ThC′	3.04×10^{-7}s	8.78
		213	Po	4.2×10^{-6}s	8.35
		214	RaC′	1.64×10^{-4}s	7.68
		215	AcA	1.83×10^{-3}s	7.36
		216	ThA	0.158 s	6.775
		218	RaA	3.05 min	5.998
85	At	217	At	0.018 s	7.05
		219	At	0.9 min	6.27
86	Rn	218	Rn	0.019 s	7.13
		219	An	3.92 s	6.813 (82 %) 6.547 (13 %) 6.419 (5 %)
		220	Tn	51.5 s	6.282
		222	Rn	3.823 d	5.486
87	Fr	221	Fr	4.8 min	6.33 (84 %) 7.12 (16 %)
		223	AcK	22 min	5.34 (4×10^{-3} %)
88	Ra	222	Ra	38 s	6.55
		223	AcX	11.68 d	...5.742 (10.5 %) 5.172(50 %)5.602(24 %) 5.534 (10.3 %) ...
		224	ThX	3.64 d	5.681 (95 %) 5.445 (4.9 %)
88	Ra	226	Ra	1622 y	4.777 (94 %) 4.590 (5.7 %)
89	Ac	225	Ac	10 d	5.818 (54 %) 5.782 (28 %) 5.721 (9.5 %)
		227	Ac	21.6 y	4.94 (1.2 %)
90	Th	226	Th	30.9 min	6.33 (79 %) 6.22 (19 %) 6.10 (1.7 %)
		227	RdAc	18.17 d	6.033 (23 %)...5.976 (24 %)...5.755(21 %) 5.710 (13.7 %)
		228	RaTh	1910 y	5.421 (71 %) 5.338 (28 %)
		229	Th	7340 y	5.02 (10 %) 4.94 (20 %) 4.85 (70 %)
		230	Th	8×10^{4}y	5.682 (76 %) 4.615 (24 %)
		232	Th	1.39×10^{10}y	4.007 (76 %) 3.942 (24 %)
91	Pa	231	Pa	3.43×10^{4} y	5.046 (10 %) 5.017 (23 %) 5.001 (24 %).. 4.938 (22 %) ... 4.722 (11 %)

Table A.II (continued)

Z	Element	A	Name	Lifetime*)	Energy (MeV) and intensity of main lines (percents)
92	U	230	U	20.8 d	5.884 (67.2 %) 5,813 (32.1 %)
		232	U	74 y	5.318 (68 %) 5,261 (32 %)
		233	U	1.62×10^5 y	4.816 (83.5 %) 4.773 (14.9 %) 4.717(1.6 %)
		234	UII	2.48×10^5 y	4.768 (72 %) 4.717 (28 %)
		235	AcU	7.1×10^2 y	4.559 (6.7 %) 4.520 (2.7 %) 4.370 (25 %) 4.354 (35 %) 4.333 (14 %) 4.318 (8 %) 4.117 (5.8 %)
		236	U	2.39×10^7 y	4.5 (73 %) 4.450 (27 %)
		238	UI	4.51×10^9 y	4.195 (77 %) 4.145 (23 %)
93	Np	237	Np	2.2×10^6 y	4.872 (3.1 %) 4.816 (3.5 %) 4.787 (53 %) 4.767 (29 %) 4.713 (1.7 %) 4.674 (33 %) 4.644 (6 %)
94	Pu	236	Pu	2.85 y	5.763 (69 %) 5.716 (31 %) 5.610(0.18 %)
		238	Pu	86.4 y	5.495 (72 %) 5.452 (28 %) 5.352 (0.09 %)
		239	Pu	24360 y	5.147 (72.54 %) 5.134(16.8 %)5.096(10.7 %)
		240	Pu	6,580 y	5.162 (76 %) 5.118 (24 %) 5.014 (0.1 %)
		242	Pu	3.79×10^5y	4.898 (76 %) 4.858 (24 %)
95	Am	241	Am	458 y	5.534 (0.35 %) 5.5 (0.23 %) 5.477 (85 %) 5.435 (12.6 %) 5.378 (1.7 %)
96	Cm	241	Cm	35 d	5.95 (0.96 %)
		242	Cm	162. 5 d	6.110 (73.7 %) 6.066 (26.3 %) 5.965 (0.035 %)
		244	Cm	17.9 y	5.801 (76.7 %) 5.759 (23.3 %) 5.66 (0.017 %)
98	Cf	252	Cf	22 y	6.112 (82 %) 6.069 (15 %)

*) y = years, d = days, min = minutes, s = seconds

APPENDIX A III

ANALYSIS OF SOME CIRCUITS USED IN CHARGE SENSITIVE PREAMPLIFIERS

Some particular amplifier circuit configurations such as cascode [B 30], cascode with bootstrapped plate load [C 5] and voltage amplifier stage with bootstrapped plate load [B 27] are frequently used in the charge sensitive preamplifiers for semiconductor nuclear particle detectors.

It is well known that the cascode stage is used at the input because it has a pentode amplification with triode noise, but the "classical" circuit does not achieve sufficiently high amplification. For this reason the dynamic capacity at the input of such a preamplifier is quite small; it is well suited for small capacity detectors [B 44].

The cascode with bootstrapped plate load permits us to obtain high input dynamic capacity. This is the most frequently used circuit in the charge sensitive preamplifiers for semiconductor detectors. Such preamplifiers usually consist of two units; besides the charge sensitive unit, they have a voltage amplifier unit; in the last one an amplifier stage with bootstrapped plate load [B 27] can be also included.

The theory of low-noise amplifier design has been adequately treated elsewhere [B 30], [F 5], [G 7], [T 17], but when designing or analysing such circuits it is necessary to know the equations for its gain, output impedance and transient response.

In this Appendix we give the relations to be used in the analysis and synthesis of the particular circuits mentioned above, for linear working conditions (small signals). The formulas for gain, output impedance and transient response will be presented on the basis of equivalent circuit analysis. Circuits with open feedback capacitive loop will be considered. First the amplifier stage with bootstrapped load will be analysed, since it is simpler and allows us to plot easily the relations obtained.

A.III.1. AMPLIFIER WITH BOOTSTRAPPED PLATE LOAD

The equivalent circuit of a bootstrapped load amplifier (Fig. A III.1 *a*) is shown in Fig. A III.1*b*. The following notations are used:

R_{i1}, R_{i2} = internal (plate) resistances of the first and second tube (T_1, T_2).

μ_1, μ_2 = amplification factors of the first and second tube (T_1, T_2),

C_1 = total capacity between point A_1 and ground,

C_3 = total capacity between point K_2 and ground,

C_f = $C_{G_2 K_2}$.

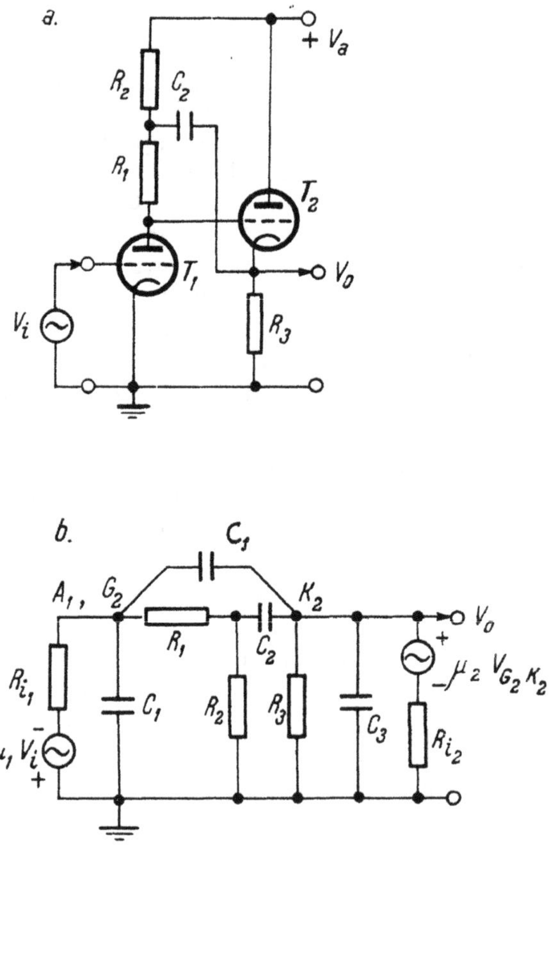

Fig. A.III.1 — Amplifier stage with bootstrapped plate load (*a*) and its equivalent circuit (*b*).

The mesh analysis of the circuit for the middle frequency range (for which all capacitances can be neglected) [V 10], gives the following relations for the gain ($\mathcal{A}_0 = U_o/V_i$) and output resistance (R_o)

$$\mathcal{A}_0 = -\frac{\mu_1 (\mu_2 R_1 + R_{i_2})}{(\mu_2 + 1)\, R_1 + R_{i_1} + R_{i_2} + \frac{R_{i_2}}{R}(R_1 + R_{i_1})}, \qquad \text{(A.III.1)}$$

$$\frac{1}{R_o} = \frac{1}{R} + \frac{(\mu_2 + 1)\, R_1 + R_{i_1} + R_{i_2}}{R_{i_2}(R_1 + R_{i_1})}, \qquad \text{(A.III.2)}$$

where

$$\frac{1}{R_2} + \frac{1}{R_3} = \frac{1}{R}. \qquad \text{(A.III.3)}$$

There are two important practical cases

1. T_1 — pentode, T_2 — triode,
2. identical tubes (double triode).

In the first case, assuming that the inequalities

$$R_{i_1} \gg R_{i_2}, \quad R_1 \quad \text{and} \quad \mu_2 \gg 1, \qquad \text{(A.III.4)}$$

are valid, Eqs. (A.III.1) and (A.III.2) become

$$\frac{|\mathcal{A}_0|}{\mu_1} \simeq \frac{1}{1 + \frac{R_{i_1}}{\mu_2 R_1}\left(1 + \frac{R_{i_2}}{R}\right)}, \qquad \text{(A.III.5)}$$

$$\frac{R_o}{R_{i_2}} \simeq \frac{1}{\frac{\mu_2 R_1}{R_{i_1}} + \frac{R_{i_2}}{R} + 1}. \qquad \text{(A.III.6)}$$

These relations are plotted in Figs. A.III.2 and A.III.3, respectively.

In this case, the gain obtained is very high because the first tube has a very large μ_1, An inspection of Fig. A.III.2, shows that the maximum value of the gain $|\mathcal{A}_0| = \mu_1$ can be obtained the easier, the greater the values of the resistances R_1 and R and the slope $S = \frac{\mu_2}{R_{i_2}}$ of the tube T_2. Fig. A.III.3 shows also that the output resistance of this circuit configuration is generally high (comparable to R_{i_2}) and tends towards the output resistance of a cathode-follower only for

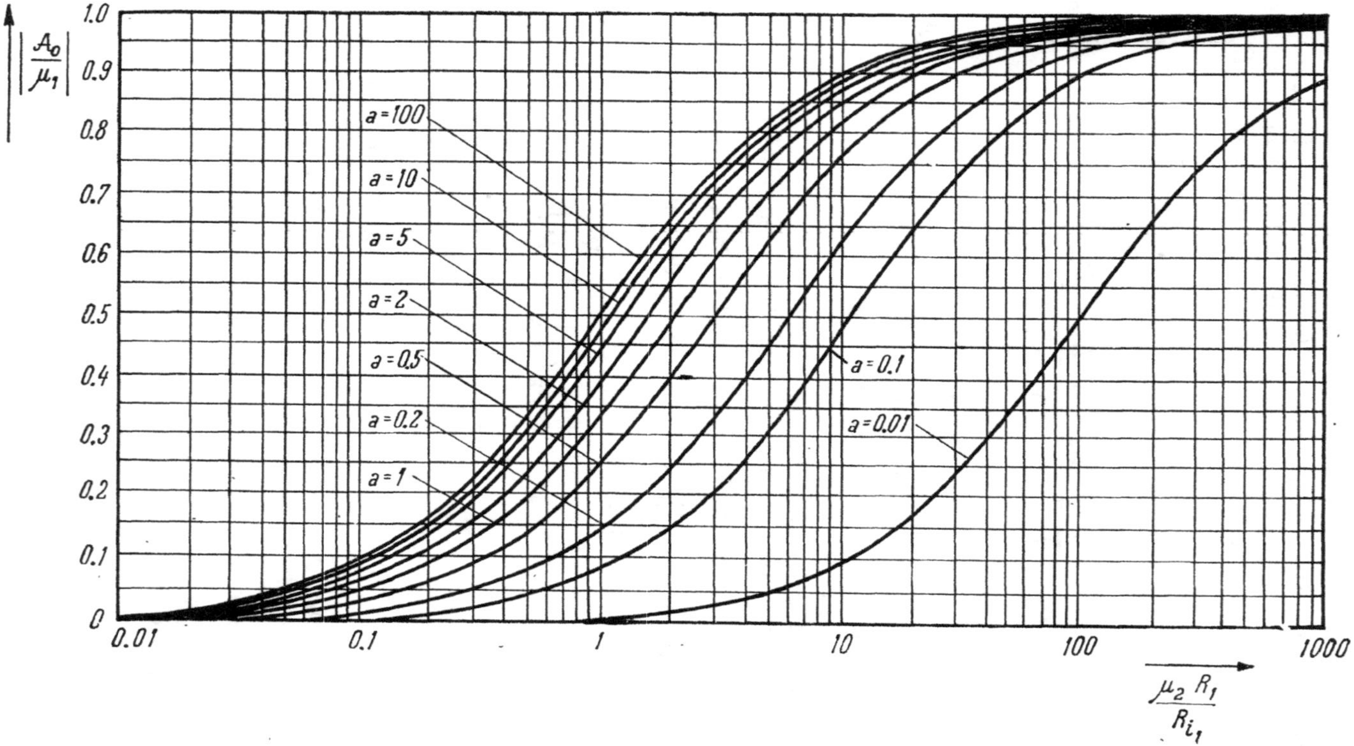

Fig. A.III.2 — The normalized amplification of the amplifier with bootstrapped plate load, having a pentode as a first tube and a triode as a second one ($a = R/R_{i_2}$)

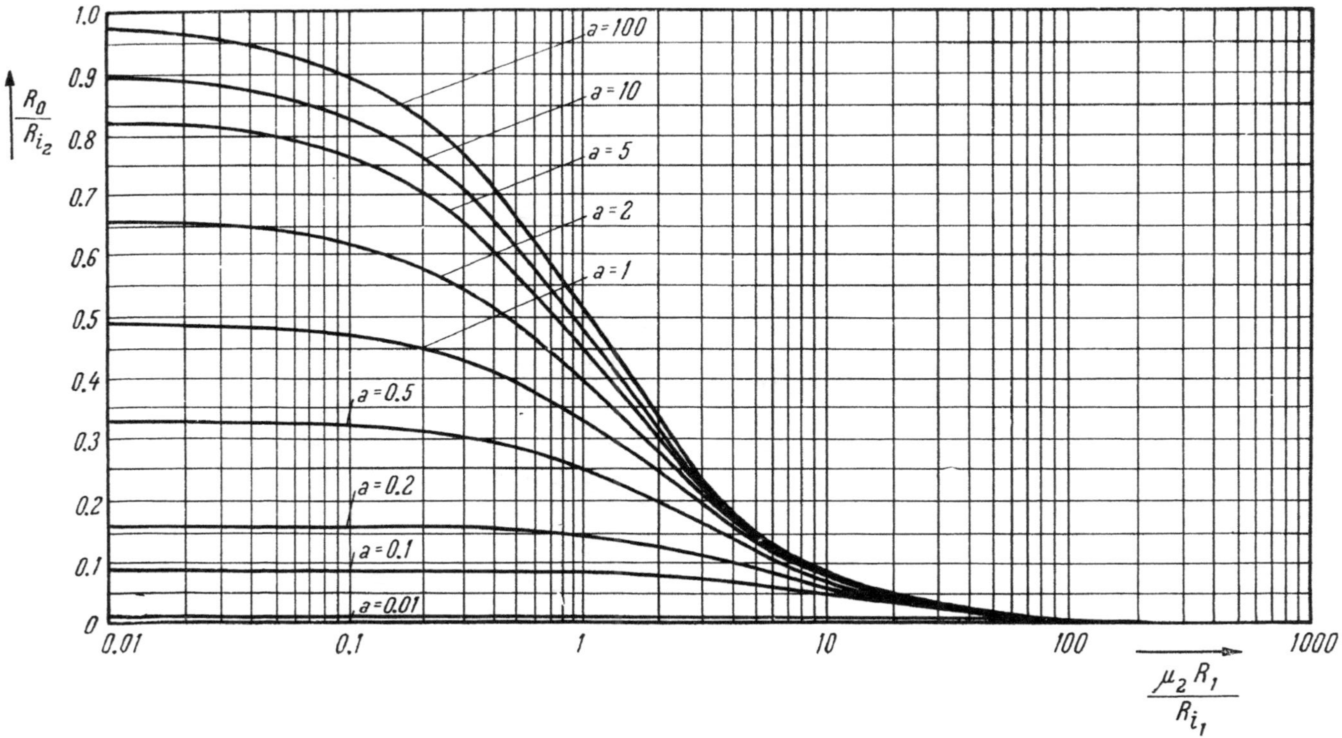

Fig. A.III.3 — The normalized output resistance of the amplifier with bootstrapped plate load, having a pentode as first tube and a triode as a second one ($a = R/R_{i_2}$).

the large values of the factor $\mu_2 R_1/R_{i_1}$ which are not usually met with in pulse amplifiers.

In the second case, (double triode) one can write

$$R_{i_1} = R_{i_2} = R_i \text{ and } \mu_1 = \mu_2 = \mu \gg 1, \tag{A.III.7}$$

and the normalized relation for the gain and output impedance becomes

$$\frac{|\mathcal{A}_0|}{\mu} \simeq \frac{1}{1 + \dfrac{R_i}{R}\,\dfrac{\dfrac{R}{R_i} + \dfrac{1}{\mu}\,\dfrac{\mu R_1}{R_i} + 1}{\dfrac{\mu R_1}{R_i} + 1}}, \tag{A.III.8}$$

$$\frac{R_o}{R_i} = \frac{1}{\dfrac{\mu R_1 + 2R_i}{R_1 + R_i} + \dfrac{R_i}{R}}. \tag{A.III.9}$$

The curves shown in Fig. A.III. 4 *a*, *b*, *c*, are calculated on the basis of equation (A.III.8) for three values of μ (20 ; 50 ; 100). For intermediate values of μ, interpolations can be easily performed, the more so as the differences between the family of curves plotted for different values of the parameter μ are high, mainly on the final part and for the small value of the ratio $a = R/R_i$. The last cases point out the advantage of high μ vacuum tubes.

The plot of the normalized output resistance against circuit parameters is shown in Fig. A.III. 5 for $\mu = 20$.

The curves for other values of μ (up to 100) differ only slightly from these and are, particularly on the final part, usually of no interest for pulse amplifiers. When μ is high, the asymptotic value obtained for $R_1 \gg R_i$ (and consequently $\mu R_1 \gg 2R_i$) may be easier obtained

$$\frac{1}{R_o} = S + \frac{1}{R}. \tag{A.III.10}$$

The maximum value of the output resistance in the case of identical tubes is

$$R_{oM} = \frac{1}{2} R_i, \tag{A.III.11}$$

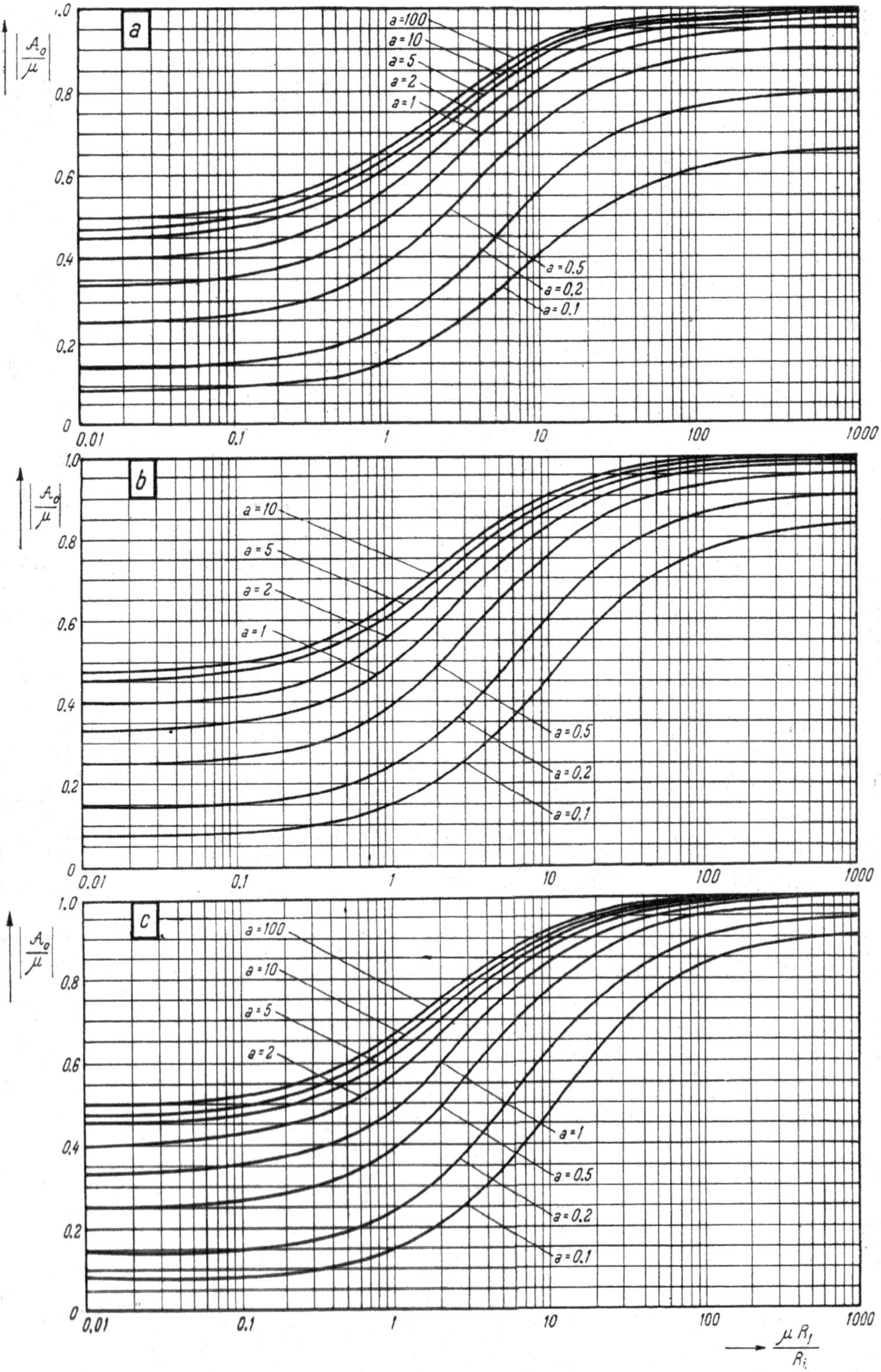

Fig. A.III.4 — The normalized amplification of the amplifier with bootstrapped plate load equiped with double triode: *a*) $\mu = 20$; *b*) $\mu = 50$; *c*) $\mu = 100$; ($a = R/R_i$).

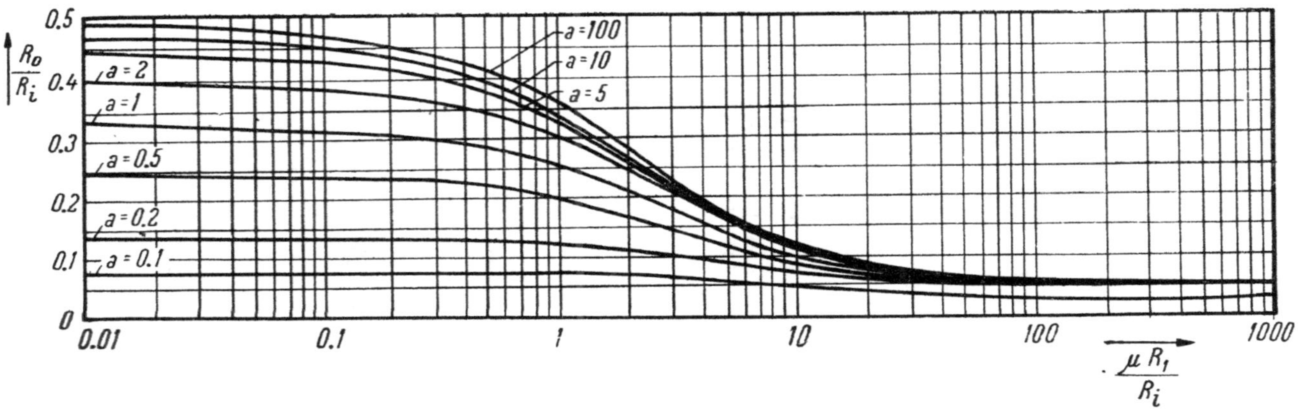

Fig. A.III.5 — The normalized output resistance of the amplifier with bootstrapped plate load, equiped with double triode : $\mu = 20$; ($a = R/R_i$).

as distinguished from the preceeding case (Fig. A.III.3) in which

$$R_{oM} = R_{i_2}. \qquad \text{(A.III.12)}$$

This value is obtained for

$$\frac{\mu_2 R_1}{R_{i_1}} \ll 1; \quad \frac{R}{R_{i_2}} \gg 1. \qquad \text{(A.III.13)}$$

The lowering of the output resistance when the plate load resistance R_1 of the first tube increases, is a characteristic of this configuration. This happens so because the voltage transfer ratio of the positive fedback $R_{i_1}/(R_{i_1} + R_1)$ decreases.

Using the mesh analysis of the equivalent circuit shown in Fig. A.III.1 *b* for the high frequency range (where the effect of coupling capacity C_2 is negligible) and the inverse Laplace transform, one can obtain [V 10] the following relation for the initial part of the transient response (edge response)

$$f(\xi) = 1 - \frac{m-n}{m-1} \exp\left(-\frac{\xi}{m}\right) + \frac{1-n}{m-1} \exp(-\xi), \qquad \text{(A.III.14)}$$

where

$$\xi = \frac{B + \sqrt{B^2 - 4C}}{2C} t = at, \qquad \text{(A.III.15)}$$

$$m = \frac{B + \sqrt{B^2 - 4C}}{B - \sqrt{B^2 - 4C}}, \qquad \text{(A.III.16)}$$

$$n = \frac{B + \sqrt{B^2 - 4C}}{2C} A_r, \qquad \text{(A.III.17)}$$

$$B = \frac{\left[(\mu+1)R_1 + R_{i_2} + R_1 \frac{R_{i_2}}{R}\right]\tau_1 + \left(R_{i_2} + R_{i_1} + \frac{R_{i_2} R_{i_1}}{R}\right)\tau_f + \frac{R_{i_2}}{R}(R_1 + R_{i_1})\tau_3}{(\mu_2 + 1) R_1 + R_{i_2} + R_{i_1} + \frac{R_{i_2}}{R}(R_1 + R_{i_1})}, \qquad \text{(A.III.18)}$$

$$C = \frac{R_{i_1} R_{i_2} R_1 (C_1 C_f + C_1 C_3 + C_f C_3)}{(\mu_2 + 1) R_1 + R_{i_1} + R_{i_2} + \frac{R_{i_2}}{R} (R_1 + R_{i_1})}; \qquad A_f = \frac{R_{i_2} \cdot \tau_f}{\mu_2 R_1 + R_{i_2}} \tag{A.III.19}$$

$$\tau_1 = C_1 R_{i_1}\ ; \ \tau_f = C_f R_1; \quad \tau_3 = C_3 R \tag{A.III.20}$$

Equation (A.III.14) can be generally used, but in most practical cases ($n < 1$ and $m \gg 1$), this equation becomes

$$f(\xi) \simeq 1 - \exp\left(-\frac{\xi}{m}\right). \tag{A.III.21}$$

The corresponding rise time (0.1—0.9) is

$$t_r = 2.2 \frac{m}{a}, \tag{A.III.22}$$

and taking into account (A.III.15) and (A.III.16) we obtain

$$t_r = \frac{4.4C}{B - \sqrt{B^2 - 4C}}. \tag{A.III.23}$$

When $C \ll B^2$, equation (A.III.23) can again be simplified

$$t_r \simeq 2.2B \simeq 2.2 \frac{|\mathcal{A}_0|}{\mu_1} \tau_1, \tag{A.III.24}$$

that is the rise time is comparable to that of the basic amplifier circuit configuration when the plate load resistance equals the internal resistance of the tube T_1.

The final part of the transient response (flat top response) is obtained from equivalent circuit analysis in the low frequency range (when the capacities C_1, C_f and C_3 in Fig. A.III.1*b* are neglected). In this way one obtains

$$F(t) = \frac{N}{M} + \left(1 - \frac{N}{M}\right) \exp\left(-\frac{M}{\tau} t\right), \tag{A.III.25}$$

where

$$\tau = C_2 R_2, \tag{A.III.26}$$

$$M = \frac{\left(\mu_2 + 1 + \frac{R_{i_2}}{R_3}\right)(R_1 + R_{i_1} + R_2)}{(\mu_2 + 1)R_1 + R_{i_1} + R_{i_2} + \frac{R_{i_2}}{R}(R_{i_1} + R_1)}, \tag{A.III.27}$$

$$N = \frac{\mu_2(R_1 + R_2)}{\mu_2 R_1 + R_{i_2}}. \tag{A.III.28}$$

It may be observed from an examination of equation (A.III.25) that when $N/M = 1$, a total compensation at low frequencies is obtained indifferently of the value C_2. It is worth mentioning that frequently $N \neq M$ and it is necessary to consider the time constant $\frac{\tau}{M}$ when the calculation of the coupling capacity C_2 is performed.

A.III.2. CASCODE AMPLIFIER

Using the same procedure as in the case of the amplifier with bootstrapped plate load, similar relations are obtained [P 9], [V 10] for the cascode amplifier (Fig. A. III. 6a).

The gain at the middle of the frequency range

$$\mathcal{A}_0 = -\frac{(\mu_2 + 1)\,\mu_1 R_a}{(\mu_2 + 1)R_{i_1} + R_{i_2} + R_a}, \tag{A.III.29}$$

and when

$$\frac{R_{i_2} + R_a}{\mu_2 + 1} \ll R_{i_1}, \tag{A.III.30}$$

the well-known relation is obtained

$$\mathcal{A}_0 = -S_1 R_a, \tag{A.III.31}$$

as if it were a pentode with slope S_1 and a plate resistance R_a.

The output resistance can be calculated with a method used by Butler which is based on the fact that the gain is halved when a load resistance equal to the output resistance is connected across the amplifier output. In this way one obtains

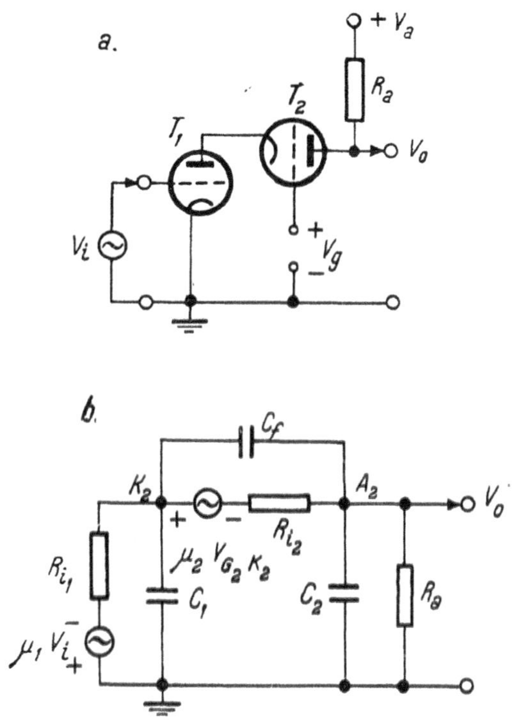

Fig. A.III.6 — Cascode amplifier configuration (*a*) and its equivalent circuit (*b*).

$$\frac{R_o}{R_a} = \frac{(\mu_2 + 1)\, R_{i_1} + R_{i_2}}{(\mu_2 + 1)\, R_{i_1} + R_{i_2} + R_a}. \tag{A.III.32}$$

The initial part (edge response) of the transient response function is similar to that of the amplifier with bootstrapped plate load

$$f(\xi) = 1 - \frac{m-n}{m-1} \exp\left(-\frac{\xi}{m}\right) + \frac{1-n}{m-1} \exp(-\xi), \tag{A.III.33}$$

where

$$\xi = at = \frac{B + \sqrt{B^2 - 4C}}{2C}\, t, \tag{A.III.34}$$

$$m = \frac{B + \sqrt{B^2 - 4C}}{B - \sqrt{B^2 - 4C}}, \tag{A.III.35}$$

$$n = \frac{B + \sqrt{B^2 - 4C}}{2C}\, A_f, \tag{A.III.36}$$

$$A_f = \frac{\tau_f}{\mu_2 + 1}, \tag{A.III.37}$$

$$B = \frac{[(\mu_2 + 1) R_{i_1} + R_{i_2}]\, \tau_2 + (R_{i_2} + R_a)\, \tau_1 + (R_{i_1} + R_a)\, \tau_f}{(\mu_2 + 1) R_{i_1} + R_{i_2} + R_a}, \tag{A.III.38}$$

$$C = \frac{R_{i_2} \tau_1 \tau_2 + R_a \tau_1 \tau_f + R_{i_2} \tau_2 \tau_f}{(\mu_2 + 1) R_{i_1} + R_{i_2} + R_a} \tag{A.III.39}$$

$$\tau_1 = C_1 R_{i_1}; \quad \tau_f = C_f R_{i_2}; \quad \tau_2 = C_2 R_a . \tag{A.III.40}$$

A careful examination of equation (A.III.33) shows that the n-values ($n < 1$) affect the rise time expressed in $\frac{\xi}{m}$ units only for small values of m ($m > 1$). As m increases this influence decreases, becoming negligible for $m \gg 10$. Then, for $m \gg 10$ the transient response can be well approximated by the relation

$$f\left(\frac{\xi}{m}\right) \simeq 1 - \exp\left(-\frac{\xi}{m}\right), \tag{A.III.41}$$

and its rise time (0.1—0.9) is

$$t_r = \frac{2.2\, m}{a} = \frac{4.4\, C}{B - \sqrt{B^2 - 4C}} \tag{A.III.42}$$

By neglecting the capacity C_f, equations (A.III.38) and (A.III.39) become

$$B = \frac{[(\mu_2 + 1) R_{i_1} + R_{i_2}]\, \tau_2 + (R_{i_2} + R_a)\, \tau_1}{(\mu_2 + 1) R_{i_1} + R_{i_2} + R_a}, \tag{A.III.43}$$

$$C = \frac{R_{i_2} \tau_1 \tau_2}{(\mu_2 + 1) R_{i_1} + R_{i_2} + R_a}, \tag{A.III.44}$$

and in the case of identical tubes ($R_{i_1} = R_{i_2} = R_i$) and ($\mu_1 = \mu_2 = \mu$) when in addition $R_a \ll R_i$ and $\mu \gg 1$, equations (A.III.43) and (A.III.44) can be much simplified

$$B = \tau_2 + \frac{\tau_1}{\mu}, \tag{A.III.45}$$

$$C = \tau_2 \frac{\tau_1}{\mu_2}. \tag{A.III.46}$$

As it was to be expected, these relations show that the effect of the capacity C_1 is smaller than that of C_2, because it is connected at a point of small impedance (the input impedance of grounded-grid configuration). In this case the well-known relation is obtained

$$t_r = 2.2\ \tau_2 = 2.2\ R_a C_2 \tag{A.III.47}$$

the rise time of the transient response being the same as that of a basic amplifier circuit configuration.

A.III.3. CASCODE AMPLIFIER WITH BOOTSTRAPPED PLATE LOAD

If the plate load of a cascode amplifier is connected in a bootstrapped mode (Fig. A.III.7*a*), the gain of this amplifier can be greatly increased [C 5]. The equivalent circuit of this configuration is shown in Fig. A.III.7*b*. From this circuit the following relation for the gain in the middle frequency range can be obtained [P 9]

$$\mathcal{A}_0 = -\frac{\mu_1(\mu_2+1)(\mu_3 R_1 + R_{i_3})}{(\mu_2+1)R_{i_1} + (\mu_3+1)R_1 + R_{i_2} + R_{i_3} + \frac{R_{i3}}{R}[(\mu_2+1)R_{i_1} + R_{i_2} + R_1]} \tag{A.III.48}$$

where

$$\frac{1}{R_2} + \frac{1}{R_3} = \frac{1}{R}. \tag{A.III.49}$$

For practical purposes the case of a circuit with identical tubes T_2 and T_3 (double triode) and low noise tube T_1 is of particular interest. In this case, by denoting

$$R_{i_2} = R_{i_3} = R_i;\ \mu_2 = \mu_3 = \mu, \tag{A.III.50}$$

and, if

$$R_i \ll \mu R_1 \, ; \; R_1 \ll \mu R_{i_1} , \tag{A.III.51}$$

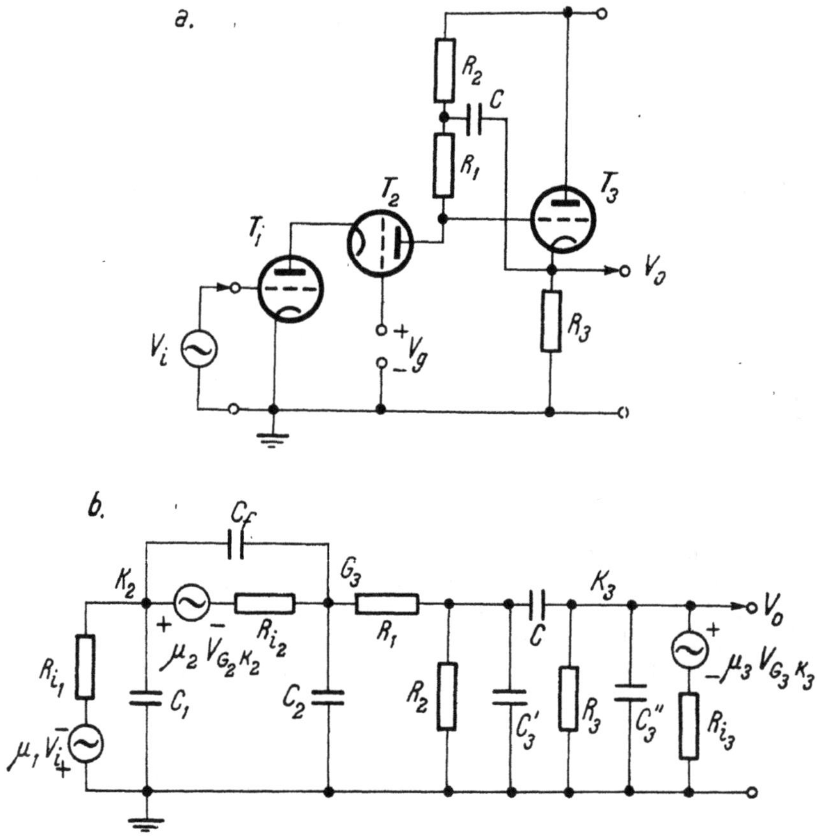

Fig. A.III.7 — Cascode amplifier with bootstrapped plate load (*a*) and its equivalent circuit (*b*).

a simplified formula for the gain is obtained

$$\mathcal{A}_0 = - S_1 R_1 \frac{\mu}{1 + \dfrac{R_i}{R} + \dfrac{R_1}{R_{i_1}}} \, , \tag{A.III.52}$$

If (A.III.52) is compared with (A.III.31) one can see that the bootstrapped connection of the plate load brings a gain multiplication of the order of μ.

The output resistance is

$$R_o = \frac{R_{i_3}\,[(\mu_1+1)\,R_{i_1} + R_{i_2} + R_1]}{(\mu_2+1)\,R_{i_1} + (\mu_3+1)\,R_1 + R_{i_2} + R_{i_3} + \dfrac{R_{i_3}}{R}\,[(\mu_2+1)\,R_{i_1} + R_{i_2} + R_1]}, \tag{A.III.53}$$

but when conditions (A.III.50) and (A.III.51) are fulfilled

$$\frac{1}{R_o} = \frac{1}{R} + \frac{1}{R_i}\left(1 + \frac{R_1}{R_{i_1}}\right), \tag{A.III.54}$$

that is, the output resistance is greater than that of a cathode follower with the same tube and the cathode load R_1.

The edge response can be calculated by taking the inverse Laplace transform of the equation

$$\frac{\mathcal{A}(p)}{p\mathcal{A}_0} = \frac{1 + A_f p}{p\,(1 + Bp + Cp^2 + Dp^3)}, \tag{A.III.55}$$

where

$$A_f = \frac{\tau_f}{\mu_2}, \tag{A.III.56}$$

$$\begin{aligned} B = {} & \left\{(\mu_2+1)\,R_{i_1} + (\mu_3+1)\,R_1 + R_{i_2} + R_{i_3} + \frac{R_{i_3}}{R}[(\mu_2+1)R_{i_1} + R_1 + R_{i_2}]\right\}^{-1} \times \\ & \times \left\{\left[(\mu_3+1)\,R_1 + R_{i_2} + R_{i_3} + \frac{R_{i_3}}{R}\,(R_1 + R_{i_1} + R_{i_2})\right]\tau_1 + \right. \\ & + \left\{\left[(\mu_2+1)\,\frac{R_{i_1}}{R_1} + \frac{R_{i_2}}{R_1}\right](\mu_3 R_1 + R_{i_3}) + \left(1 + \frac{R_{i_3}}{R}\right)[(\mu_2+1)\,R_{i_1} + R_{i_2}]\right\}\tau_2 + \\ & + \frac{R_{i_3}}{R}\,[(\mu_2+1)\,R_{i_1} + R + R_{i_2}]\,\tau_3 + \left[(\mu_3+1)\,R_1 + R_{i_1} + R_{i_3} + \right. \\ & \left. \left. + \frac{R_{i_3}}{R}\,(R_{i_1} + R_1)\right]\tau_f\right\}, \end{aligned} \tag{A.III.57}$$

$$C = \left\{(\mu_2+1) R_{i_1} + (\mu_3+1) R_1 + R_{i_2} + R_{i_3} + \frac{R_{i_3}}{R}[(\mu_2+1) R_{i_1} + R_1 + R_{i_2}]\right\}^{-1} \times$$

$$\times \left\{\left[(\mu_3 R_1 + R_{i_3}) \frac{R_{i_2}}{R_1} + (R + R_{i_3}) \frac{R_{i_2}}{R}\right] \tau_1\tau_2 + \frac{R_{i_3}}{R} (R_1 + R_{i_2}) \tau_1\tau_3 + \right.$$

$$+ \left[(\mu_3 + 1) R_1 + R_{i_3} + \frac{R_{i_3}}{R} R_1\right] \tau_1\tau_f + \frac{R_{i_3}}{R} [(\mu_2 + 1) R_{i_1} + R_{i_2}] \tau_2\tau_3 +$$

$$\left. + \left[\frac{R_{i_1}}{R_1} (\mu_3 R_1 + R_{i_3}) + \frac{R_{i_1}}{R} (R_{i_3} + R)\right] \tau_2\tau_f + \frac{R_{i_3}}{R} (R_1 + R_{i_1}) \tau_3\tau_f\right\}, \qquad \text{(A.III.58)}$$

$$D = \left[\frac{R_{i_3}}{R} (R_{i_2} \tau_1\tau_2\tau_3 + R_{i_1}\tau_2\tau_3\tau_f + R_1\tau_3\tau_f\tau_1)\right] \times$$

$$\times \left\{(\mu_2+1) R_{i_1} + (\mu_3+1) R_1 + R_{i_2} + R_{i_3} + \frac{R_{i_3}}{R} [(\mu_2 + 1) R_{i_1} + R_1 + R_{i_2}]\right\}^{-1}, \qquad \text{(A.III.59)}$$

$$\tau_1 = C_1 R_{i_1};\; \tau_2 = C_2 R_1;\; \tau_3 = C_3 R;\; \tau_f = C_f R_{i_2}. \qquad \text{(A.III.60)}$$

This form of calculation is very laborious, but a simplified formula can be obtained, observing that in equation (A.III.57) the time constant τ_2 has a coefficient of at least an order of magnitude greater than that of τ_1, τ_f and τ_3. This was to be expected because both capacitances C_1 and C_3 are connected at points of small impedances (input impedance of grounded grid amplifier and output impedance of a cathode follower, respectively). By taking $\tau_1 = \tau_3 = \tau_f = 0$ in equations (A.III.56) — (A.III.59), we obtain

$$A_f = C = D = 0, \qquad \text{(A.III.61)}$$

$$B = \left\{\left[(\mu_2 + 1)\frac{R_{i_1}}{R_1} + \frac{R_{i_2}}{R_1}\right](\mu_3 R_1 + R_{i_3}) + \left(1 + \frac{R_{i_3}}{R}\right)[(\mu_2+1) R_{i_1} + R_{i_2}]\right\} \times$$

$$\times \left\{ (\mu_2+1) R_{i_1} + (\mu_3+1) R_1 + R_{i_2} + R_{i_3} + \frac{R_{i_3}}{R} [(\mu_2+1) R_{i_1} + R_1 + R_{i_2}]\right\}^{-1}. \qquad \text{(A.III.62)}$$

The rise time (0.1—0.9) of the corresponding transient response is

$$t_r = 2.2\,B \simeq 2.2\,\frac{|\mathcal{A}_0|}{\mu_1}\left(1 + \frac{R_{i_1}}{\mu_3 R_1}\right)\tau_2, \tag{A.III.63}$$

hence greater than that of a basic amplifier configuration with the anode circuit time constant τ_2.

The flat top response is given by

$$F(t) = \frac{M}{N} + \frac{N - M}{N}\,\exp\left(-\frac{N}{\tau}\,t\right), \tag{A.III.64}$$

where

$$\tau = CR_2, \tag{A.III.65}$$

$$= \mu_3\,\frac{M(R_1 + R_2)}{\mu_3 R_1 + R_{i_3}} \tag{A.III.66}$$

$$N = \frac{|\mathcal{A}_0|}{\mu_1}\cdot\frac{\left(\mu_3 + 1 + \dfrac{R_{i_3}}{R_3}\right)[(\mu_2 + 1)\,R_{i_1} + R_{i_2} + R_1 + R_2]}{(\mu_2 + 1)\,(\mu_3 R_1 + R_{i_3})} \tag{A.III.67}$$

As in the case of an amplifier with bootstrapped plate load, for $N = M$ one obtains $F(t) = 1$, indifferently of the value of the time constant.

REFERENCES

A1. ADAMS J. R., Nucleonics, **22**, *8*, 112 (1964).
A2. AIRAPETIANT A. V., RIVKIN S. M., J. T. F., **27**, *1*, 95 (1957).
A3. AJDACIC V.S., LALOVIC B. I., PETROVIC B.P., *Electronique Nucléaire*, Colloque—Paris, 1962, C.E.A. Paris, 1963, p. 181.
A4. ALBERIGI A., QUARANTA A., MARTINI M., OTTAVIANI G., ZANARINI G., Nucl. Instr. Meth., **35**, 93 (1965).
A5. ALKHAZOV G. D., KOMAR A. P., VOROB'EV A. A., Nucl. Instr. Meth., **48**, 1 (1967).
A6. AMSEL G., BARUCH P., SMULKOWSKI O., Nucl. Instr. Meth., **8**, 93 (1960).
A7. AMSEL G., BOSSHARD R., ZAJDT C., I.E.E.E. Trans., **NS—14,** 1 (1967).
A8. ANDERSON-LINDSTROEM G., ZAUSIG B., Nucl. Instr. Meth., **40**, 277 (1966).
A9. ANDERSON-LINDSTROEM G., Nucl. Instr. Meth., **56**, 309 (1967).
A10. ANDREWS P. T., *Proc. Symp. Nucl. Instr., Harwell*, 1961, p. 93.
A11. ANTMAN S. O. W., LANDIS D. A., PEHL R. H., Nucl. Instr. Meth., **40**, 272 (1966).
A12. ARBEL A. F., *N.A.S.—N.R.C. Pub., 1184*, Washington, 1964, p. 79.
A13. ARMANTROUT G., U.C.R.L. — 14263, Livermore (1965).
A14. ARMANTROUT G., I.E.E.E. Trans., **NS—13**, 328 (1966).
A15. ARON W. A., HOFFMAN B. G., WILLIAMS F. C., A.E.C.U. — 663 (U.C.R.L. — 121), Berkeley (1960).
A16. ASTRÖM B., JOHANSSON N. G. E., WICKENBERG G., *Nuclear Electronics. Proc. Conf. Belgrade, 1961*, vol. 1, I.A.E.A. Vienna, 1962, p. 535.
A17. AWCOCK M. L., YOUNG D. C., A.E.R.E.—R 4710, Harwell (1964).
B1. BABCOCK R. V., DAVIS R. E., RUBY S. L., SUN K. H., WOLLEY E. D., Nucleonics, **17**, 116 (1959).
B2. BABCOCK R. V., I.R.E. Trans., **NS—8**, *1*, 98 (1961).
B3. BABUSKIN A. A., MEDNIKOV A. K., P.T.E., *1*, 105 (1965).
B4. BĂDĂNOIU M., CONSTANTINESCU S., Rev. Roum. Phys., **11**, *7*, 559 (1966).
B5. BAILY N. A., MAYER J. W., CRAINGER R. J., I.R.E. Trans., **NS—9**, *1*, 91 (1962).
B6. BALDINGER E., FRANZEN W., *Advances in Electronics*, vol. 8, Academic Press, New York, 1956, p. 256.
B7. BALDINGER E., CZAJA W., Nucl. Instr. Meth., **10**, 237 (1961).
B8. BALDINGER E., CZAJA W., GUTMANN J., *Proc. Symp. Nucl. Instr., Harwell*, 1961, p. 98.
B9. BALLAND J. C., PIGNERET J., SAMUELI J. J., SARAZIN A., Nucl. Instr. Meth., **37**, 276 (1965).
B10. BALLAND J. C., GOYOT M., PIGNERET J., SAMUELI J. J., L'Onde El., **46**, 472, 844 (1966).
B11. BALLAND J. C., PIGNERET J., SAMUELI J. J., SARAZIN A., LYCEN--6771, Lyon (1967).
B12. BALLAND J. C., *Thèse*, Lyon, 1967.
B13. BENVENISTE J., BOOTH R., MITCHELL A. C., Nucl. Instr. Meth., **12** (1961).
B14. BERG R. E., KASHY E., Nucl. Instr. Meth., **39**, *1*, 169 (1965).
B15. BERGSTRÖM I., DOMEIJ B., Nucl. Instr. Meth., **43**, 146 (1966).

B16. Bertrand F. E., Peelle R. W., Love T. A., Fox R. J., Hill N. W., Todd H. A., I.E.E.E. Trans., **NS–13**, *3*, 279 (1966).

B17. Beskrovnii I. M., Izv. A.N. S.S.S.R. Ser. Fiz., **27**, *7*, 956 (1963).

B18. Bethe A. H., Rev. Mod. Phys., **22**, 213 (1960).

B19. Bilaniuk O. M., Nucl. Instr. Meth., **13**, 906 (1961).

B20. Bilaniuk O. M., Marsh B. B., *Nuclear Electronics. Proc. Conf. Belgrade, 1961*, vol. 3, I.A.E.A. Vienna, 1962, p. 231.

B21. Bilger H. R., Nucl. Instr. Meth., **40**, 54 (1966).

B22. Blalock T. V., I.E.E.E. Trans., **NS–11**, *3*, 365 (1964).

B23. Blalock T. V., Pierce J. F., *N.A.S. – N.R.C. Pub., 1184*, Washington, 1964, p. 204.

B24. Blalock T. V., Rev. Sci. Instr., **36**, *10*, 1448 (1965).

B25. Blalock T. V., I.E.E.E. Trans., **NS–13**, *3*, 457 (1966).

B26. Blankenship J. L., Borkowski C. J., I.R.E. Trans., **NS–7**, *2–3*, 190 (1960).

B27. Blankenship J. L., Borkowski C. J., I.R.E. Trans., **NS–8**, *1*, 17 (1961).

B28. Blankenship J. L., Borkowski C. J., I.R.E. Trans., **NS–9**, *3*, 181 (1962).

B29. Blankenship J. L., Borkowski C. J., Fox R. J., *Nuclear Electronics. Proc. Conf. Belgrade, 1961*, vol. 1, I.A.E.A. Vienna, 1962, p. 379.

B30. Blankenship J. L., I.E.E.E. Trans., **NS–11**, *3*, 373 (1964).

B31. Blankenship J. L., Nowlin C. H., I.E.E.E. Trans., **NS–13**, *3*, 495 (1966).

B32. Blank D., Casanovas J., Chapuis A. M., Soudain G., Nucl. Instr. Meth., **37**, 90 (1965).

B33. Blatt F. J., *Theory of Mobility of Electrons in Solids*, Academic Press Inc., New York, 1957.

B34. Bloch F., Z. Phys., **81**, 363 (1933).

B35. Bock R., Duhm H. H., Melzer W., Pühlhofer F., Stadler B., Nucl. Instr. Meth., **41**, 190 (1966).

B36. Bok J., Cosnac de B., Noël J. P., Schuttler R., *Nuclear Electronics. Proc. Conf. Belgrade, 1961*, vol. 1, I.A.E.A. Vienna, 1962, p. 447.

B37. Borisov M., Marinov M., *Nuclear Electronics. Proc. Conf. Belgrade, 1961*, vol. 1, I.A.E.A. Vienna, 1962, p. 363.

B38. Bosshard R., L'Onde El., **46**, 804 (1966).

B39. Braid T. H., Heinrich J. I., *Nuclear Electronics. Proc. Conf. Belgrade, 1961*, vol. 1, I.A.E.A. Vienna, 1962, p. 447.

B40. Bridgers H. E., Scaff J. F., Shive I. N., Biondi F. J., *Transistor Technology*, D. Van Nostrand, New York, 1958.

B41. Broadhurst J. H., Pyle G. J. B., Nucl. Instr. Meth., **48**, 117 (1967).

B42. Bromley D. A., Sachs M. W., Anderson C. E., *Proc. Symp. Nucl. Instr., Harwell*, 1961, p. 67.

B43. Bromley D. A., *N.A.S.–N.R.C. Pub. 871*, Washington, 1961, p. 61.

B44. Brovchenco V. G., Molchanov Iu. D., P.T.E., **4**, 5 (1964).

B45. Brown W. L., I.R.E. Trans., **NS–8**, *1*, 2 (1961).

B46. Brown W. L., *Proc. Conf. Semiconductor Physics*, Dunod, Paris, 1965, p. 1313.

B47. Buck T. M., McKim F. S., J. Electrochem. Soc., *105*, 709 (1958).

B48. Buck T. M., *N.A.S.–N.R.C. Pub. 871*, Washington, 1961, p. 111.

B49. Buck T. M., Wheatley G. J., Rogers T. W., I.E.E.E. Trans., **NS–11**, *3*, 294 (1964).

C1. Canepa P. C., Malinaric P., Cambell R. B., Ostroski J., I.E.E.E. Trans., **NS–11**, *3*, 262 (1964).

C2. Cartianu Gh., *Bazele Radiotehnicii*, Ed. St. Did. Ped., Bucharest, 1962.

C3. Cavalleri G., Fabri G., Gatti E., Svelto V., Nucl. Instr. Meth., **21**, 177 (1963).

C4. CHADEERTON L. T., *Radiation Damage in Crystals*, Methuen and Comp. Ltd., London, 1965.
C5. CHASE R. L., HIGINBOTHAM W. A., MILLER G. L., I.R.E. Trans., **NS−8**, *1*, 147 (1961).
C6. CHASE R. L., I.R.E. Trans., **NS−9**, *1*, 119 (1962).
C7. CHASMAN C., JONES K. W., RISTINEN R. A., B.N.L., −9157, New York, (1965).
C8. CHRISTIE D. R., DANIELS J. M., FELSTEINER J., Nucl. Instr. Meth., **37**, 165 (1965).
C9. CHYNOWETH A. G., J. Appl. Phys., **31**, *7*, 1161 (1960).
C10. CHYNOWETH A. G., *N.A.S.−N.R.C. Pub.*, *871*, Washington, 1961, p. 171.
C11. CHWASZCZEWSKA J., DAKOWSKI M., DABROWSKI A., SOWINSKI M., STAPA M., Nucleonika, **9.**, *1*, 11 (1964).
C12. CIORĂSCU F., ONCESCU M., *Detectarea şi Măsurarea Radiaţiilor Nucleare*, Ed. Acad., Bucharest, 1964.
C13. CIPOLLA F., TARONI A., Nucl. Instr. Meth., **54**, 331 (1967).
C14. COHEN E. R., DUMOND J. W. M., Rev. Mod. Phys., **37**, *4*, 537 (1965).
C15. COLEMAN J. A., RODGERS J. W., I.E.E.E. Trans., **NS−11**, *3*, 213 (1964).
C16. COROLLER Y. LE, R 2722, C.E.A. Paris, (1963).
C17. COTTINI C., GATTI E., GIANELLI G., ROZZI G., Nuovo Cimento, Ser. X, **3**, *2*, 473 (1956).
C18. COTTINI C., GATTI E., SVELTO V., *N.A.S.−N.R.C. Pub.*, *1184*, Washington, 1964, p. 53.
C19. CROWELL C. R., SHORE H. B., LABATE E. E., J., Appl. Phys., **36**, *12*, 3824 (1965).
D1. DAKOWSKI M., SOWINSKI M., Nucleonika, **8**, *4*, 280 (1963).
D2. DEARNALEY G., I.R.E. Trans., **NS−8**, 11 (1961).
D3. DEARNALEY G., FERGUSON A.T.G., *Proc. Symp. Nucl. Instr., Harwell*, 1961, p.84.
D4. DEARNALEY G., WHITEHEAD A. B., Nucl. Instr. Meth., **12**, 205 (1961).
D5. DEARNALEY G., FERGUSON A. T. G., MORRISON G. C., I.R.E. Trans., **NS−9**, *3*, 174 (1962).
D6. DEARNALEY G., FERGUSON A. T. G., WHITEHEAD A. B., MONTAGUE J. H., *Nucl. Electronics. Proc. Conf. Belgrade, 1961*, vol. 1, I.A.E.A. Vienna, 1962, p. 523.
D7. DEARNALEY G., LEWIS I. C., *Electronique Nucléaire. Colloque — Paris*, 1962, C.E.A. Paris, 1963, p. 147.
D8. DEARNALEY G., I.E.E.E. Trans., **NS−10**, *1*, 106 (1963).
D9. DEARNALEY G., I.E.E.E. Trans., **NS−11**, *3*, 249 (1964).
D10. DEARNALEY G., Nucl. Instr., Meth., **25**, *2*, 237 (1964).
D11. DEARNALEY G., NORTHROP D. C., *Semiconductor Counters for Nuclear Radiations*, Spon Limited, London, 1964.
D12. DEARNALEY G., *Progress in Nuclear Physics*, vol. 9, Pergamon Press, 1964, p. 29.
D13. DEARNALEY G., Nucleonics, **22**, *7*, 78, (1964).
D14. DEIGHTON M. O., A.E.R.E. — R 5021, Harwell (1967).
D15. DELAUNAY B., DELAUNAY I., MEURGUES J. P., PASSERIEUX J. P., PEREY F. J., Bull. Inf. C.E.A., **91** 1 (1965).
D16. DE LYSER H., ZIEMBA F. P., ANTWERP W. R. VAN., I.E.E.E. — Trans., **NS−12**, *1*, 265 (1965).
D17. DESHANDE R. Y., Nucl. Instr. Meth., **46**, 255 (1967).
D18. DOBRETOV L. N., *Atomnaia Fizika*, Fizmatghiz, Moscow, 1960.
D19. DODGE W. R., DOMEN S. R., HIRSHFELD A. T., HOPPES D. D., I.E.E.E. Trans., **NS−11**, *3*, 238 (1964).
D20. DODGE W. R., COLEMAN J. A., DOMEN S. R., WHITTAKER J. K., Rev. Sci. Instr., **37**, *9*, 1151 (1966).

D21. Donovan P. F., I.E.E.E., Trans., **NS—13**, *3*, 21 (1966).
D22. Drăgănescu M., Mircea A., *Circuite cu tranzistoare*, Ed. tehnică, Bucharest, 1961.
D23. Drăgănescu M., *Procese Electronice în Dispozitive Semiconductoare*, Ed. Acad., Bucharest, 1962.
D24. Drexler G., Perzl F., Nucl. Instr. Meth., **48**, 332 (1967).
D25. Dubrau H. J., Nucl. Instr. Meth., **15**, 77 (1962).
D26. Dubrau H. J., Nachrichtentechnik, **4**, 147 (1962).
D27. Dummur I., George G., Gunnersen E. M., Hitchcock A., *Nuclear Electronics. Proc. Conf. Belgrade, 1961*, vol. 1, I.A.E.A. Vienna, 1962, p. 501.
D28. Dunlap W. C., *An Introduction to Semiconductors*, John Wiley and Sons, New York, 1957.
E1. Eisler P. L., Nucl. Instr. Meth., **44**, 253 (1966).
E2. Eisler P. L., Nucl. Instr. Meth., **48**, 103 (1967).
E3. Elad E., Nakamura M., U.C.R.L. = 17149, Berkeley (1966).
E4. Elad E., Nakamura M., U.C.R.L. = 17632, Berkeley (1966).
E5. Elliott J. N., Nucl. Instr., **12**, 60 (1961).
E6. Emerson S. T., Simpson W. D., Legg J. C., Phillips C. C., Nucl. Instr. Meth., **52**, 229 (1967).
E7. Emery F. E., Rabson T. A., I.E.E.E. Trans., **NS—13**, *1*, 48 (1966).
E8. Emmer T. L., I.R.E. Trans., **NS—8**, *1*, 140 (1961).
E9. Emmer T. L., I.R.E. Trans., **NS—9**, *3*, 305 (1962).
E10. Engelkemeir D., Nucl. Instr. Meth., **48**, *2*, 335 (1967).
E11. Engler H. D., Nuckleonik, **2**, *6*, 215 (1960).
E12. Ewan G. T., Tavendale A. J., Canad. J. Phys., **42**, 2286 (1964).
E13. Ewan G. T., Malm H. L., Fowler I. L., A.E.C.L. — 2607, Chalk River (1966).
E14. Ewan G. T., Graham R. L., Mac Kenzie I. K., I.E.E.E. Trans., **NS—13**, *3*, 297 (1966).
E15. Ewing R. J., I.R.E. Trans., **NS—9**, *3*, 207 (1962).
F1. Fabri G., Gatti E., Svelto V., Nucl. Instr. Meth., **15**, 237 (1962).
F2. Fabri G., Gatti E., Svelto V., Phys. Rev., **131**, *1* ,134 (1963).
F3. Fabri G., Gatti E., Svelto V., *N.A.S. — N.R.C. Pub. 1184*, Washington, 1964, p. 49.
F4. Fabri G., Svelto V., Nucl. Instr. Meth., **35**, 83 (1965).
F5. Fairstein E., I.R.E. Trans., **NS—8**, *1*, 129 (1961).
F6. Fairstein E., Hahn J., Nucleonics, **23**, *7*, 56 ; *9*, 81 ; *11*, 50 ; (1965) **24**, *1*, 54 ; *3*, 68 (1966).
F7. Fairstein E., I.E.E.E. Trans., **NS—13**, *1*, 596 (1966).
F8. Falk K., Tove P. A., Madakbats M., Nucl. Instr. Meth., **34**, 157, (1965).
F9. Fano V., Ann. Rev. Nucl. Sci., **13**, 1 (1963).
F10. Fefilov B. V., P.T.E., **5**, 121 (1964).
F11. Fenyves E., Haiman O., *Die Physikalischen Grundlagen der Kernstrahlungs Messungen*, Akademiai Kiadó, Budapest, 1965.
F12. Forgue V., Kahn S., Nucl. Instr. Meth., **48**, *1*, 93 (1967).
F13. Fox R. J., Borkowski C. J., I.R.E. Trans., **NS—9**, *3*, 213 (1962).
F14. Freeman J. M., Jenkin J. G., Nucl. Instr. Meth., **43**, 269 (1966).
F15. Freycenon J., Rosny G. D., L'Onde El., **47**, **488**, 1363 (1967).
F16. Friedland S. S., Mayer J. W., Wiggins J. S., Nucleonics, **18**, 54 (1960).
F17. Friedland S. S., Mayer J. W., Denny J. H., Keywell F., Rev. Sci. Instr., **31**, 74 (1960).

F18. Friedland S. S., Ziemba F. P., Olson R. M., Lyser H. De, I.R.E. Trans., **NS—9**, *3*, 391 (1962).
F19. Friedland S. S., Katzenstein H. S., Ziemba F. P., I.R.E. Trans., **NS—10**, *1*, 190 (1963).
G1. Gärtner R., Zfk. Ph.A.12, Rossendorf, 172 (1963).
G2. Gemmell D. S., I.E.E.E. Trans., **NS—11**, *3*, 409 (1964).
G3. Gibbons P. E., *Electronique Nucléaire. Colloque — Paris, 1962*, C.E.A. Paris, 1963, p. 243.
G4. Gibbons P. E., Owen R. B., A.E.R.E. — M 1602 Harwell (1965).
G5. Gibbons P. E., A.E.R.E. — R 5206, Harwell (1966).
G6. Gibson W. M., I.E.E.E. Trans., **NS—13**, *3*, 162 (1966).
G7. Gillespie A. B., *Signal, Noise and Resolution in Nuclear Counter Amplifiers*, Pergamon Press, London, 1953.
G8. Gîdea S., Protopopescu M., Drimer D., *Metalurgia fizică a semiconductoarelor*, Ed. Acad., Bucharest, 1963.
G9. Glotin Ph., Bull. Inf. Sci. Tech., C.E.A., **98**, 1 (1965).
G10. Gobeli G. W., Phys. Rev., **103**, 275 (1957).
G11. Goldstein B., J. Appl. Phys., **36**, *12*, 3853 (1965).
G12. Goldsworthy W. W., I.R.E. Trans., **NS—10**, *1*, 61 (1963).
G13. Goldsworthy W. W., U.C.R.L. — 16838, Berkeley (1966).
G14. Golovin B. M., Osipenco B. P., Sidorov A. I., P—510, I.U.C.N. Dubna (1960).
G15. Golovin B. M., Osipenco B. P., Sidorov A. I., P.T.E., **6**, 5 (1961).
G16. Goodman C. D., Ball J. B., *Proc. Symp. Nucl. Instr., Harwell*, 1961, p. 163.
G17. Gorodetzky S., Port M., Graff I., Nucl. Instr. Meth., **15**, 183 (1962).
G18. Gossick B., Mayer J., Rev. Sci. Instr., **27**, 407 (1956).
G19. Gossik B., *Potential Barriers in Semiconductors*, Academic Press, New York, 1964.
G20. Goulding F. S., Hansen W. L., Nucl. Instr. Meth., **12**, 249 (1961).
G21. Goulding F. S., I.E.E.E. Trans., **NS—11**, *3*, 177 (1964).
G22. Goulding F. S., Landis D. A., Cerny J., Pell R. H., I.E.E.E. Trans., **NS—11**, *3*, 388 (1964).
G23. Goulding F. S., Hansen W. L., I.E.E.E. Trans., **NS—11**, *3*, 286 (1964).
G24. Goulding F. S., U.C.R.L. — 11302, 18 (1964).
G25. Goulding F. S., Landis D., *N.A.S. — N.R.C. Pub. 1184*, Washington, 1964, p. 61.
G26. Goulding F. S., Landis D., *N.A.S. — N.R.C. Pub. 1184*, Washington, 1964, p. 124.
G27. Goulding F. S., *N.A.S. — N.R.C. Pub. 1184*, Washington, 1964, p. 59.
G28. Goulding F. S., U.C.R.L. — 16231 (1965).
G29. Goulding F. S., Jarrett B. V., U.C.R.L. — 16480 (1966).
G30. Graham R. L., Mac Kenzie I. K., Ewan G. T., I.E.E.E. Trans., **NS—13**, *1*, 72, (1966).
G31. Graniger R. I., Mayer J. W., Oliver J. W., I.R.E. Trans., **NS—8**, *1*, 116 (1961).
G32. Gunn J. B., Solid State Electronics, **7**, *10*, 739 (1964).
H1. Hahn J., Mayer R., S.R.E. Trans., **NS—9**, *4*, 20 (1962).
H2. Haitz R. H., Smits F. M., I.E.E.E. Trans., **NS—13**, *3*, 198 (1966).
H3. Halbert M. L., *Nuclear Electronics. Proc. Conf. Belgrade, 1961*, vol. 1, I.A.E.A. Vienna, 1962, p. 379.
H4. Hansen N. J., I.R.E. Trans., **NS—9**, *3*, 217 (1962).
H5. Hansen W. L., Goulding F. S., *Nuclear Electronics. Proc. Conf. Belgrade, 1961*, vol. 1, I.A.E.A. Vienna, 1962, p. 583.
H6. Hansen W. L., Jarrett B. V., U.C.R.L., 11589 (1964).
H7. Hansen W. L., Goulding F. S., I.E.E.E. Trans., **NS—11**, *3*, 388 (1964).

H8. Hatch K. E., U.C.R.L. —7947 (1965).
H9. Heath R. L., Nucl. Instr. Meth., **43**, 209 (1966).
H10. Heath R. L., Black W. W., Cline J. E., I.E.E.E. Trans., **NS—13**, *3*, 445 (1966).
H11. Heerschap M., Coninck R. De, *Nuclear Electronics. Proc. Conf. Belgrade, 1961*, vol. 1, I.A.E.A. Vienna, 1962, p. 513.
H12. Heiland G., Usp. Fiz. Nauk, **82**, *2*, 325 (1964).
H13. Hick H., Rumpold K., Weinzierl P., *Electronique Nucléaire. Colloque — Paris, 1962*, C.E.A. Paris, 1963, p. 229.
H14. Hofker W. K., et al., I.E.E.E. Trans., **NS—13**, *3*, 208 (1966).
H15. Hollander J. M., U.C.R.L. — 16307 (1965).
H16. Hooton B. W., A.E.R.E. — R 4921, Harwell (1965).
H17. Hotz G. C., Mathiesen J. M., Hurley J. P., Nucl. Instr. Meth., **37**, 93 (1965).
H18. Huth G. C., Trice J. B., Mc Kinney R. A., Rev. Sci. Instr., **9**, 1220 (1964).
H19. Huth G. C., Trice J. B., Shannon J. A., Mc Kinney R. A., I.E.E.E. Trans., **NS—12**, *1*, 275 (1965).
H20. Huth G. C., Trice J. B., Mc Kinney R. A., Locker R. J., General Electric, Report N.Y.O. 3246, 5 (1965).
H21. Huth G. C., Locker R. J., McKinney R. A., General Electric, Report N.Y.O. 3246, 6 (1965).
H22. Huth G. C., I.E.E.E. Trans., **NS—13**, *1*, 36 (1966).
I1. Inskeep C. N., Eidson W. W., Lasalle R. A., I.R.E. Trans., **NS—9**, *3*, 167 (1962).
I2. Inskeep C. N., Eidson W. W., *Electronique Nucléaire. Colloque — Paris, 1962*, C.E.A. Paris, 163 (1963).
I3. Ioffe A. F., *Fizika poluprovodnikov*, Izd. Akad. Nauk S.S.S.R., Moscow, 1957.
J1. Jackson R. W., Webb P. P., Williams R. L., I.R.E. Trans., **NS—8**, *1*, 29 (1961).
J2. Jen C. K., Proc. I.R.E., **29**, 345 (1941).
J3. Jen C. K., Proc. I.R.E., **29**, 464 (1941).
J4. Jones A. R., A.E.C.L. — 2252, Chalk River (1965).
J5. Jorgensen M. H., Meyer N. J., Schmidt J. J., Preprint Technical University of Denmark, Copenhagen (1965).
J6. JoynerW. T., Schmitt N. Y., Neiler J. H., Silva R. J., I.R.E. Trans., **NS—8**, *1*, 54 (1961).
K1. Kalbitzer S., Bader R., Melzer W., Stumpfi W., Nucl. Instr. Meth., **54**, 323 (1967).
K2. Kalinin A. I., **Ph. A. 12**, Zfk Rossendorf, 219 (1963).
K3. Kandiah K., *N.A.S. — N.R.C. Pub. 1184*, Washington, 1964, p. 65.
K4. Kaufmann C., **Ph. A. 12**, Zfk Rossendorf,, 10 (1963).
K5. Katzenstein H. S., Alexander R. M., *Electronique Nucléaire, Colloque — Paris, 1962*, C.E.A. Paris, 1963, p. 129.
K6. Katzenstein H. S., I.E.E.E. Trans., **NS—13**, 527 (1966).
K7. Klein S. S., Hulstman L., Blok J., Nucl. Instr. Meth., **54**, 190 (1967).
K8. Klema E., I.E.E.E. Trans., **NS—12**, *1*, 288 (1965).
K9. Klemma E. D., Pleasonton F., Schmitt H. W., Unik J., I.E.E.E. Trans., **NS—13**, *1*, 43 (1966).
K10. Klingersmith R. W., I.R.E. Trans., **NS—8**, *1*, 112 (1961).
K11. Kobayashi T., Takayanagi S., Nucl. Instr. Meth., **44**, 145 (1966).
K12. Kock L., Messier J., Kerns O., I.R.E. Trans., **NS—8**, *1*, 83 (1961).
K13. Kock L., Messier J., Valin J., *Nuclear Electronics. Proc. Conf. Belgrade, 1961*, vol. 1, I.A.E.A. Vienna, 1962, p. 465.
K14. Kohler T. R., *N.A.S. — N.R.C. Pub. 871*, Washington, 1961, p. 193.
K15. Konijn J., Lauber A., Tollander B., Nucl. Instr. Meth., **36**, *2*, 302 (1965).

K16. Kraner H. W., Sovka I. A., Breckenridge R. W., Nucl. Instr., **36**, 328 (1965).
K17. Krulisch A. H., Axtmann R. C., I.E.E.E. Trans., **NS—14**, *4*, 58 (1967).
K18. Kuchly J. M. et al., Nucl. Instr. Meth., **47**, *1*, 148 (1967).
K19. Kuhlmann W. R., Lauterjung K. H., Schimmer B., Nucl. Instr. Meth.,**40**, 109 (1966).
K20. Kurtz A. D., Yee R., J. Appl. Phys., **31**, 303 (1960).
L1. Lalovic B. I., Ajdacic V. S., *Nuclear Electronics. Proc. Conf. Belgrade, 1961*, vol. 1 I.A.E.A. Vienna, 1962, p. 543.
L2. Lalovic B. I., Ajdacic V. S., *Electronique Nucléaire. Colloque — Paris, 1962*, C.E.A. Paris, 1963, p. 171.
L3. Landau L., J. Phys. U.S.S.R., **8**, 201 (1944).
L4. Landau L. D., Lifşiţ E., *Mecanica, cuantică* Ed. tehnică, Bucharest, 1965.
L5. Langmann H. J., Meyer O., Nucl. Instr., **30**, 135 (1964).
L6. Langmann H. J., Meyer O., Nucl. Instr. Meth., **44**, 55 (1966).
L7. Lecomte J. L., *Electronique Nucléaire. Colloque — Paris, 1962*, C.E.A. Paris, 1963, p. 341.
L8. Levich V. G., *Kurs teoreticheskoi fiziki*, Fizmatghiz, Moscow, 1962.
L9. Levin S. N., *Quantum Physics of Electronics*, Macmillan Comp., New York, 1965.
L10. Libert J., Nucl. Instr., **35**, *1*, 148 (1965).
L11. Lindmayer I., Solid State Electronics, **9**, 225 (1966).
L12. Llacer I. L., I.E.E.E. Trans., **NS—11**, *3*, 221 (1964).
L13. Lothrop R. P., Smith H. E., U.C.R.L. 16190 (1965).
L14. Love T. A., Murray R. B., I.R.E. Trans., **NS—8**, 1, 91 (1961).
L15. Love T. A., Murray R. B., Manning J. H., Todd H. A., *Nuclear Electronics. Proc. Conf. Belgrade, 1961*, vol. 1, I.A.E.A. Vienna, 1962, p. 415.
L16. Ludwig E. J., Rev. Sci. Instr., **36**, *8*, 1175 (1965).
L17. Ludwig E. J., Gibson W. M., Hood J. S., I.E.E.E. Trans., **NS—12**, *1*, 247 (1965).
M1. MacCabee H. D., Raju M. R., Nucl. Instr., **37**, 176 (1965).
M2. Madden T. C., Gibson W. M., I.E.E.E. Trans., **NS—11**, *3*, 254 (1964).
M3. Malm H. L., Tavendale A. J., Fowler I. L., Canad. J. Physics, **43**, 1173 (1965).
M4. Malm H. L., Fowler I. L., I.E.E.E. Trans., **NS—13**, *1*, 62 (1966).
M5. Malm H. L., I.E.E.E., Trans., **NS—13**, *3*, 285 (1966).
M6. Mann H. M., Haslett J. W., Janarek F. J., I.R.E. Trans., **NS—9**, *4*, 43 (1962).
M7. Mann H. M., Yntema L. J., I.E.E.E. Trans., **NS—11**, 201 (1964).
M8. Mann H. M., Sherman I. S., I.E.E.E. Trans., **NS—11**, *3*, 271 (1964).
M9. Mann H. M., I.E.E.E. Trans., **NS—12**, *2*, **88**, (1965).
M10. Mann H. M., Bilger H. R., Sherman I. S., I.E.E.E. Trans., **NS—13**, *3*, 252 (1966).
M11. Marcazzan M. G., Merzari F., Tovolini F., I. R.E. Trans., **NS—9**, *3*, 234 (1962).
M12. Martin F. W., King W. I., Harrison S., I.E.E.E. Trans., **NS—11**, *3*, 280 (1964).
M13. Martin F. W., Harrison S., King W. J., I.E.E.E. Trans., **NS—13**, *1* (1966).
M14. Maslova L. V., Matveev O. A., Rivkin S. M., Strokan N. B., Tarhin D. V., Hozov V. G., Izv. A.N.S.S.S.R. Ser. Fiz., **26**, *12* (1962).
M15. Mayer J. W., Rev. Sci Instr., **27**, 407 (1956).
M16. Mayer J. W., J. Appl. Phys., **30**, *12*, 1937 (1959).
M17. Mayer J. W., I.R.E. Trans., **NS—7**, *2—3*, 178 (1960).
M18. Mayer J. W., *N.A.S. — N.R.C. Pub. 871*, Washington, 1961, p. 8.
M19. Mayer J. W., I.R.E. Trans., **NS—9**, *3*, 124 (1962).
M20. Mayer J. W., Nucleonics, **20** (1962).
M21. Mayer J. W., Baily N. A., Dunlap H. L., *Nuclear Electronics. Proc. Conf. Belgrade, 1961*, vol. 1, I.A.E.A. Vienna, 1962, p. 567.

M22. Mayer J. W., J. Appl. Phys., **33**, 2894 (1962).
M23. Mayer J. W., *Electronique Nucléaire. Colloque — Paris, 1962,* C.E.A. Paris, 1963, p. 129.
M24. Mayer J. W., Baron, R., Marsh O. J., Phys., Rev., **137**, 286 (1965).
M25. Mayer J. W., Nucl. Instr. Meth., **43**, 55 (1966).
M26. Mayer J. W., J. Appl. Phys., **38**, 296 (1967).
M27. Mayer J. W., Marsh O. J., Mankarious R., Bower R., J. Appl. Phys., **38** (1967).
M28. McKay K. G., Phys., Rev., **76**, 1537 (1949).
M29. McKay K. G., McAfee K. B., Phys. Rev., **91**, *5*, 1079 (1953).
M30. McKenzie J. M., Ewan G. T., I.R.E. Trans., **NS—8**, *1*, 50 (1961).
M31. McKenzie J. M., Waugh I. B. S., I.R.E. Trans., **NS—7**, *2—3*, 195 (1960).
M32. McKenzie J. M., Donovan P. F., Gynn A. C., Nucl. Instr. Meth., **54**, 147 (1967).
M33. Messier J., Coroller Y. Le., Flores J. M., I.E.E.E. Trans., **NS—11**, *3*, 276 (1964).
M34. Meyer H., I.E.E.E. Trans., **NS—13**, 180 (1966).
M35. Meyer O., Langmann H. J., Nucl. Instr. Meth., **34**, 77 (1965).
M36. Meyer O., Langmann H. J., Nucl. Instr. Meth., **36**, 119 (1966).
M37. Meyer O., Haushahn G., Nucl. Instr. Meth., **56**, 177 (1967).
M38. Miehe J. A., Ostetag E., Coche A., L'Onde El., **46**, *472*, 801 (1966).
M39. Miller G. L., Brown W. L., Donovan P. F., MacKintosh J. M., I.R.E. Trans., **NS—7**, *2—3*, 185 (1960).
M40. Miller G. L., *N.A.S. — N.R.C. Pub. 871*, Washington, 1961, p. 19.
M41. Miller G. L., Foreman B. M., Yuan L.C.L., Donovan P. F., Gibson W. M., I.R.E. Trans., **NS—8**, *1*, 73 (1961).
M42. Miller G. L., Gibson W. M., Donovan P. F., Ann. Rev. Nucl. Sci., **12**, 189 (1962).
M43. Miller G. L., Gibson W. M., *Nuclear Electronics. Proc. Conf. Belgrade, 1961*, vol. 1, I.A.E.A. Vienna, 1962, p. 477.
M44. Miller G. L., Pate B. D., Wagner S., I.R.E. Trans., **NS—101**, 220 (1963).
M45. Moak C. D., Dabbs J. W. T., Walker W. W., Rev. Sci. Instr., **37**, *9*, 1131 (1966).
M46. Moll J. L., *Physics of Semiconductors*, McGraw-Hill, New York, 1964.
M47. Monca ster M. E., Northrop D. C., Raines J., Nucl. Instr. Meth., **22**, 157 (1963).
M48. Monteith L. K., Rev. Sci. Instr., **35**, *3*, 388 (1964).
M49. Motoc C., *Acţiunea Radiaţiilor Nucleare asupra Corpului Solid*, Ed. Acad., Bucharest, 1964.
N1. Nanavati R. P., *An Introduction to Semiconductor Electronics*, McGraw-Hill, New York, 1963.
N2. Norbeck E., Carlson R. C., *N.A.S. — N.R.C. Pub. 1184*, Washington, 1964, p. 42.
N3. Norgate G., McIntyre R. J., I.E.E.E. Trans., **NS—11**, *3*, 291 (1964).
N4. Nowlin C. H., Blankenship S. L., Rev. Sci. Instr., **36**, *12*, 1830 (1965).
N5. Nussbaum A., *Semiconductor Device Physics*, Prentice Hall, London, 1962.
N6. Nygaard K., Nucl. Instr. Meth., **54**, *1*, 98 (1967).
P1. Papadopoulos L., Nucl. Instr. Meth., **41**, 241 (1966).
P2. Parkinson J., Rev. Sci. Instr., **32**, 1136 (1961).
P3. Pehl R. H., Landis D. A., Goulding F. S., I.E.E.E. Trans., **NS—13**, *3*, 274 (1966).
P4. Pell E. M., J. Appl. Phys., **31**, *2*, 291 (1960).
P5. Pell E. M., *N.A.S. — N.R.C. Pub. 871*, Washington, 136 (1961).
P6. Pigneret J., Samueli J. J., Sarazin A., Genre R., LYCEN — 6514, Lyon (1965).
P7. Pigneret J., Samueli J. J., Sarazin A., I.E.E.E. Trans., **NS—13**, *3*, 306 (1966).
P8. Pisarevski A. N., Sosin L. D., Firsov E. I., P.T.E., **6**, 14—20 (1961).
P9. Poenaru D. N., Vîlcov N., Automatica şi Electronica, **9**, *5*, 228 (1965).
P10. Poenaru D. N., Vîlcov N., Nucl. Instr. Meth., **36**, 52 (1965).

P11. Poenaru D. N., Vîlcov N., Automatica şi Electronica, **11**, 203 (1967).
P12. Poenaru D. N., Vîlcov N., Rev. Roum. Phys., **12**, 243 (1967).
P13. Poenaru D. N., Vîlcov N., Automatica şi Electronica, **10**, 5 (1967).
P14. Poenaru D. N., Nucl. Instr. Meth., **54**, 229 (1967).
P15. Poenaru D. N., Nucl. Instr. Meth., **54**, 242 (1967).
P16. Poenaru D. N., I.E.E.E. Trans., **NS—14**, *5*, 1 (1967).
R1. Radeka V., *N.A.S. — N.R.C. Pub. 1184*, Washington, 1964, p. 70.
R2. Radeka V., I.E.E.E. Trans., **NS—11**, *3*, 358 (1964).
R3. Radeka V., Chase R. L., I.E.E.E. Trans., **NS—13**, *3*, 477 (1966).
R4. Ramo S., Proc. I.R.E., **27**, 584 (1939).
R5. Raju M. R., Aceto H., Richman C., Nucl. Instr. Meth., **37**, 152 (1965).
R6. Raymo C. T., Mayer J. W., I.R.E. Trans., **NS—8**, *1*, 157 (1961).
R7. Rivkin S. M., Matveev O. A., Strokan M. B., Husaikov A. H., *Conference on Semiconductor Detectors, Dubna, 26—28 Oct. 1964.*
R8. Roosbroeck W. Van, Phys. Rev., **135**, 5A, A 1702 (1965).
S1. Sakai E., *Nuclear Electronics, Proc. Conf. Belgrade, 1961*, vol. 1, I.A.E.A. Vienna, 1962, p. 551.
S2. Salai R., *Electronique Nucléaire. Colloque — Paris, 1962*, C.E.A. Paris, 1963, p. 205.
S3. Schiff L. I., *Quantum Mechanics*, McGraw-Hill, New York, 1955.
S4. Schmitt H. W., Dabbs S. W. T., Miller P. D., *Physics and Chemistry of Fission*, I.A.E.A. Vienna, 1965, p. 517.
S5. Schmitt H. W., Gibson W. M., Neiler J. H., Walter F. J., Thomas T. D., *Physics and Chemistry of Fission*, I.A.E.A. Vienna, 1965, p. 531.
S6. Schmitt H. W., Pleasonton F., Nucl. Instr. Meth., **40**, 204 (1966).
S7. Scott E. E., I.E.E.E. Trans., **NS—11**, *3*, 206 (1964).
S8. Sergiescu V., *Introducere în Fizica Solidului*, Ed. Tehnică, Bucharest, 1956.
S9. Shannon J. A., Trice J. B., Huth G. C., I.E.E.E. Trans., **NS—12**, *1*, 281 (1965).
S10. Sharpe J., *Nuclear Radiation Measurement*, Temple Press, London, 1960.
S11. Shockley W., Bell. Syst. Techn. J., **28**, *3*, 435 (1949).
S12. Shockley W., Read W. T., Phys. Rev., **87**, *5*, 835 (1952).
S13. Shockley W., Proc. I.R.E., **40**, 1365 (1952).
S14. Shockley W., *Electrons and Holes in Semiconductors*, D. van Nostrand Company, New York, 1954.
S15. Shockley W., Solid State Electronics, **2**, *1*, 35 (1961).
S16. Siegbahn K., *Alpha-, Beta- and Gamma-ray Spectroscopy*, North-Holland Publ. Comp., Amsterdam, 1965.
S17. Siffert P., Rougeot H., Coche A., *Electronique Nucléaire. Colloque — Paris, 1962*, C.E.A. Paris, 1963, p. 155.
S18. Siffert P., Coche A., I.E.E.E. Trans., **NS—11**, *3*, 244 (1964).
S19. Siffert P., Coche A., Hibou F., I.E.E.E. Trans., **NS—13**, *3*, 225 (1966).
S20. Siffert P., Coche A., I.E.E.E. Trans., **NS—13**, **757** (1966).
S21. Simpson O., *Proc. Symp. Nucl. Instr., Harwell*, 1961, p. 75.
S22. Smith K. F., Cline J. E., I.E.E.E. Trans., **NS—13**, *3*, 468 (1966).
S23. Smith R. A., *Semiconductors*, University Press, Cambridge, 1959.
S24. Sona A., Svelto V., Energia Nucleare, **9**, *12*, 694 (1962).
S25. Spătaru Al., *Teoria Transmisiunii Informaţiei*, Ed. tehnică, Bucharest, 1965.
S26. Starodubtev S. V., Romanov A. M., *Prokhozhdenie Zariazhenyk Chastit cherez Veshchestvo*, Izd. A. N. Uzbekskoi S.S.R., Tashkent, 1962.
S27. Stojnic M. D., Slavic J. A., *Electronique Nucléaire. Colloque — Paris, 1962*, C.E.A., Paris, 1963, p. 189.
S28. Sternheimer R. M., Phys. Rev., **115**, 137 (1959).

S29. Sternheimer R. M., Phys. Rev., **117**, 485 (1960).
S30. Sternheimer R. M., Phys. Rev., **118**, 1045 (1960).
S31. Strauss M. G., Larsen R. N., Sifter L. L., I.E.E.E. Trans., **NS—13**, *3*, 265 (1966).
T1. Tanguy J., L'Onde El., **46**, 785 (1966).
T2. Tavendale A. J., *Electronique Nucléaire. Colloque — Paris, 1962*, C.E.A. Paris, 1963, p. 235.
T3. Tavendale A. J., I.E.E.E. Trans., **NS—11**, *3*, 191 (1964).
T4. Tavendale A. J., I.E.E.E. Trans., **NS—12**, *1*, 255 (1965).
T5. Tavendale A. J., Rev. Sci. Instr., **36**, *8*, 1275 (1965).
T6. Tavendale A. J., Nucl. Instr. Meth., **36**, 325 (1965).
T7. Tavendale A. J., I.E.E.E. Trans., **NS—13**, *3*, 315 (1966).
T8. Taylor J. M., *Semiconductor Particle Detectors*, Butterworth, London, **1963**.
T9. Taylor K. N. R., Brit. J. Appl. Phys., **13**, *5*, 242 (1962).
T10. Takaki R., Perkins M., Tuzzolino A., I.R.E. Trans., **NS—8**, *1*, 64 (1961).
T11. Tepper L., Miller Gl., Kyvia T., I.E.E.E. Trans., **NS—11**, *3*, 431 (1964).
T12. Tove P. A., Falk K., Nucl. Instr. Meth,. **12**, 278 (1961).
T13. Tove P. A., Falk K., Nucl. Instr. Meth,. **29**, 66 (1964).
T14. Tove P. A., Seibt W., Nucl. Instr. Meth., **51**, 261 (1967).
T15. Tove P. A., Seibt W., Leitz W., Nucl. Instr. Meth., **51**, 304 (1967).
T16. Tsukuda M., Nucl. Instr. Meth., **14**, 241 (1961).
T17. Tsukuda M., I.R.E. Trans., **NS—9**, *4*, 63 (1962).
T18. Tuzzolino A. J., Perkins M. A., Kristoff J., Nucl. Instr. Meth., **37**, *2*, 204 (1965).
T19. Tuzzolino A. J., Kristoff J., Perkins M. A., Nucl. Instr. Meth., **36**, *1*, 73 (1965).
T20. Ţiţeica Ş., *Elemente de Mecanică Statistică*, Ed. tehnică, Bucharest, **1956**.
V1. Valentin F., *Nuclear Electronics. Proc. Conf. Belgrade, 1961*, I.A.E.A. Vienna, **3**, 249 (1962).
V2. Van Der Dores De Bye J. A. W., Philips Research Reports, **16**, 85 (1961).
V3. Van Putten J. D., Van Der Velde J. C., I.R.E. Trans., **NS—8**, *1*, 124 (1961).
V4. Van Der Ziel, Proc. I.R.E. **50**, 1808 (1962).
V5. Vavilov V. S., *Deistvie izlucheniya na poluprovodniki*. Fizmatghiz, Moscow, 1963.
V6. Vavilov V. S., Chukichev M. V., Uvarov E. F., *Conference on semiconductor Detectors, Dubna, 26—28 Oct. 1964*.
V7. Vitouski N. A., Maleev P. I., Matveev O. A., Rivkin S. M., Tarkhin D.V., P.T.E., **2**, 82 (1961).
V8. Vîlcov N., St. Cerc. Fiz., **11**, *5*, 715 (1962).
V9. Vîlcov N., Rev. Roum. Phys., **7**, *4*, 397 (1962).
V10. Vîlcov N., Poenaru D. N., Automatica şi Electronica, **9**, *3*, 107 (1965).
W1. Wahl H., Nucl. Instr. Meth., **25**, *2*, 274 (1964).
W2. Wahl R., Industries Atomiques, *9—10*, 89 (1962).
W3. Wainio K. M., Knoll G. F., Nucl. Instr. Meth., **44**, 213 (1966).
W4. Walter F. J., Dabbs J. W. T., Roberts L. D., Rev. Sci. Instr., **31**, *7*, 756 (1960).
W5. Walter F. J., *N.A.S. — N.R.C. Pub. 871*, Washington, 1961, p. 1.
W6. Walter F. J., Dabbs J. W. T., Roberts L. D., I.R.E. Trans., **NS—8**, *1*, 79 (1961).
W7. Walter F. J., I.E.E.E. Trans., **NS—11**, *3*, 232 (1964).
W8. Walter F. J., Boshart R. R., I.E.E.E. Trans., **NS—13**, *3*, 189 (1966).
W9. Warble K. V., Gri N. I., I.E.E.E. Trans., **NS—13**, *1*, 346 (1966).

W10. WAUGH J. B.S., Nucl. Instr. Meth., **8**, 257 (1960).
W11. WEGNER H. E., I.R.E. Trans., **NS—8**, *1*, 103 (1961).
W12. WEGNER H. E., *N.A.S. — N.R.C. Pub. 871*, Washington, 1961, p. 182.
W13. WEGNER H. E., *Nuclear Electronics. Proc. Conf. Belgrade, 1961*, vol. 1, I.A.E.A. Vienna, 1962, p. 425.
W14. WEGNER H. E., ERGINSOY C., GIBSON W. M., I.E.E.E. Trans., **NS—12**, *1*, 240 (1965).
W15. WEISS W. L., I.R.E. Trans., **NS—10**, *1*, 202 (1963).
W16. WERTHEIM G. K., *N.A.S.—N.R.C. Pub. 871*, Washington, 1961, p. 128.
W17. WILLIAMS R. L., WEBB P. P., I.R.E. Trans., **NS—8**, *1*, 35 (1961).
W18. WILLIAMS R. L., *N.A.S. — N.R.C. Pub. 871*, Washington, 1961, p. 28.
W19. WILLIAMS R. L., *N.A.S. — N.R.C. Pub. 871*, Washington, 1961, p. 182.
W20. WILLIAMS R. L., WEBB P. P., I.R.E. Trans., **NS—9**, *3*, 160 (1962).
W21. WILLIANSON C. F., BOUJOT J. P., PICARD J., C.E.A. Paris, 3—3042 (1966).
W22. WILSON R. R., Phys. Rev., 60, 749 (1941).
Z1. ZAJDE C., L'Onde El., **46**, *472*, 818 (1966).
Z2. ZIEMBA E. P., PELT G., RYAN G., WANG L., ALEXANDER R., I.R.E. Trans., **NS—9**, *3*, 155 (1962).
Z3. ZUBRITKI L. A., POPOV A. I., SOROKIN P. V., SAMOILOV V. F., *Nuclear Electronics. Proc. Conf. Belgrade, 1961*, vol. 1, I.A.E.A. Vienna, 1962, p. 591.
Z4. ZURMÜHLE R. W., Nucl. Instr., 36, 168 (1965).
1 * * * *Dispozitive electronice şi Circuite pentru Aplicarea Izotopilor*. Ed. tehnică, Bucharest, 1965.
2 * * * *Commercially available semiconductor detectors and preamplifiers*, Nucleonics, **22**, *5*, 62 (1964).
3 * * * *American Institute of Physics Handbook*, McGraw-Hill, New York, 1963.
4 * * * *Poluprovodnikovye schetshiki izlucheniya*, Gosatomizdat, Moscow, 1962.

www.ingramcontent.com/pod-product-compliance
Lightning Source LLC
LaVergne TN
LVHW020630300926
849161LV00004B/658
* 9 7 8 0 8 2 0 6 0 1 0 3 8 *